Product Design and Manufacture

JOHN R. LINDBECK
Professor Emeritus
Engineering Technology
Western Michigan University

CONTRIBUTING AUTHOR:
ROBERT M. WYGANT
Professor of Industrial Engineering
Western Michigan University

PRENTICE HALL, Englewood Cliffs, New Jersey 07632

Library of Congress Cataloging-in-Publication Data

Lindbeck, John R.
    Product design and manufacture / John R. Lindbeck ; contributing
author, Robert M. Wygant.
        p.   cm.
    Includes bibliographical references and index.
    ISBN 0-13-034257-2
    1. Design, Industrial.   2. Concurrent engineering.   3. Production
planning.   I. Wygant, Robert M.   II. Title
    TS171.L57   1994
    658.5'752—dc20   94-20422
                                                                              CIP

Editorial/production supervision: Lynda Griffiths, TKM Productions
Cover design: Tom Nery
Manufacturing buyer: Lori Bulwin

Cover credits: 1937 Elgin Skylark: Courtesy of Schwinn Bicycle
Center. Table: Courtesy of Peter Czuk Woodworking and Design.
Poster design: Diverse Directions, Women's Faculty Exhibition,
Dept. of Art, Western Michigan University; Design: Tricia Hennessey.

© 1995 by Prentice-Hall, Inc.
A Simon & Schuster Company
Englewood Cliffs, New Jersey 07632

Printed in the United States of America

10   9   8   7   6   5   4   3   2   1

ISBN 0-13-034257-2

PRENTICE-HALL INTERNATIONAL (UK) LIMITED, *London*
PRENTICE-HALL OF AUSTRALIA PTY. LIMITED, *Sydney*
PRENTICE-HALL CANADA INC., *Toronto*
PRENTICE-HALL HISPANOAMERICANA, S.A., *Mexico*
PRENTICE-HALL OF INDIA PRIVATE LIMITED, *New Delhi*
PRENTICE-HALL OF JAPAN, INC., *Tokyo*
SIMON & SCHUSTER ASIA PTE. LTD., *Singapore*
EDITORA PRENTICE-HALL DO BRASIL, LTDA., *Rio de Janeiro*

Contents

Preface

There is a prevailing general impression that the quality of U.S. products and the efficiency with which they are made has deteriorated somewhat in recent years. To be sure, there are many examples of American product excellence to be found at home and abroad, but the fact remains that the situation could be improved. Much has been written about this predicament, and it has been the subject of numerous seminars and symposia. Some critics blame the Japanese for their selfishness, self-centeredness and self-righteousness, and at once applaud them for their dedication to optimum standards in design and production. Others maintain that the Europeans and the Japanese deny an open market to U.S. goods. Also brought to charge are the consumers for their acceptance of shoddy American goods; the industrial managers and designers for their demonstrated concern for making money rather than quality products; the chief executive officers, whose salaries and bonuses too often are a function of the whims of boards of directors rather than being linked to corporate productivity; the educators for failing to provide state-of-the-art programs; and the production workers for their indifference to their work. Everyone is responsible and no one is responsible.

Customer complaints, product recalls, liability suits, and the industrial degradation of the environment are issues that appear in the press with depressing regularity. The fact of consumer distrust and loss of confidence in U.S. manufacturers and their artifacts cannot be denied. This becomes the challenge for all those who design and make the things used by humans, and the challenge is being addressed. Essential to the resolution of this problem is to design and manufacture better things through improved methodology.

One purpose of this book is to introduce engineering, technology, and design students to the methods of concurrent or simultaneous design, commonly known as *design for manufacturing*, where all aspects of product design and planning are considered as a totality. Significant issues that must be addressed early in the design process include proper material and process selection; product safety, maintainability, producibility, quality, reliability and usability; human factors; legal issues; and product aesthetics. Ideally, teams of specialists join together to create products that are as close to perfect as they can become. No longer does a designer plan an artifact and then deliver it to production engineer who must figure out an economic way to make the thing. And then, further down the line, the safety and maintenance people consider the design from their special viewpoints. To reiterate, all people involved with product design and manufacture must be involved at the outset of the planning effort. This approach sounds very logical and it is, but it was not always that way. If this nation is to regain its reputation as a provider of high-quality products, it must reorient its design procedures to incorporate the important issues just described. A second reason for this book is to provide an examination of historical antecedents, information, and data on product design theory and procedures.

This book begins with an examination of the background for design as an inquiry into the genesis of artifact planning and making. Here, it can be seen that people have always developed the things they use in direct response to need, and as a reflection of the prevailing technology and aesthetics. From this base, attention is directed to design theory, looking first at the fundamental functional, material, and visual requirements of a product, and then moving to an examination of the approaches utilized by industrial, engineering, graphic, and craft designers. Considerations of design materials follow, as an introduction to the nature and diversity of the substances of which products are made.

Manufacturing processes are presented as an overview of the means by which materials are transformed into usable products, recognizing the vast array of processes available, and introducing some rationale for their selection. The theory of human factors in design is developed as a means of guiding product-planning methods by recognizing the strengths, limitations, and characteristics of humans and their behavior. Logically, the chapter on applied ergonomics follows with a series of illustrations and examples of human needs as directives to quality product design. Included is information on serviceability, security, safety, usability, vandalism, the physically disadvantaged, and the environment.

The subject of computers in design and manufacturing is presented to illustrate the uses of these remarkable tools in product design and manufacture. Computers enhance human effort and make it more efficient in resolving the many problems in modern production and design. Design for producibility is treated next by describing this important element in manufacturing efficiency and illustrating the methods for achieving it. The subject of design for graphics and packaging is a significant adjunct to product planning and is based on a background of printing reproduction technology. The methods of graphic design and packaging techniques are described and exemplified. The important legal issues are introduced next to alert persons associated with product design and manufacture to the many related problems of protecting intellectual property and preventing product liability claims. The book closes with a list of design resources, including books, periodicals, visual materials, and design centers of value to the practitioner as well as the learner and teacher.

In summary, the primary aim in writing this book is not to prepare design experts but instead to provide a background for beginners in the field to give greater meaning to the many advanced courses in the technical program, and to help fledgling designers to make more intelligent design decisions.

ACKNOWLEDGMENTS

Writing this textbook has been made more enjoyable and satisfying because of the cooperation of those many industrial consultants, foreign trade representatives, engineers, designers, technicians, and media specialists who gave freely of their time, information, data, and graphics. Their kindness and pleasant conversations were helpful and valuable. Those persons deserving special mention are:

My wife Marilyn, for her patience and competence as my word processor

Dr. David Gregg, for his CAD skills and his forbearance in preparing excellent charts and drawings

Dr. Robert Wygant, for contributing Chapter 5, Human Factors in Design

Dr. Arvon Byle, Dr. Ralph Tanner, Mr. John Vargo, and Mr. John Walker for kindly consenting to review portions of the manuscript, and for their helpful suggestions

Mr. Ronald Tanis, for his generous assistance with the material on patents, trademarks, copyrights, and trade secrets

Dr. Jerry Hemmye, for providing the interesting engineering design example

Ms. Tricia Hennessy, for her spectacular section on graphics design

Mr. Scott Spink, for his fine photography

Mr. Dennis Smith, for the innovative material on tiling construction

Mr. Bruce Sienkowski, for the informative narrative on the development of the Eli chair

And to the countless students of the 34 years of my tenure, who listened, inquired, challenged, experimented, designed, built, discussed, and learned.

It was truly a pleasure to work with everyone.

CHAPTER 1 | Background for Design

Humans have always been confronted with a need for well-conceived, functional artifacts; from primitive societies to the contemporary, there are numerous examples of such products. Though the techniques for producing these articles have changed appreciably, the reasons for their existence and the methods of planning for them have changed very little. Each period in history had its own styles, technology, and material limitations, but the rationale for product design was always one of form developed in response to need.

Beginning with the primitive humans, the specialization and technological progress, coupled with the prevailing cultural-aesthetic tastes, formed the background for the styles and periods of furniture and other articles. It is important to realize that each design was developed during and for a specific period in history, and each was controlled by the technical, material, and aesthetic requirements of that day. Only through an acceptance of this fact can the designs of all periods of history be truly understood and appreciated.

THE PRIMITIVE

The story of human progress toward our present civilization is illustrated with examples of success in developing suitable tools with which to build shelters, till the soil, hunt, fight, and produce. Without them, societies would not have evolved or endured. Before humans were toolmakers, they were tool users. The stick was a club for hunting, a hoe for tilling, or a spear for fighting. These simple wooden tools were

1

made more serviceable by hardening and shaping them over a campfire. The humans used stones in their natural shapes before putting sharp edges on them by chipping and flaking. These crude eoliths (or early stones) served as aids in cutting, scraping, and grubbing. They were all-purpose tools that provided a measure of assistance in acquiring the necessities for existence. (See Figure 1.1.)

In subsequent anthropological periods, humans improved their tools. In the Paleolithic (Old Stone) Age, tools of stone, wood, root, hide, ivory, horn, and bone were fashioned. To the simple "blow-of-the-fist" stone was added a wooden handle fastened with strips of hide or root, producing a more effective hammer, ax, or hoe. The Neolithic (New Stone) Age was characterized by polished stone artifacts of a wider variety and a higher level of development. Here, humans also learned to weave and make pottery—an achievement that made possible the production of containers

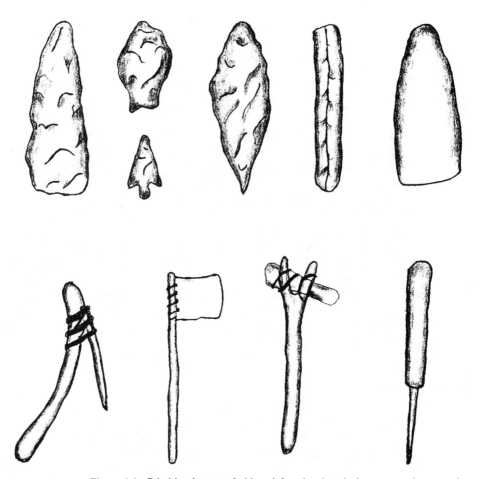

Figure 1.1 Primitive humans fashioned functional tools from stone, bone, and wood, with little decoration or embellishment.

other than the natural gourds and shells. The wheel was invented, along with pulleys, levers, looms, and ladders.

Metal appeared relatively recently in human history. Gold, copper, bronze, and iron were used some 6,000 years ago by the first civilizations of Mesopotamia, Egypt, and the Orient. Exactly how they came upon it is only surmised; possibly the heat from some primeval campfire caused surrounding copper-bearing rocks to release the metal in a molten form. Copper, being very soft, was easily worked with the crudest of tools and could be work-hardened by hammering to produce more durable edges. The working cycle of annealing (or softening) and hammering to shape produced a range of refined implements. With the discovery of methods of combining tin or zinc with the copper to form bronze and brass, humans had harder, tougher metals with which to work. The story of iron is essentially the same but begins somewhat later, around 1000 B.C. Though it was found in greater abundance than nonferrous metals, perhaps its use was postponed because it was more difficult to secure in forms suitable for working. The availability of metal, and the advancement of skills and techniques for using it, were tremendous steps forward in toolmaking. Being able to fashion tools by means of hammering, casting, or bending was far superior to the more restrictive chipping and flaking methods necessary in stoneworking. Humans had sharper, stronger, more lasting implements, which figured strongly in the further development of civilization.

Aesthetically, many of the tools of primitive humans displayed an admirable simplicity and a directness. When they devised a utilitarian tool, no conscious effort was made to make it beautiful as well. That these primitive designs are pleasing lends substance to the contention that even if form does not always follow function, at least it offers a point of departure in design. Modern humans could do well to consider this approach.

For reasons not entirely clear, humans gradually began to decorate the things they made. Bone handles were carved to resemble human and animal forms; clay vessels were painted to depict familiar incidents, as were the walls of caves; and woven materials were dyed with juices and minerals. Perhaps these artistic ventures had their bases in religious beliefs, or perhaps they were manifestations of some innate aesthetic desire. In any event, they were not realized until the essentials for physical survival were assured.

EGYPT

A study of the furniture of the Egyptian Period (ca. 4000–300 B.C.) is enlightening because it reveals the considerable influence on crafters in subsequent eras, and illustrates the antiquity of furniture shapes and structures. These ancients, working with simple bronze and stone tools, reached a technical standard in woodwork not equaled in Europe until the time of the Renaissance. (See Figure 1.2.) Egyptian designs predated the classic lines of the Greeks and Romans, and Greek masters were schooled in technique and design by the Egyptians. Thus, the craft of these people was of great significance to future generations.

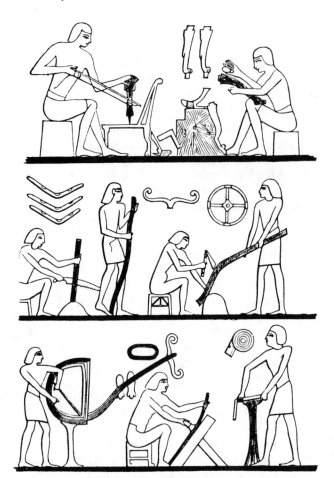

Figure 1.2 Ancient Egyptian crafters used rudimentary tools to create magnificent artifacts.

Some of the most impressive collections of furniture and artifacts come from the Eighteenth Dynasty (ca. 1575–1310 B.C.). These were the times of the great rulers such as Akhanaten and Tutenkhamen. Evidences of Egyptian magnificent technical achievements are seen in the pyramids, temples, and immense sculptures that are today found along the Nile River. But grace of line and sensitivity of form are not attributes generally associated with Egyptian furniture. Many of the regal chairs and beds were decorated to excess, structurally awkward, and uncomfortable.

Interestingly enough, some of the most important furniture construction techniques were developed to meet serious material shortages. The ingenious scarf joints, as well as intricate pegging and fitting, were necessary because of a dearth of suitable timber. For the same reason, inferior woods were frequently employed as foundations, and more precious ones—such as ebony—were adhered to them. Thus was born the art of veneering. Many of the wood-joining methods in use today have their origin with the Egyptian carpenter. The mortise and tenon, dowel, and dovetail joints exemplify the high level of technical skill reached in this era. (See Figure 1.3.)

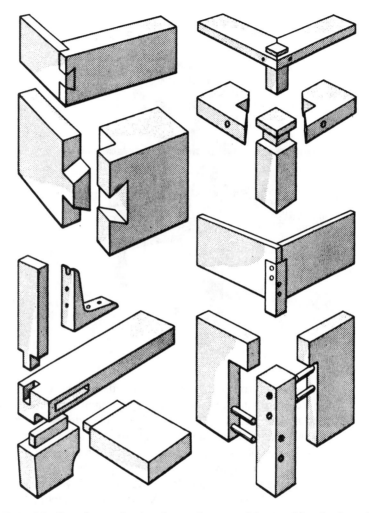

Figure 1.3 Egyptian woodworkers invented a range of clever and functional wood joint systems, many of which are still used.

Fish and animal glues were developed, and tools were greatly improved. The bow lathe was invented and used, and was probably based on the bow drill. Lathes similar to these are still being used in some parts of the world.

Fortunately, many ancient Egyptian furniture pieces have survived their burial in tombs and drifting sands. An examination of these actual objects, or a study of the paintings that adorn tombs and buildings, provides much information about the skills, materials, and processes used by artisans of the period. This furniture was built primarily of wood, with seats of laced leather thongs or reeds. Some of the oldest furniture featured simulated ox legs carved in wood or ivory, a practice testifying to the significance of agriculture in ancient Egypt. The lion-leg style gradually replaced

Figure 1.4 An Egyptian chair from the Eighteenth Dynasty. Note the applications of some of the wood joints shown in Figure 1.3. Reproduction by the author.

these—an approach that perhaps paralleled the emergence of Egypt as a great empire and that acted as a mute testimonial to leonine courage. Despite the numerous examples of intricately carved backs and leg structure, some of the furniture has an aura of classic simplicity about it. Such a piece, shown in Figure 1.4, is a reproduction of an original chair housed in a museum in Florence, Italy. Of special interest is the attention to proper technique; the crafter made good use of the triangle to strengthen the back and legs, and the employment of interlocking wood joints is impressive. The resultant chair is simple, clean, technically correct, and quite pleasing in appearance. Indeed, it has an almost contemporary look about it.

GREECE AND ROME

Probably no single era of human existence has contributed as much to the history of humankind as the Golden Age of Greece (ca. 450–404 B.C.). People today continue to feel the impact of this civilization, as shown in the following points:

1. Modern philosophy, science, mathematics, and politics abound in concepts of Greek origin.
2. The Hippocratic oath is still recited by doctors of medicine.
3. Athletes today strive for that zenith of all sporting events—the Olympic Games.

4. The classic form and structure of Greek art influences current painting, architecture, sculpture, literature, and drama.

This was the age of Pericles, the ruler who encouraged and led a people possessing critical, inquiring spirits to create art forms after the manner of idealism and humanism. The great and familiar Greek figures lived during this period—Sophocles, Aeschylus, Plato, Socrates, Aristotle, Herodotus, and others. The *classic style*, developed and nurtured in Greece and transported to Rome, was rooted in Egyptian art and design. Many skills and techniques were learned from the Egyptians, as were the principles of the arch and vault. But it is said that while the Egyptians created from technical knowledge, the Greeks learned to use their hearts and their eyes. The classic spirit of Greece—so evident in their beautiful, sensitive pottery, sculpture, and architecture—has influenced and endured, bearing witness to this claim. Gradually, the inherited naturalistic and ritualistic sources of design were disregarded in favor of a search for ideal forms, a mission from which emerged the Golden Section. (See Chapter 2.)

From this search for beauty evolved the classic Greek form, embodying the spirit of the times and characterized by simplicity and clear arrangement. This ideal is reflected in the Parthenon, which featured gently tapering columns and a sensitively proportioned rectangular structure. Examples of this search for correct organization were also extended to politics, drama, and philosophy. Plato, in his *Republic*, set forth the requirements for sound government and the necessary qualifications for statesmen. In Aristotle's *Poetics* can be found the neatly circumscribed rules for writing a good tragedy, and Aristotle's writings on logic establish rules for sound and orderly thinking.

Whereas the climatic conditions and the tombs of Egypt preserved examples of its furniture, little of that of the Greek Period survived because of time and conflict. Most of the impressions of furniture forms come from vase paintings or relief carvings on gravestones or buildings. The classic chair, or klismos (shown in Figure 1.5), is based on the relief sculpture from the tombstone of Hegeso (ca. 420 B.C.), and with reasonable accuracy represents the form, joinery methods, and materials used in originals. The strikingly beautiful curve of the back and leg structure reflects the Greek concern for ideal form. The joinery methods are of Egyptian origin, but the debt stops there, for the chair configuration is visual splendor. It is not, however, without flaws; the splayed leg structure is weak by any measure, and the absence of a rail connecting the legs adds to the technical deficiencies. But Greek designers would not compromise in the search for the ideal. If err they must, they would sacrifice structural integrity to preserve grace of form. With regard to this chair, they might well say, "Don't sit on it—admire it." Homer said that the klismos was fit only for goddesses.

Stone and bronze were used in this period, and turned wooden members appeared in furniture. Much garden furniture was of stone or worked metal, for the obvious reason of durability. Decoration was used widely, but not obtrusively; it was aesthetic in principle and used harmoniously to emphasize structure and form.

Figure 1.5 A classic Greek klismos (ca. 450 B.C), sensitive and elegant, but with its slender, splayed legs, technically at fault. Reproduction by the author.

Contrary to the Greek aesthetic, the works of the Roman Period (ca. 600 B.C.–400 A.D.) centered about an appeal to the senses and self-gratification. Their indiscriminate use of ornamentation was a common practice that has plagued society ever since. Typical motifs included the reintroduction of animal forms in leg structures. Crafters of that period, however, preferred to fashion furniture supports after the hind legs, rather than the forelegs, because it gave them the freer, concave form they found so appealing. Many pieces of furniture were of bronze alone, or this metal in combination with marble. Some of their most sensitive works were the graceful, well-conceived, and functional bronze chariots. The range of Roman design motifs ultimately gave way to the rigid patterns of the Gothic style.

THE VIKINGS

A brief digression in this somewhat systematic treatment of design history might be both interesting and valuable. The preceding unit dealt with design in the Roman Period. This section treats a parallel chronologic period, but one occurring in another part of the civilized world and under different social conditions.

The exploits of the Vikings constitute a historic event familiar to all. These Northmen—Danes, Swedes, and Norwegians—had a passion for daring adventure, and raided and colonized the coasts of Britain and Northern Europe, as well as Russia and Turkey, for almost three centuries (800–1000 A.D.). These were a courageous people living in a hostile climate where good land was scarce, and they had to learn to adjust to and compensate for the physical conditions that surrounded them. Their ingenuity, creativity, and resourcefulness were the equal of their courage, as evidenced by the ships and implements yielded by grave findings in recent years. Typical of these is the Viking ship representation in Figure 1.6, which offers tangible evidence of the Viking technical and aesthetic design abilities. To a people dominated by a harsh northern sea, ships were indispensable to communication and travel. A great invention, the keel, made them masters of the craft of shipbuilding and sailing. It is not known precisely how or when they developed the keel, but the significance of the achievement is clear. These graceful ships could be made broad and flat to permit travel in shallow, rocky waters. On the open seas, they were more stable and easier to propel but yet sufficiently elastic to yield to the pressure of the waves. The sight of these splendid ships, in full sail with armor shields lining the gunwales, must have evoked a feeling of admiration as well as fear in the intended victims of a coastal raid.

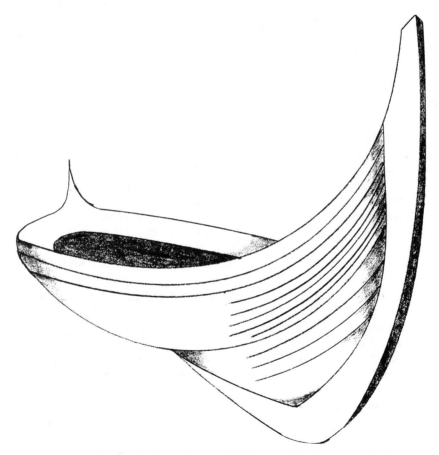

Figure 1.6 The graceful lines of a typical Viking ship. The intersecting planking members create a pleasing structural pattern.

One village on the coast of Normandy erected an altar and sign inscribed, "From the fury of the Vikings protect us, oh Lord."

Their sturdy utensils and furniture were generally of wood with occasional metal fastenings. Seats often served a dual purpose as storage chests—a practical solution for cramped quarters. Joints were pegged and seats were of leather thongs, and carved or embossed decorations are found on many of these articles. The ships, sledges, and carts often featured carved details, sometimes as pure decoration and at other times to depict some dramatic epic tale. The gold and silver jewelry and sword handles were also ornamented with filigree, granulations, or carved figures, skillfully and sensitively executed.

The distinctions between Roman and Viking designs are obvious: different kinds of products reflecting different materials and methods, emanating from a dissimilar social and cultural milieu. The Romans were influenced by the Greeks, but added their own appointments as dictates of a confused, emerging cultural mode. The Vikings, emerging from geographic isolation, could not fail to be affected by the contacts with new peoples. But the traditions and needs of life in the North were sufficient to maintain a design outlook necessary to sustain a simple and arduous mode of living.

GOTHIC

The Gothic period in art and architecture (ca. 1100–1500) was born of an era of artistic sterility and religious fervor. Following the decline of the Roman Empire, there ensued a stagnant time known in history as the Dark Ages. The church played a dominant role in the lives of the people, generally discouraging individual thought, initiative, and creativity. In this setting, and at a time when the Crusaders were returning from foreign soil with fresh ideas, the *Gothic style* developed. The leader in European literature, education, philosophy, and craftsmanship was France. She gave impetus to the movement, and it reached its zenith in her hands.

Where the Greeks had sought bodily perfection and the Romans sensual pleasure, the Gothic era was characterized by the demotion of the body for the advancement of the soul. Life generally was dedicated to and centered around the church. The soul of Gothic architecture is verticality and a variety of pointed arches, ribbed vaulting, and flying buttresses that soar upward, accentuating height at the expense of breadth. These factors are typical of the magnificent Gothic cathedral (see Figure 1.7), which evolved from the resolution of rather imposing engineering problems. The development and perfection of the ribbed vault lessened the ceiling weight, thereby permitting more slender supporting columns. Localizing support by means of the flying buttresses allowed greater window space in thinner walls, which made possible the inclusion of impressive stained glass windows as integral parts of the imposing effect. Even the soaring grace of the pointed arch was not possible until suitable means for bearing these tremendous pressures were found. It might be said,

Figure 1.7   A work of Gothic art—the Cathedral of Notre Dame. One of the finest examples of early Gothic architecture, work was begun in 1163 and completed some 150 years later. Courtesy of the French Government Tourist Office.

then, that Gothic architecture evolved from the needs of the church for solutions to structural problems and from artistic inspiration.

The furniture of the time naturally reflected this atmosphere of religious fervor. The chair in Figure 1.8 exemplifies the manifestation of this ecclesiastical influence. Note the perpendicular lines, heavy and solid, with carved ornamentations borrowed from the Gothic arch. Even the principal materials, oak and wrought iron, illustrate this quality of endurance and substantiality.

Interestingly enough, the *Renaissance Italians* assigned the name *Gothic* (from the Goths) out of contempt because they felt that only the barbarians from the North could have developed such an extravagant art. Be that as it may, the Gothic design was distinctive in both spirit and execution. Also, it was one of the most splendid styles the world has ever known, and remained dominant in Europe for over 400 years.

Figure 1.8 French Gothic chair, oak, fifteenth century. The Metropolitan Museum of Art. Gift of J. Pierpont Morgan, 1916.

RENAISSANCE

The Renaissance (rebirth) (ca. 1400–1700) is significant in the progress of design because it marks the transition from the medieval to the modern. This revival of classical learning and art began in Italy in the sixteenth century and rapidly spread throughout the whole of Europe. With respect to design, it was antiquarian and revivalistic, which was the process of adapting classic ornamentation and applying it with little restraint. But when one considers the conditions and atmosphere that gave birth to the Gothic style, it is understandable that public sentiment should swing away from an era of religious fervor to one less restrictive and severe.

Typical Renaissance furniture (see Figure 1.9) showed classical inspiration but was lavishly decorated with carving or inlays of Greek and Roman ornamental motifs: the acanthus, lotus, egg-and-dart, bead, fret, and rosette. It varied in form from country to country, reflecting a rejection of the Gothic structure, turning instead to horizontal lines, palatial sizes and effects, beautiful proportions, and comparatively restrained design. Walnut was the predominate material in Italy, Spain, and France, whereas oak was used in England and Holland.

Figure 1.9 Italian Renaissance folding armchair, carved walnut frame and red velvet back and cushion, sixteenth century. All rights reserved, The Metropolitan Museum of Art. Bequest of Benjamin Altman, 1913.

Baroque is a term used to denote the period of greater artistic freedom that succeeded the regularity and restraint of the Renaissance. At first marked by formal, stately, but somewhat fantastic structure, baroque later became a profusion of meaningless rock-work, shells, and scrolls massed together. This furniture, more playful and piquant in character, became known as *rococo*. A typical baroque piece appears in Figure 1.10. The Renaissance artists displayed a marked versatility in the scope of their work. They accepted practically any task that involved the application of aesthetic elements, even the utilitarian objects such as silver dishes and battle armor. These were objects of both utility and interesting form.

EIGHTEENTH CENTURY

The eighteenth century is appropriately described as The Golden Age of Furniture. It was an age that saw the development of the *Georgian* group in England: Chippendale, Hepplewhite, Sheraton, and the Adams brothers. The Provincial, Directoire, and Regency styles prevailed in France. The United States produced the Colonial and Duncan Phyfe. The pieces in the English group were elaborate and often quite overdone, but many were graceful and beautiful. The *Provincial* pieces—French, Italian, and American—were quietly dignified and unassuming, with a noticeable absence of lavish decoration. It is in these Provincial styles that one sees a marked departure from the ornate pretentiousness of the traditional eighteenth-century English and French designs.

The Chippendale-style chair (ca. 1740–1749) typifies the traditional English pieces. (See Figure 1.11.) Thomas Chippendale, the first craftsman not a reigning monarch to give his name to a furniture style, was inspired by preceding English,

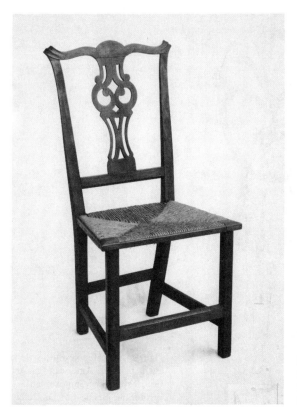

Figure 1.11 Chippendale style side chair, New England (ca. 1760). Courtesy of The Smithsonian Institution. Gift of Mrs. Charlotte Ellis Danforth.

French, and Chinese furniture forms. He depended heavily on carving for decoration, delicate in some pieces, bold and lavish in others; occasionally, however, he sacrificed comfort for appearance.

George Hepplewhite's name became synonymous with grace and elegance during this period, a factor borne out in the striking oval and shield-back motif. (See Figure 1.12.) He was a London cabinetmaker and original designer whose break with tradition did much toward the development of a simpler, more appealing style. His work is characterized by lightness and an emphasis on the pure beauty of line. Other names associated with the harmony of this Georgian period are Robert and James Adams, and Thomas Sheraton. Sheraton was impressive in his use of pure and beautiful perpendicular lines. When he did employ a curve, it became a graceful sweep. The Adams brothers' interpretation of classic forms resulted in motifs quite different from any that had been used prior to that time, even though their pieces were sometimes structurally impractical. Of the Provincial styles, the French is perhaps the best known. It is characterized by restraint, and by simple cabriole legs and gently curved stretchers, with a marked absence of intricate decoration.

Figure 1.12 Hepplewhite-style side
chair, painted, American (ca. 1800).
Courtesy of The Smithsonian
Institution. Bequest of Dr. John
Watt, Jr.

COLONIAL AMERICA

Generally, the early colonists built in America as they had built in Europe, for they
brought mental images or sketches, and some pieces, of styles typical of their
homelands. In architecture, predominant influences were the Flemish as the ances-
tor of the well-known "saltbox" and the English medieval which was the forerunner
of the still popular "Cape Cod." In furniture, the English Jacobean and Elizabethan,
and the French Louis XIV traditions prevailed, though with some modifications. The
colonists also brought the skills and tools of Europe, so that America in these early
days became a veritable melting pot for furniture, even as it has since proved to be a
melting pot for humankind. It is difficult to consider the very early Colonial as a
definitive, identifiable style because of the profusion of periods transported across
the ocean. Only because the restrictions of materials and time were accepted as a
challenge did Colonial emerge as a style—simple, sturdy, practical, sometimes crude,
yet not unattractive, and seldom imitating materials or using them unnaturally.

The Shakers, an American religious sect deriving from the Quakers, appeared
first in New York state during the Revolutionary War and spread all over the east
establishing semi-monastic, communal villages. Religion dominated their daily life,

simplicity was the first requirement in all activities, and hard work was not only an economic but a moral obligation. Their furniture reflected this philosophy. They were excellent crafters of pieces characterized by simple, clean, severely functional lines, a complete absence of decoration, and strong, honest, and sensible wooden structures. (See Figure 1.13.) Chairs typically featured turned legs and rails, and slat backs of pine, cherry, maple, and oak.

Figure 1.13 Shaker side chair, American (ca. 1860-1900). Courtesy of The Smithsonian Institution.

The Windsor chair, named for the Windsor Castle (Figure 1.14), is reminiscent of the furniture of Europe, yet with modifications paralleling the requirements of life in the United States. Absent are intricate carvings and inlays, embellishments that detract from the article's usefulness. Instead, the crafters gradually departed from the imitated styles, working toward forms that were lighter, easier, and cheaper to produce. They invented and adapted, developing all manner of innovations from rocking chairs to clapboard siding.

Wood was the predominant construction material because of its abundant supply. It was used for everything from tools and implements to machines. Tin, pewter, some gold and silver, and iron were used as they became more plentiful. Simple tools and household articles were formed as functional, durable examples of colonial crafting.

Figure 1.14 Windsor-style rocking chair, American (ca. 1800). Courtesy of The Smithsonian Institution. Gift of Miss Gertrude D. Ritter Webster.

NINETEENTH CENTURY

First and foremost, the nineteenth century was the age of the Industrial Revolution. With new fortunes, new machinery, and new skills and processes came new opportunities for design. This was the era of great expansion and growth in the railroads and new industries, and a renewed faith in the future. The result was that more people could now afford to be more concerned with fashion and design. More money meant more furniture, more and better homes, and more time to attend to such matters of taste.

Duncan Phyfe was America's first great furniture designer. His work of the early nineteenth century was influenced by the Georgian and French designers, but he gave these a modified and distinctive touch. Most characteristic of Phyfe's detail is the lyre, which he incorporated into many structural members. The chair (Figure 1.15) displays a simple motif in the back, while the uprights slope down and gently flow toward the seat rail. The penchant for rich ornamentation in much of his work, however, identifies him with the aesthetic abuses that plagued the century.

Mention was made of this era of growth and expansion. With it came the mass

production of heretofore costly and scarce items. In 1838, William Compton patented the fancy-weaving loom. In 1844, the first practical wallpaper printing machines were imported from England. Also in 1844, Erastus B. Bigelow invented a power loom for ingrain rugs and, later, methods for loom production of tapestry carpets. New mechanical methods of producing furniture and appliances came along, and designers went wild in experimenting with fancily conceived decorations. Flowers, fruit, and other elements of nature were used as motifs, while turned spindles became integral parts of wooden furniture. With mass production methods, more and more people had access to "tasteful" carpets, tables, and furnishings.

Applied ornamentation embellished everything from chairs to machines and lamps. Designers and crafters appear to have attempted to outdo one another in wildly decorated objects of every description. The chair in Figure 1.16 exemplifies the extreme aberrations of taste that persisted at the time. This cast iron chair structure features a wild array of sinuous vines and twigs and leaves. Early sewing machine and sink bases were similarly done. (See Figure 1.17.)

This, then, was the state of design and taste in the nineteenth century. Buildings, machinery, and furniture were made to meet certain requirements and then a bit of "art" was added to meet the public demand. Victorian styling was popular, as typified by the Belter chair in Figure 1.18. John H. Belter was an American cabinet-

Figure 1.16 Spring office chair, American (ca. 1850). The cast-iron base supports springs that are attached to the seat, allowing the chair to recline and then snap back into place. The steel back is elaborately painted. Courtesy of Detroit Institute of Arts.

maker noted for his Late Empire (Victorian) furniture. Facades lifted from books of historical styles were plastered on buildings, and absurd ornaments were applied to all manner of manufactured products. But a deep dissatisfaction regarding these abuses was felt by artist and manufacturer alike—a feeling that was bound to spread and alter the situation. It is difficult to set chronologically any period of design or style because of the ebb and flow of changing thought patterns, but newer, refreshing concepts of modern design began toward the end of the nineteenth century—a logical reaction to an age of aesthetic opulence.

The events relative to art, architecture, and design since that period appear to have had a significant bearing on the look of things today. The movements, the styles, the individual pioneers, and the experimental groups of that time made significant contributions to what is commonly referred to as *contemporary design*. These were the beginnings of design for today.

Figure 1.17   Wilcox and Gibbs sewing machine, New York (ca. 1870), featuring a molded wood top and ornamental iron structure. From the collections of Henry Ford Museum & Greenfield Village.

Contemporary design is a term encompassing those artifacts that are planned and developed to meet the current needs of humans. Accordingly, it is difficult, and perhaps improper, to consider contemporary design as a definite period or style. If products are designed from the standpoint of function, materials, and appearance, it is inconceivable that they will ever be anything but contemporary. Even though modern forms had their roots in past concepts and movements, humans continually develop new techniques, materials, and ideas that directly influence the forms that the products assume.

Four movements or groups, that shaped modern products were the Arts and

Figure 1.18 Belter side chair, Victorian style, American (ca. 1850). Courtesy of The Smithsonian Institution. Bequest of Miss Annie McCardle.

Crafts Movement, Art Nouveau, de Stijl, and Bauhaus. There were others, but these serve to illustrate how artists, crafters, and designers—recognizing the significance of their work—pioneered and contributed to a direction and a rationale for product design. They are described here, followed by an examination of modern concepts of product design. The impact of modern technology on the field of design is introduced, as well as the development of the place of the designer in industry. The purpose is to develop an awareness of the evolutionary process that led to the present state of product design.

ARTS AND CRAFTS MOVEMENT

Crafters of the Industrial Revolution faced a real challenge, for they had no aesthetic rationale to guide their actions. The objects formerly made by hand were now being mass manufactured, except that often the attempt was made to have them appear to be handcrafted. This situation did not arise as a conscious effort to deceive, but because people did not yet know how to design for the machine. The traditional handmade products in which the crafter took so much pride were poorly adapted to the machine, and a complete reorientation to the machine became necessary. Designers faced the same problem.

Especially in England, there was a feeling of unhappiness about the general decline in craft that resulted from the Industrial Revolution. The critics and artists were disgusted with the cheap, machine-made imitations of the work of craft guilds, which at one time had individual meaning and nobility. It was this situation that in 1861 compelled William Morris and a group of colleagues to form a company devoted to the improvement of the art of decoration. Under the name Morris, Marshall, Faulkner and Company, Fine Art Workmen in Painting, Carving, Furniture and Metals, the firm had as its mission the production of well-designed, well-made artifacts for the home. The group dedicated itself to the reform of the arts and crafts toward a replacement of the tawdry results of mass production by inspired and meaningful handwork. By returning to pre-industrial conditions and abolishing machinery, Morris and his followers hoped for a regeneration of crafting. Though this was an unrealizable goal, it did foster a taste for the simple and the homespun. It further served to cause crafters to reconsider their tasks and arrive at some new and refreshing solutions to the problems of designing home furnishings.

Not all of the work results of the Arts and Crafts group were examples of functional simplicity. There is a sharp contrast between the patterned Morris chair (Figure 1.19) and the clean and direct Heal cupboard (Figure 1.20). In many cases, directness was the key, but decoration was rationalized in pieces that were hand-crafted, for this was honest manual work.

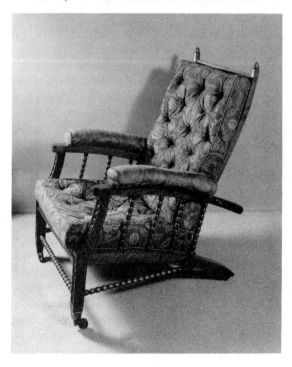

Figure 1.19 Furniture styling of the Arts and Crafts Movement. This adjustable-back chair was designed by William Morris. Courtesy of Victoria and Albert Museum, London.

Figure 1.20 Cupboard by Sir Ambrose Heal (ca. 1850). An excellent example of the simplicity and directness attained in some of the arts and crafts pieces. Courtesy of Heal's, London.

The Arts and Crafts Exhibition Society, which Morris helped to found in 1888 sparked a movement that was to have worldwide repercussions. This gave inspiration and direction to the Art Nouveau Movement in Belgium and France, and the Jugendstil and Werkbund in Germany. Effects of the exhibits were also felt in the United States. An American, Gustav Stickley, established a furniture factory in which he practiced the handicraft traditions preached by Morris. It was Stickley who was in some measure responsible for the development of the plain, sturdy, four-square furniture known as *Mission*. (See Figure 1.21.) These simple and rigid pieces of unadorned, walnut stained oak originated in California, based on Spanish Mission styles, and were popular from about 1895 to 1910. The style included tables, chairs, lamps, settees, rockers, and stools that were durable, honest, and inexpensive. They also were very producible, being comprised of rectangular parts that were easy to saw, finish, and assemble. The Mission furniture styles persist to this day, where manufacturers are producing attractive pieces lighter in both color and structure. Though the proponents of the Arts and Crafts Movement failed in their efforts to force a return to the handicraft tradition, their influence was far-reaching and lasting.

For example, the Deutsche Werkbund (German work band or group) was founded in Munich in 1907 and was influenced by the work of Morris and colleagues. Unlike the Arts and Crafts Movement, which championed the handmade artifacts, its aim was to enhance artifact design through a cooperative effort between art and industry and the handicrafts—a philosophy that rather anticipated the later work of

Figure 1.21 Mission-style chair, oak with woven rush seat (ca. 1900). Other pieces include tables, rockers, desks, and settees. Later chairs had wooden seats.

the Bauhaus. While advocating a reconsideration of the handicrafts as a means of raising product quality, the Werkbund recognized that industrial production was a fact of life that had to be met and with which to deal. The experimental involvement of this and other early groups is typified by the ingenious work of the German, Michael Thonet. In 1856, he developed a process of steam-bending rods of solid beechwood into long curved workpieces, which he used to make the famous bentwood furniture. He also experimented with producing chair parts from bundles of thick veneer strips saturated with glue and then heated and bent over prepared molds. These methods made it possible to eliminate intricate carved and fabricated joints, which led to the first mass production of standardized furniture by the Austrian firm of Gebruder Thonet, the Thonet Brothers.

ART NOUVEAU

With its beginnings at the turn of the century and flourishing in France and Belgium, Art Nouveau (New Art) was perhaps the first conscious attempt to develop a nonhistorical style in art and architecture. This approach was typified by an expressive and forceful use of freely flowing lines, often originating with natural plant forms. Those associated with this movement—such as Gaillard, Van de Velde,

Guimard, Mackintosh, and Tiffany—utilized this whiplash curve in the design of everything from furniture and glassware to posters and buildings. Charles Rennie Mackintosh was a Scottish architect and a brilliant original designer. His severe oak dining chair (Figure 1.22) characterizes his work with its elongated form and restrained ornament. Gaillard's work (Figure 1.23) was usually of symmetrical elements organized to flow freely into one another, thereby forming a graceful unit. This same motif was obtained in the glass work of American Louis C. Tiffany, who perfected a process for working glasses of various colors, while hot, into an iridescent

Figure 1.22 Art Nouveau style dining chair by Charles Rennie Mackintosh, oak and black silk (ca. 1904). Courtesy of Victoria and Albert Museum, London.

Figure 1.23 This curvilinear Art Nouveau side chair (ca. 1900), by Eugene Gaillard, features a sensitive and gentle carved structure. Note the opening at the back, which serves as a hand hold.

composition of striking beauty. He gave the name *favrile* to this special process by which the glass became a form suggestive of organic abstractions.

The sinuous forms characteristic of this style did not derive from the requirements of structure, but were instead a result of ornamental ideas imposed on a variety of materials. Because it was essentially an "antimovement" against historical style and because of a preoccupation with applied decoration, Art Nouveau never succeeded as a permanent style in design. It was an interesting interlude between the nineteenth and twentieth centuries. Its primary contribution to the matter of style was that it gave impetus to a fresher and freer approach to design, a force long stifled by tradition and sterile imitation.

DE STIJL

For the designers and architects of today, de Stijl (the Style) has special importance. This influential group of artists was formed in Holland during the First World War and included painters such as Mondrian and Van der Leck; the architect, Oud; and the very versatile Theo Van Doesburg. Many contemporary buildings and products display forms that are rooted in the work of this group.

Three fundamental principles were identifiable in the work of de Stijl—principles that were reflected in architecture, typography, painting, and furniture. In *form*, the rectangle was favored; in *composition*, occult or asymmetric balance was featured; and in *color*, red, blue, and yellow predominated. These three resulted from years of experimenting with basic aesthetic organization. Their adaptability to construction methods and the myriad new possibilities in the asymmetric balance of assembled shapes were primary reasons for the influence de Stijl had in the field of design and architecture. This group worked closely with the Bauhaus in developing what came to be known as The Machine Style in design. Their experimental design work is reflected in the armchair in Figure 1.24. The Rietveld chair is composed of basic geometric elements that result in an appearance of weightlessness by separation of the parts. Though awkward in appearance by current standards, it was a significant achievement in functional design of the de Stijl period. The key word here is

Figure 1.24 De Stijl armchair (ca. 1918), by Gerrit Rietveld, is an interesting study of intersecting linear elements, painted in black, blue, and red.

experimental, for Rietveld was attempting to reduce the function of a chair to a set of constituent rectangular elements, and organize them into usable forms.

BAUHAUS

The revolution in the shape of contemporary products and architecture was decidedly influenced by a pioneering team who experimented with new ideas, techniques, materials, and forms. These were the members of the Bauhaus (building house), an organization founded in Weimar, Germany, in 1919 by Walter Gropius. The function of the school was forthright: to create forms that symbolized the Machine Age. Trained simultaneously by both crafters and artists, the students were encouraged to exercise their imaginations boldly, yet never lose sight of their design purpose, which was the creation of products that were primarily functional yet visually correct.

One of the advanced designs that emerged from the Bauhaus was stacking furniture. They used metal tubing in the design of furniture and lamps, and textures (instead of ornamentation) to relieve the monotony of an overabundance of highly polished surfaces. They employed simple, unadorned, geometric forms, adaptable to machine production, in the design of products. An early furniture group designed by Gropius in 1913 anticipated this design approach. (See Figure 1.25.) The tubular steel side chair by Marcel Breuer was a prototype of hundreds of variations used throughout the world. Ludwig Mies van der Rohe also created bent tube chairs, but he perhaps is best known for his remarkable Barcelona chair made of bent steel bars and black leather upholstered cushions. (See Figure 1.26.) This fine chair is still being manufactured. These are classic examples of the revolutionary designs that arose from the Bauhaus experiments.

Figure 1.25 Bauhaus style furniture (ca. 1913), by Walter Gropius. Courtesy of ClassiCon America.

Figure 1.26 The elegant Barcelona lounge chair by Ludwig Mies van der Rohe. Chrome-plated steel bars and leather, 1929. The Knoll Group.

The Bauhaus Program

The impetus for these new ideas become obvious as one examines those factors that guided the work of this extraordinary group and that influenced the future industrial design profession. The Bauhaus philosophy embraced several significant precepts, which gave direction to its revolutionary and innovative curriculum. Bauhaus held that design teachers should be persons in the vanguard of their professions, enlightened practitioners, leaders not followers. The schools of design should draw their faculties from the various arts of sculpture, painting, weaving, and architecture, among others, as well as the manual arts and crafts, to achieve a complete synthesis of creative work. Students should be immersed in a creative effort combining practical manual experiences with the artistic to develop a new and modern sense of aesthetics. And finally, both students and faculty must recognize that they live in the twentieth century and that their futures lie with industry and mass production. The curriculum was designed to reflect this philosophy by providing a range of learning experiences, as shown in Figure 1.27.

The Bauhaus remained at Weimar until 1925. During this period, it developed the curriculum (Figure 1.27) for its students, assembled a very distinguished faculty (Klee, Kandinsky, Itten, Moholy-Nagy, and others), and struggled to achieve a direction for its work. Their learning program was creative and sensible, and current schools of industrial design could well emulate it. To promote their program, hostilities from all directions had to be overcome—from the art academies, from the artists, and from the public. Gradually, the people at Bauhaus succeeded in uniting creative imagination with the practical knowledge of the crafter, from which evolved a new

PRELIMINARY COURSE (6 MONTHS)

*ELEMENTARY FORM THEORY
 -CONSTRUCTION TECHNIQUES
 -MODEL BUILDING

*MATERIAL EXPERIMENTATION
 -CLAY, STONE, GLASS
 -METAL
 -WOOD
 -TEXTILES

GENERAL COURSE (3 YEARS)

*MATERIALS AND TOOLS INSTRUCTION
*STUDY OF NATURE
*ANALYSIS OF MATERIALS
*THEORY OF COLOR, SPACE, COMPOSITION
*TECHNICAL CONSTRUCTION
*GRAPHIC REPRESENTATION
*BUSINESS PRACTICES-ESTIMATING,
 CONTRACTING, ACCOUNTING

PRACTICUM (VARIABLE)

*BUILDING
 -EXPERIENCE
 -EXPERIMENTS

*DESIGN
 -BUILDING SCIENCES
 -ENGINEERING SCIENCES

Figure 1.27 The Bauhaus curriculum was organized to prepare student designers with a working knowledge of art, materials, and technique, and it influenced future design programs.

sense of functional design. The precept that it is much more difficult to design a first-rate chair than to paint a second-rate painting, and much more useful, pretty well sums up the nature of their goals.

By the time the Bauhaus moved to Dessau, Germany, in 1925, a proficient core of teachers had been trained who were at once creative artists, industrial designers, and accomplished craftspeople. This group of many nationalities approached problems with a rational simplicity, employing geometric forms and common materials with a special inventiveness. They were much more concerned with the problems of functional design than was the de Stijl group, but there is no doubt that their geometric solutions were influenced by the Dutch concepts. In like manner, the free spirit of Art Nouveau and the dedication of Werkbund permeated the work of the

Bauhaus, to the end that these designers approached problems unencumbered by the requirements of historical styles.

Gropius was the director of the Bauhaus until the rise of Nazism in Germany forced the closing of the Bauhaus in 1933. In the years following, many of its prominent designers—such as Van der Rohe, Moholy-Nagy, Breuer, Kepes, Gropius, and Albers—emigrated to the United States where they introduced and further refined the Bauhaus methods through their teaching and practice.

Further creative work was accomplished by other people in other places. French architect Charles le Corbusier designed fantastic tubular steel and leather chairs and recliners in the late 1920s. Eileen Gray, born in Ireland, did much of her design work in France during the 1920s and 1930s, and became one of the leading exponents of the revolutionary theories of design and construction. One of her most notable pieces is the elegant steel tube table in Figure 1.28. Her sensible, minimalist

Figure 1.28   Adjustable table of chrome-plated steel tubes supporting a polished plate glass top by Eileen Gray, 1927. Courtesy of ClassiCon America.

approach resulted in this innovative structure whose top is adjustable to various heights. The dining furniture by Eliel Saarinen (father of Eero) shows a Bauhaus influence and a good use of pattern. (See Figure 1.29). The desk and chair (Figure 1.30) by Raymond Hood is pure machine structure, evincing a sense of sturdiness and durability. Raymond Loewy's office suite reflects the streamlining approach that became so popular, with its sweeping curves of polished steel tubes. (See Figure 1.31.) The famous architect, Frank Lloyd Wright, designed not only his sensitive naturalistic Prairie Style homes but also innovative furniture such as the desk and chair in Figure 1.32. The several models of the chair (Figure 1.33) featured three-point supports and adjustable seats to move backward and forward, up and down, with no tools required. The back was completely reversible and self-adjusting to the user. The caster models, however, were somewhat tricky to use.

Why was the tube construction technique such a dramatic advancement in furniture making, as evidenced by the work of the aforementioned pioneers? First of all, it made possible a far more efficient method of joining components. The parts could be welded together to form clean and strong joints, with no bulky structures. Second, the tubes could be finished with chrome plating, paint, enamel or other coloring systems to provide an attractive and durable protection. Finally, the results of this system were lighter in appearance and far stronger than wooden members of

Figure 1.29 Dining room furniture by Eliel Saarinen, showing Bauhaus influence, 1929. All rights reserved, The Metropolitan Museum of Art.

Figure 1.30   This business executive office, by Raymond H. Hood, 1929, is a good example of Machine Age design.

Figure 1.31 Industrial design office, by Raymond Loewy and Lee Simpson, 1934. Note the glossy appearance of the curved chromed metal tube and glass structure which illustrate the stream-lining approach. All rights reserved, The Metropolitan Museum of Art.

Figure 1.32   Desk and chair by Frank Lloyd Wright, enamelled and brass-plated steel, and walnut (ca. 1936-1939). Courtesy of Steelcase.

similar size. The nature of wood required that joints be of substantial size to give necessary strength, as can be seen in the Egyptian joinery methods described earlier. (Refer to Figure 1.3.) This discussion provides a good example of how technological advances affect the way products are made and how they look.

It also should be noted that during the period between World War I and World

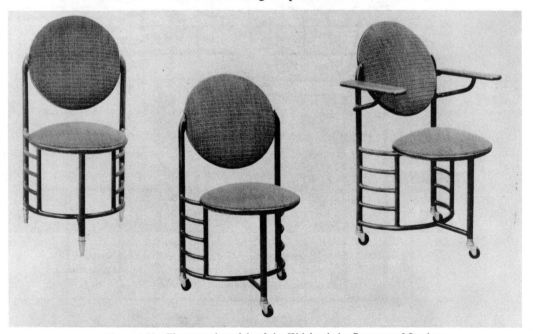

Figure 1.33   The several models of the Wright chair. Courtesy of Steelcase.

War II, the machine had a powerful impact on culture and design in both America and Europe. Artists and designers struggled to understand and accept the machine-driven world and to experiment with the direction it might give to their work. The precise geometric shapes assumed by the honest and seemingly undesigned tools and machines became for many a matter of conscious aesthetic preference. A tool was designed to perform a prescribed function and incorporated the proper materials to assure this function without any extraneous and unnecessary embellishments. The various styles and artistic movements such as de Stijl, Bauhaus, impressionism, expressionism, cubism, and Art Deco influenced artifact design, but the machine also was a driving force. Indeed, to some this was the era of Machine Art or the Machine Age, and it still gives direction to much of the design of the present or contemporary period. The movement toward a concept of modern design truly depended on new materials and techniques for working them, on a more careful consideration of the function of an artifact, and on new approaches to aesthetics.

Architecture was and is influenced by these new concerns. Wright applied his efforts to the design of a few very ordinary structures, such as the gasoline station in Figure 1.34. This interesting structure, which is located in a pleasant valley in the city

Figure 1.34 This unique Wright gasoline station is located in Cloquet, Minnesota. Courtesy of Best Oil Company.

of Cloquet, Minnesota, is both striking and functional. The famous chapel on the campus of the Massachusetts Institute of Technology is an extraordinary piece of impressionism designed by Eero Saarinen. (See Figure 1.35). The problem was to create a chapel to meet the spiritual needs of all who may choose to use it, whether it be Catholic, Protestant, Muslim, Jew, agnostic, or atheist. This goal was achieved by providing a quiet, windowless, cylindrical structure whose subdued lighting reflected from a surrounding moat causes the eyes to focus on a marble pedestal frontispiece. This contrast is further accentuated by a lovely metal screen sculpture by Harry Bertoia that hangs from ceiling to floor and heightens the impact of the lantern light above the marble pedestal or altar. (See Figure 1.36.) The effect is dramatic. The viewer steps into the quiet, dimmed chapel, the eyes take in the soft moat light and move to the marble slab, and the combined altar lights and metal sculpture cause

Figure 1.35   The auditorium and chapel (ca. 1955), by Eero Saarinen, is located on the campus of The Massachusetts Institute of Technology. The chapel appears on the right. Courtesy of The Massachusetts Institute of Technology Historical Collections.

them and the mind to raise upward. No crosses or common religious symbols are present or necessary. Saarinen used visual impact in an extraordinary way.

Figure 1.36 The MIT chapel interior, showing the pedestal and the hanging sculpture. Courtesy of The Massachusetts Institute of Technology Historical Collections.

PRINCIPLES OF MODERN DESIGN

Modern design, contemporary design, or design for today—whichever term one uses—developed steadily over the past century. This maturation was forged by the innovative work of the aforementioned pioneers to the matter of producing things for human consumption. Out of this period of evolution have emerged certain precepts that have endured and directed the course of product design. There follows a generally accepted set of guiding principles:

1. Product materials should be used honestly, taking full advantage of their unique attributes and never making them seem to be what they are not.
2. Products should be functional, useful, and expressive of their intended purposes.
3. Products should incorporate the latest advances in material science, computer applications, and process technology, and extract the most from those that are traditional and familiar.
4. Products should be derived from new combinations of form, color, and texture to enhance both their functions and appearances.
5. Products should be expressions of the prevailing aesthetic of our age.
6. Products should express the techniques used to make them, not disguising machine production as simulated handcrafting or camouflaging integral elements such as fasteners.
7. Products should be devoid of unnecessary or extraneous decoration or embellishment.

Many product designs were reflections of these principles. Following the Second World War, Scandinavian furniture burst onto the international market and captured the popular imagination for several reasons. Most important, it rejected the austerity of some functionalistic pieces, substituting instead light, gentle forms, and naturally patterned surfaces. This style took wood as its primary medium and concentrated on clean, well-structured joints and natural finishes. The very pleasant side chair by the Danish crafter Hans Wegner epitomizes the Scandinavian approach. (See Figure 1.37.) Arne Jacobsen created a delightfully minimalist chair of molded beech plywood and three bent steel legs. Borge Morgensen devised a simple, country-inspired wooden chair with a low back, turned dowel front legs and a woven rush seat. Alvar Alto, the Finn, experimented with structures of molded and bent birch plywood. The connecting feature of all such pieces was the recognition of the requirements of Scandinavian living, with the smaller rooms and the need for light, open styles. And further, the style's success was influenced by the Nordic reputation for craft technique, quality, and natural aesthetics. The sensitive works of these designers fostered a tradition that endures and is emulated today.

The classic chair by Charles and Ray Eames combines wood and metal in a technologically innovative and aesthetically honest way that reveals mechanical connections instead of hiding them. The rubber shock mounts that connect the bent chrome-plated steel rods to the walnut plywood are plainly visible, and are as integral to the design as the elegant shapes of the seat, back, and frame. (See Figure 1.38.)

Figure 1.37 The classic armchair by Hans J. Wegner, 1949. Oak and cane, 30″ high, 24⅝″ wide, 21¼″ deep. Courtesy of The Museum of Modern Art, New York. Gift of Georg Jensen, Inc.

Figure 1.38 Dining chair by Charles and Ray Eames, molded walnut plywood and plated metal rod, 1946. Courtesy of Herman Miller Inc.

The chairs in George Nelson's swaged leg group have molded plastic seats and backrests, leading to a reputation as the first plastic chair with a movable back attached with innovative plastic shock mounts. (See Figure 1.39.) The legs of the pieces are thick and straight at the top, and are then formed as swaged or tapered

Figure 1.39 Molded plastic chair with movable back and swaged steel tubing legs, by George Nelson (ca. 1958). Courtesy of Herman Miller Inc.

tubes that gently curve outward and downward at the bottom. A flexible back model is shown in Figure 1.40.

The Yugoslavian designer Dragomir Ivecevic regards design as functional sculpture, and so he designed his Proper chair to be viewed from every angle. Therefore, the viewer will not find an exposed screw, bolt or other fastener, even if

Figure 1.40 Molded plastic chair with a flexible back, of the Nelson swaged leg group. Courtesy of Herman Miller Inc.

the chair is turned upside down. (See Figure 1.41.) The arms and legs are made of cold-rolled oval steel tubing, and feature contoured corner connectors of die cast aluminum. The outer shell is polystyrene, and the inner construction includes molded, glass-reinforced polypropylene covered with a thin sheet of urethane foam. The Proper chair is true to its name—elegant enough for a formal dining room, yet with durable materials and a stackable design that make it a very practical side chair for offices, and reception and conference areas. With its sleek, sophisticated lines and contoured comfort, Ivecevic's design is a perfect marriage of form and function.

Figure 1.41 The Proper chair by Dragomir Ivecevic, steel tubing and plastic shell, 1986. Courtesy of Herman Miller Inc.

The Balerafon chair is a dynamic structure of curved thrusting arms and front legs of stainless steel tubing, with a linear undercarriage tied in to provide a sturdy support. The shape of the upholstered seat and back logically follows the contour of the metal frame, presenting a feeling of unity and grace. (See Figure 1.42.) It is a good example of applying engineering design principles to industrial design problems.

Present-day furniture design, especially chairs, can probably be best described as creative activities involving materials, technology, and a somewhat eclectic approach to experimental aesthetics. Many of the contemporary chair designs in Fig-

Figure 1.42 Balerafon chair, stainless steel frame and upholstered seating structure, 1991. Courtesy of Brueton Industries Inc. Designed by Glenn Polinsky.

ure 1.43 reveal truly innovative uses of materials and form. Their structures display a working knowledge of manufacturing processes, for they can be economically produced. This is not so with other designs. Many are wild attempts at creating unique forms, quite unsittable, structurally unsound, unattractive, and almost impossible to produce. The piece in Figure 1.44 is a case in point. It is visually unappealing, rather uncomfortable, and difficult to manufacture. Designs such as these should be avoided, but not necessarily rejected. They should be studied with the aim of extracting from them any new ideas regarding structure and material innovations. Recall the sensitive form of the Greek klismos in Figure 1.5. It is beautiful, reasonably comfortable, but ultimately weak in structure. The de Stijl armchair in Figure 1.24 displays an excellent utilization of wood joinery leading to sturdiness, but is uncomfortable and visually awkward. However, each contributes something to the progression of ideas regarding chair design. Don't sit in them—look at them and appreciate what can be learned about style and structure.

Figure 1.43 Examples of contemporary chair designs that display innovations in form and structure, good uses of materials. Note the two chairs of sturdy wood construction: one utilizes a combination of wood and metal rod, and one is made of bent plywood or steelsheet.

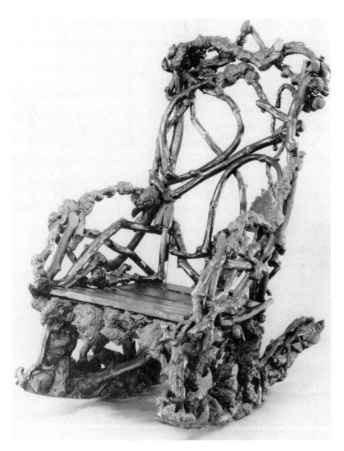

Figure 1.44 Classic early rustic-style rocking chair made of grapevines, bark, and oak (ca. 1865–1890). An experiment gone wrong; not very inviting. From the Collections of Henry Ford Museum & Greenfield Village.

QUESTIONS AND ACTIVITIES

1. Collect some illustrations of artifacts of the Greek Golden Age and compare them to the Greek klismos. Do each of these items reflect the prevailing aesthetic of idealism and humanism?

2. Prepare a report on Gothic architecture to describe the engineering concepts relative to the flying buttress, the pointed arch, and the vault. How did these three elements contribute to the appearance of the Gothic cathedral?

3. Describe the essential formal and structural differences between Gothic and Renaissance chairs.

4. Prepare sketches to compare the backs of Chippendale, Hepplewhite, Adams brothers, and Sheraton chairs.

5. How would you describe the features of the furniture, artifacts, and architecture of the American Colonial Period?

6. A chair and a sewing machine are shown in this chapter as examples of elaborate ornamental cast-iron structures. Collect some illustrations of other such pieces of this style.

7. Describe the visual differences between the works of the Art Nouveau and the de Stijl groups. Were the paintings and sculptures of these groups consonant with their styles? Show some examples.

8. It is held by many that the Bauhaus movement laid the groundwork for what was to become industrial design. What evidences can you find to support this claim?

9. Select a product, such as the telephone or the radio, and prepare sketches to describe how their appearances have changed from their inception to the present. What role has technology had in influencing these changes?

10. Prepare a report on the Machine Age design in the United States.

11. How would you describe the Scandinavian style of furniture?

12. How would you define the term *Contemporary Design*?

CHAPTER 2

Design Theory

As suggested in Chapter 1, humankind always has been involved with fashioning artifacts to improve life and to ensure its continued progression. Remarkable as they were, these early works generally lacked an analytical dimension because to the primitive, design was essentially a matter of trial and error.

At some further period in history, humans began to direct this process of designing as they observed and experienced the requirements of a tool. For example, an implement to fell a tree had to have a sharpened edge in order to slice through the wood fibers. Field use led to the addition of a handle to this blow-of-the-fist stone ax because its superior performance was experienced, although its physics probably was not known. Thus, people continued for eons inventing baskets, pottery, wheels, and bow drills, finally discovering fire and metals and using this new technology in the further improvement of their artifacts.

It would appear that the first notable effort at systematizing the design process came from Aristotle and the important theories he established in his writings on metaphysics. (See Chapter 11: Adler, 1978.) Aristotle, the philosopher of Periclean Athens, was searching for answers to the fundamental questions of the causes and the principles of the known universe. He rejected the prevailing theories of leading thinkers because none was sufficiently analytical and none came to grips with the question of the existence of things everywhere. He theorized that for every artifact there were four reasons or causes that gave rise to its existence. Take, for example, a wine jug made of clay. The exemplification of these causes would run as follows:

1. The *material cause* refers to the substance of which the object is made. One reason for the jug's existence is the clay from which it is fashioned. It answers the question, What is it going to be made of?
2. The *formal cause* refers to the shape or configuration assumed by the object material. This particular container is distinguished from any other, or from a shapeless lump of clay, because of its form or shape. The question answered is, What is it that is being made?
3. The *efficient cause* refers to the maker or prime mover whose skills caused the object to be produced. The efficient cause of the jug is the potter who transformed a mental image of a jug into a physical reality—in this instance, one made of clay. This answers the question, Who made it and how?
4. Finally, and most important perhaps, is the *final cause*, the purpose, end, or function to which the object was brought into existence, answering the question, What is it being made for?

The final cause of the wine jug was the potter's desire to create a container to hold wine. The functional requirement of the end product is significant because it determines the form the jug will take in order to contain wine. Furthermore, there is a significant relationship between material and form because clay can be easily fashioned in a manner to strongly suggest a containment form to meet a specific function. Clay could be worked in this way; stone could not.

Since Aristotle first propounded this theory some 2,500 years ago, there have been many occasions for designers to develop and refine theories and give them currency. Obviously, the kinds of products humans came to know and the methods by which they were made have been changing constantly, but the basic theory underlying their existence has changed little. The scheme developed by Robert Scott relates to Aristotle's. (See Chapter 11: Scott, 1951.)

Scott also used the term *cause*, except that his list included the *first cause* (need or purpose); the *formal cause* (shape or contour); the *material cause* (substance); and the *technical cause* (methods employed to shape the substance)—quite similar to Aristotle, but different. Other theorists have offered schemes for examining product requirements, as well as providing methods that would guide their development. This appears to be the significant feature of all these systems, as they contrast with Aristotle's. Whereas his was concerned primarily with providing a rational explanation for objective existence, theorists subsequently have been concerned with perfecting schemes that would direct human effort in designing things efficiently in light of current technology. Designers may use different terminologies, but their intent is clearly the same—namely, to provide a set of considerations to guide the systematic process of creating new products and artifacts.

DESIGN DEFINED

One of the most difficult tasks in any treatment of design theory is to arrive at a clear, workable, and acceptable definition of the term. A review of past and current literature reveals a plethora of suggested descriptions:

> Design is the quest for simplicity and order.
>
> Explicit in the term *design* are the concepts of order and organization.
>
> Design is the process of inventing artifacts that display a new physical order, organization, and form in response to function.
>
> Design is a conscious and intuitive effort directed toward the ordering of the functional, material, and visual requirements of a problem.
>
> Design is a statement of order and organization. Its goal is unity. It must hold together. It is an expression of the human ubiquitous quest for order.
>
> Design implies intention, meaning, and purpose.
>
> The planning and patterning of any act toward a desired, foreseeable end constitutes the design process.
>
> Designing is creative problem solving.

The common thread that connects these definitions is the assumption and expectation of order and organization. Further, these thoughts suggest that the predisposing factor in all design is an expressed human need for some product, and that the effort results in something useful. Consequently, the working definition to be employed in this book is that *design is the conscious, human process of planning physical things that display a new form in response to some predetermined need*. Further, this activity implies a creative, purposeful, systematic, innovative, and analytical approach to a problem—key events that distinguish serious design from mere idle speculation.

GUIDING THE DESIGN PROCESS

In the act of designing, primary consideration must be given to the needs of the user, involving the special functional, material, and visual requirements of the problem. These human factor needs translate as a series of subsets pertaining to purpose or use, physical substance, and appearance. Human factors are important in designing and are treated in Chapter 5. The obvious aim is to create a product that not only works and is durable but also looks nice. Good design demands this attention to the several aspects of the problem at hand.

As an example, see Figure 2.1. Industrial casters are rolling supports for furniture or equipment that is frequently moved from place to place. Many are rather dull visually, even though they may work well. However, the wheel material so often is wood, brittle plastic, or hard rubber, which wears and flattens in time so that the caster is inoperable. Others are so slick and slippery that they slide aimlessly rather than rolling smoothly and providing convenient movement without scratching the floor. The example shown is a chrome-plated, twin-wheel model that offers increased mobility on both hard and carpeted floors in medical facilities, for instance. The soft plastic tires move over moderate obstacles and into elevators quietly and with ease. The unit has a safety brake and also features a ball-bearing support to permit easy and sure swiveling to change travel direction. The gently sloping hood protects the axle from hair, string, and other debris.

Figure 2.1 This elegant and functional industrial caster displays a sensitive organization of visual elements. Note the nice contrast of material tones and the smooth tire and hub interface. Courtesy of Jilson Casters, Incorporated.

This caster also is a classic formal display of unity and variety, where the smooth transition from one visual element to another is quite remarkable. For example, the eye moves very gently from the threaded stem to the hooded fender and then to the wheel. There is no rough transition between the hub of this wheel and the firm, skid-proof plastic that forms the tire. The elements are well balanced and form a visually satisfactory whole by appearing that they do, in fact, belong together. Functionally and aesthetically, this is an excellent example of integrity in modern product design. It looks like it belongs in a hospital. It also serves to introduce the ramifications of the three design requirements, as developed more fully in the following paragraphs.

Functional Requirements

A product must fit the purpose or need for which it is intended. In other words, the well-designed article works as it is supposed to work. It functions. It is usable. (The factor of usability is treated more fully in Chapter 6.) The appliance that is difficult and awkward to use, the tool that fails to perform as intended, and the chair that neither adequately nor comfortably supports the human frame, are examples of poor design.

One of the better examples of a functional product is the common, homely little punch-type can opener seen in Figure 2.2. The tool was developed to cut a safe opening (one that would not sliver the lips) in the top of the flat top steel drink can (before the days of pull-tabs), and to be effective, reliable, and inexpensive. Manufactured by a simple two-step shearing and bending operation, it will open cans forever, even when it is rusted and worn. Tool convenience is extended by including a bottle-cap lifter opposite from the cutting end.

Figure 2.2   The familiar flat-top can opener is the penultimate utilitarian tool— inexpensive, durable, easy to use, easy to make, and very reliable.

Considerable attention must given to this matter of functionality. When planning a desk chair, for instance, the designer must either consult a reference containing anthropometric data or otherwise secure the bodily dimensions of those for whom the chair is intended, and analyze its purpose. For the chair to be functionally correct, it must fit the human frame and permit proper use. The gentle forms of the side chair in Figure 2.3, sensitively created by a clever combination of bent wood and steel rod, support the human frame comfortably and safely. Other chairs have other functions and must be designed for their special uses, such as those for typists, dentists, drafters, and milkers. Those readers who have ever hand-milked a cow in a rural barn will appreciate the efficiency of the three-legged milking stool. Barn floors are uneven and often dung-encrusted, and such a tri-structure stool will always be planted securely without wobbling. This barn stool always works, and exemplifies an approach to structural stability worthy of consideration in all design problems.

Tools must be usable. A good example is the hacksaw, shown in Figure 2.4. This saw has a number of features that make it work better than others. For one, the blade is positioned so that cuts can be made flush to a surface—a boon to a plumber who has to saw a rusty pipe that lies flat against a wall. Provision also is made for storing extra blades in the tubular handle. Note that the front grip and handle provide a safe, comfortable, and convenient shape for holding the instrument in use. These attri-

Figure 2.3   Bent metal and bent wood are thoughtfully combined to create this pleasant side chair. Courtesy of Casala-Werke, Germany.

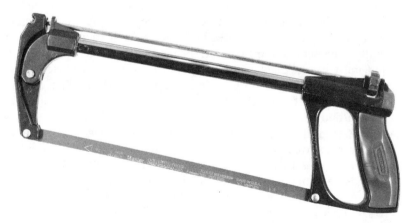

Figure 2.4   Usability, safety, and appearance are evident in this well-designed hacksaw.

butes, coupled with an obvious attention to appearance, make it an ideal metalworking tool.

   The clever L-clamp lends itself to efficient and positive workpiece holding. (See Figure 2.5.) An easy push on the tough, forged clamping arm forces the jaws into contact with the work, friction locks them in place, and a few turns on the screw completes the holding operation. It is interesting to note the good design of many hand tools. Through hundreds of years of use and modification, these vernacular artifacts have been perfected for convenient and effective operation. Their shortcomings generally lie with an imprudent use of materials, and perhaps faulty fabrication, all to the end of reducing costs. Someone once wrote, appropriately, that the bitterness of poor design remains long after the sweetness of cheapness has passed.

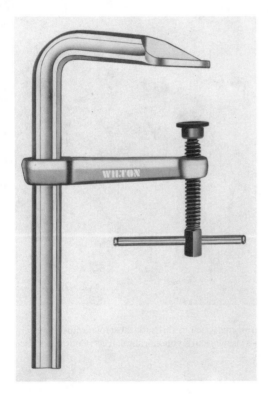

Figure 2.5 This rapid-acting clamp is durable, reliable, effective, and easy to use. Courtesy of Wilton Corporation.

   Medical devices (another class of tools) also are planned with considerable attention to functional needs. The elegantly simple stapler (Figure 2.6) is in general use to join flaps of human skin following surgery. It is easy to hold and direct the staple tip to the incision area, and a gentle squeeze of the handle cleverly bends the staple to effect the fastening. A clearly visible staple counter constantly alerts the surgeon to the number of fasteners remaining in the instrument. Form has followed function to the creation of a usable and attractive tool. The sad part is that it is a throw-away item for hygienic reasons, but perhaps for good cause.

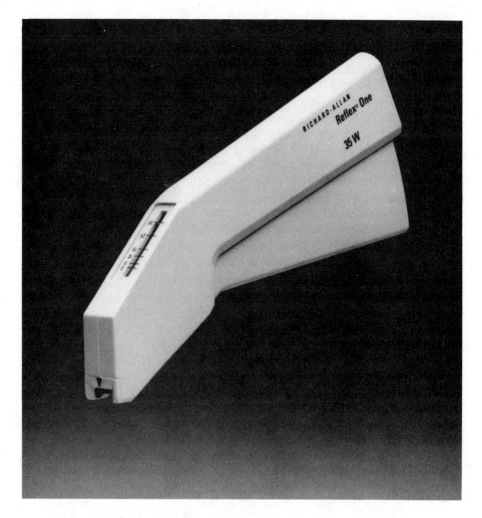

Figure 2.6 Medical instruments, such as this surgical stapler, must be essentially functional tools. This one is visually and ergonomically correct. Courtesy of Richard-Allan Medical Industries.

Designers can be guilty of allowing aesthetics to interfere with function. The two nutcrackers in Figure 2.7 are a case in point. One is an elegant piece of metal sculpture, formally sensitive but lacking in the ability to crack nuts. It does not work very well. The other common shape is designed primarily to shell nuts, and is a very efficient tool. The message is to let function guide the design effort. The term, *form follows function*, has become a verbal icon for designers over the years, and while it may be creatively stifling if embraced too fervently, it does have some merit.

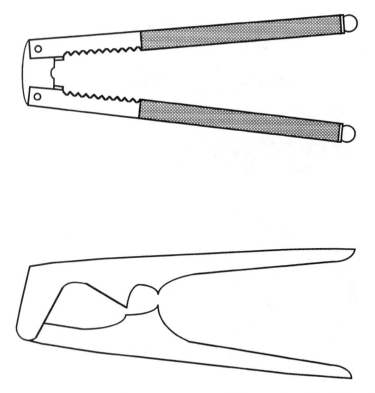

Figure 2.7   The nutcracker at the bottom is gracefully suggestive of function but is nonperforming. The one at the top is plain and uninspiring, but extremely functional.

However, there lies a certain danger in oversimplifying or overemphasizing the role of function in design. As a response to a given need, an article may be perfectly adequate from the functional standpoint, but fail to be appealing to the senses. For example, a can or a drinking glass may work perfectly well as a container for flowers, but neither is hardly suitable for the dining room table. A fine ceramic vase or a brass holder (Figure 2.8) would be far more visually pleasing. This is simply to say that functional sufficiency is no guarantee of good or appropriate design.

In any number of useful products, function does not dictate form, it merely indicates form in a general, logical way. The designer must select from a variety of possible solutions, each of which may be functionally and materially correct, that form that is aesthetically most satisfactory. That person does not unnecessarily embellish, elaborate, imitate, or enrich, but instead refines, simplifies, and perfects. The key to the wise and proper employment of this element of function is in recognizing that the question, What is it to be used for? must be coupled with, Where and how is it to be used? Significantly, this leads directly back to the necessity of clearly defining the design problem.

Figure 2.8 This deep-etched brass vase is a very appropriate flower holder, visually as well as functionally. Design by the author.

Material Requirements

The project or product should reflect a simple, direct, and practical use of the substance of which it is made. The designer should achieve maximum benefit from a minimum amount of judiciously selected materials. If this element of wise utilization is present, the structure of the product will be sound and will be as strong as necessary without any waste of materials or excess bulk.

As will be discussed in Chapter 3, the creators of products must select from woods, metals, plastics, and ceramics in order to cause their artifacts to assume some shape and to work as they should. Wood is warm, pleasing, soft, insulating, easily worked, and combustible, but not as strong as most metals and plastics. Metal is durable, fire proof, tough, harder to work than wood, and can be fabricated by a broad range of processes, but it rusts and corrodes and is noisy. Plastic is easily formed, insulating, subject to temperature limitations, quiet, break resistant, and can be compounded to meet almost any product requirement. Ceramic materials are noisy, brittle, good insulators, easy to form while in a plastic state, and can withstand high temperatures, but difficult to form or modify after they have been fired or set.

A shortcoming often leveled at industrial designers is that they are insufficiently schooled in the science of materials and their processes. This can be remedied by the curriculum directors of design schools so that design education can be complete.

Barring this, designers should include materials technologists on their design teams and consult them frequently. The bedside table shown in Figure 2.9 is a type commonly found in motels and hotels. This particular piece is an example of the work of a poorly prepared vernacular designer who lacked a knowledge of materials. It is misconstructed of a poor-quality particle board covered with an inexpensive, thin, simulated-wood, plastic-coated paper never intended to withstand the normal wear and abuse it will experience at the hands of room cleaners and guests. Note that the corner has eroded through repeated knocks and bumps and is unsightly. A competent, well-educated industrial designer would not make such a mistake. A more durable construction material is solid wood or particle board covered with a tough, high-quality plastic laminate sheet, preferably a solid color and not a simulated grain pattern. Such cost-cutting measures are uneconomical in the long run, for such tables soon must be discarded and replaced at a higher cost.

Figure 2.9 A designer more knowledgeable in material science would not have miscast cheap wood fiberboard, covered with flimsy wood-grain plastic paper, as the material of choice for commercial furniture such as this side table.

Questions of integrity and respect for materials might well be raised when a softwood is stained to imitate hardwood, or when plastic is finished to simulate wood or marble. In like manner, applying unnecessary decorations or enrichment to alter the appearance of a material is in most cases questionable. Each of the many materials, both new and old, has some inherent qualities that should be exploited to the fullest. Only through a mature attitude toward them can one discover what they can be properly made to do. Metals can be bent, folded, and formed, as can plywood, plastic, and some other materials. Some can be soldered, welded, riveted, and glued,

while others cannot. A working knowledge of material limitations manifests itself in
many ways. Because of the softness of pewter, surface ornamentation is kept at a
minimum or disregarded completely. Because a plastic becomes pliable when heated,
the compression, vacuum, or manual forming methods are far preferable to cutting,
squaring, and cementing methods. Materials should not be subjected to wild experimentation; they should be studied and tried, and they should be used for their own
intrinsic qualities.

A designer beginning to visualize the form of an artifact also should understand
the relationships between material, process, and structure. A common lever-action
corkscrew used to remove the corks from wine bottles is a good illustration. (See
Figure 2.10.) The device as shown on the left reveals the front and side views of the
die-cast zinc remover housing. Note that the legs are properly straight, heavy, and
durable. Shown in the inset is a lesser model, where the designer attempted to make
the legs more visually pleasing by tapering them to reduce the mass. This aesthetic

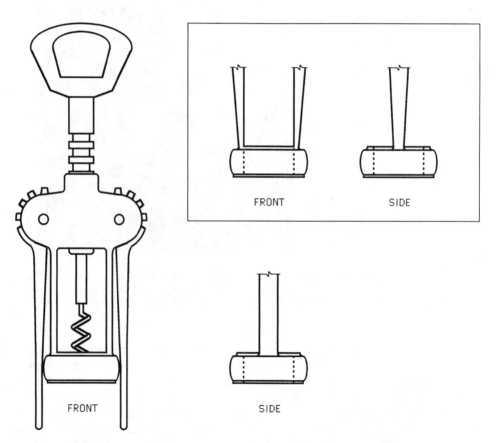

Figure 2.10   The corkscrew above is much stronger than the model shown in the
inset, where the strength of the die-cast section was sacrificed for style.

venture also reduced the strength of the legs at the point of highest stress, as can be seen, with the result that the tool broke at that point. Product failure was caused by a designer not knowing that narrow die-cast parts can be very brittle in section, and that the product form must account for this. The visualized form must be right for the purpose of the object and must grow out of the qualities of the material and its process. Form and material and process always have this interdependence—a factor that is further described in Chapter 4. When in doubt, consult with a manufacturing engineer.

This interdependence also is illustrated by examining the article in Figure 2.11, a container for watering plants. The light, easily formed plastic material ends the problem of breakage and rusting. The handle is functional in that it fits the hand comfortably, makes carrying and pouring easier by offering alternative holding positions, and is easy to fill. This graceful container is a great improvement over its traditional metal counterpart.

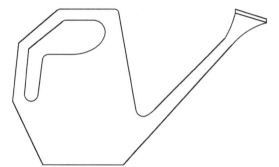

Figure 2.11 Easily formed and durable plastic is the proper material for this convenient and attractive watering can.

A piece of heavy equipment has numerous special materials requirements. The plan for a dump box and its mechanisms require the particular qualifications of a design engineer. The model shown in Figure 2.12 demands that tough, durable, reliable steels be used in its construction. The box must bear heavy loads, is subjected to considerable field abuse, and must withstand high stresses. The hydraulic hoisting mechanism must be capable of lifting the box and its load reliably and safely. Knowing the peculiar functions of a product and applying the necessary analyses will lead to some very logical materials choices.

Too many materials or contrasting forms employed in one object can result in a "busy" appearance. In such situations, the wide variety in form and material seriously impairs the unified appearance that must be present in objects. In short, the several elements of an article should appear to belong together, and not as exemplified in Figure 2.13. The somewhat bizarre lamp has too many conflicting formal, textural, and material elements for visual comfort. There is too much going on. The key point is to provide enough contrast to relieve monotony, but not enough to disturb unity.

Figure 2.12 The structure and mechanisms for this heavy duty dump box require tough, durable steels. Material matches function in this engineering design example. Courtesy of Peterbilt Motors Co.

Figure 2.13 Too many contrasting materials and forms can be visually disturbing, as is apparent in this odd lamp.

Visual Requirements

A project or product should have a pleasing appearance to the beholder. Simply stated, this requirement translates as a concern for those factors that figure into the visual correctness of a thing. It deals with aesthetics, or the philosophic study dealing with the nature of the beautiful and with judgments concerning beauty, with showing good taste, and with a sensitivity to appearance. Humans respond more positively to pleasing rather than ugly objects. This is probably the most difficult of any of the design requirements, for what one person views as beautiful may be visually unacceptable to another. Some individuals like buildings that display a clean, crisp, rectangular organization of forms, while others prefer gently curving structures quite similar to what a sculptor would do. Taste, or a person's likes and dislikes, cannot be dictated. There is no mathematically positive right or wrong, inasmuch as human aesthetic responses emerge through nature, nurture, and experience. This elusive quality of taste may be expressed as: "For the sort of people who like this sort of thing, this is just the sort of thing that that sort of people like." While the formal aspect of an object is elusive, the designer must reckon with it in order to achieve product success. It should be pointed out here that engineering designers generally are less concerned with appearance than are industrial designers. This is not a criticism, but an observation. A technical person designing an automotive disc brake does not really care what it looks like. Appearance in this instance is not part of the product specifications. It must only work, and work perfectly.

Most design or art textbooks have a chapter or two devoted to *the elements and principles of design*. In six words, this sums up the context of the term *visual requirements*. It has to do with proper balance, correct proportion, compatible colors and textures, and structure. It means that through practice and the cultivation of a discriminating eye, the designer will be able to recognize what is worthwhile and what is not, and apply this attribute to creative tasks. A person's ability to use this knowledge rests with an understanding of the visual symbols with which one works in the process of designing.

Humans are, by nature, creatures of organization, and as such find existence and progress difficult amidst chaos. Forward movement requires the measured step, the action based on logic and decision. People are creatures of order because they are surrounded by it. There is order in the measure of time that gives control to the day, to the rhythm of the seasons, and to work and play, all of which contribute to the human penchant for propriety and organization. This factor is both a plague and a source of immeasurable enjoyment.

Although there are few rigid rules regarding how one should respond to this matter of order, people still feel compelled to attend to it. It is no accident, therefore, that one's designs also reflect this feeling for order. The several definitions of design presented earlier generally reflect this concern for order. Each physical unit in the universe is comprised of lines and shapes, forms, colors, and textures organized in such a manner as to create functional and pleasing objects. Without some measure of organization or order, a contemporary living room would be nothing more than a jumbled tangle of shapes and forms. Instead, because of some indefinable law of order, design features can be arranged in such a manner as to build a visually pleasing totality. It is an examination of these arrangements and principles—*the visual organization theory*—that follows.

Design elements.   Designers communicate ideas by manipulating visual symbols in much the same fashion as they use letters in expressing a written language. This language of vision has four basic symbols that are used to graphically represent design ideas: lines, planes, forms, and surface qualities. These are quite properly called *elements of design* because they are the irreducible components of the more complex two- and three-dimensional design objects, and as such are the fundamental building blocks of all structures.

A *line* is the path generated by a point moving through space. Because it is unidimensional, direction is its most significant property. A line is an expression of continuity between two points. Lines carefully drawn and controlled are the connecting links between a mental image and a resultant physical shape. When properly spaced and joined, they establish surfaces and determine form. (See Figure 2.14.) Note that the several recurring lines give definition to the attractive and functional interchangeable garden rake and hoe parts. All outlines, contours, shapes, openings, appointments, and plane intersections are established by lines.

Whether lines are straight or curved, their role in directing attention and determining form is significant. Such lines serve to suggest emotional feelings. Vertical lines are strong, dignified, and aggressive. Horizontal lines are passive. Angular lines show motion. Each curved line expresses beauty in its grace and elegance, though the feeling it creates is not necessarily one of strength. Curved lines arcing outward produce a sense of fullness and charm, whereas those that curve inward create a feeling of poverty and emptiness. However, in combination, such curves can reflect the dignity of the candlestick in Figure 2.15, or of a lamp or a soaring arch. Straight lines, in turn, represent or suggest strength, vitality, stability, and security. Direction also plays a part here. A vertical line is noble and in balance, as suggested in the towering strength of a tall tree. Diagonals convey movement and, when used alone and unsupported, they convey a sense of falling. This becomes a challenge to one's sense of gravity, and from this emerges a feeling of lightening-like, or broken lines. In a sense, these lines lie in opposition to one another, and they tend to create discord or a harsh effect, as with the bandsaw blades in Figure 2.16. However, the dissonant lines of the blades are quite functional, for they provide sharp cutting edges that can tear through the fibers of a piece of pine or oak. Such an academic discussion of linework could continue, but the point has been made. Lines serve to convey feelings and to determine basic shapes, and an appreciation of this will contribute much to design success.

The second class of design elements are *shapes, planes*, or *surfaces*, which are created when lines are joined, closed, or comingled. For example, the forests of wooden telephone poles add little to the appearance of urban America, although many of the rural electric power transmission structures are aesthetic marvels of engineering design. Those ungainly posts with timber cross-members, generally

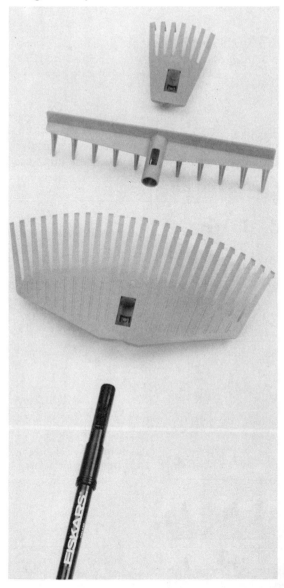

Figure 2.14 Repeated forms were used to create the rake structure, resulting in a pleasing rhythmic pattern. The parts are interchangeable. Courtesy of Design Forum Finland.

tilted and askew, are uniquely ugly—a pollution upon our cities—better they should be buried, and many are, thankfully. Barring that, those created some years ago by Henry Dreyfuss and Associates for the Edison Electric Institute demonstrate that above-ground utility poles can be as aesthetically pleasing as they are structurally sound. Note the simple yet elegant line creations that define the shapes of the structures in Figure 2.17. Their construction materials include wood, metal, plastic and concrete to good advantage. Note that these "sculptures" are quite adequately line-described visually in two dimensions, with their vertical members, arced trusses, circles, diagonals, and bold insulators.

Figure 2.15 This elegant teak candle-stick is comprised of graceful curved lines. Design by the author.

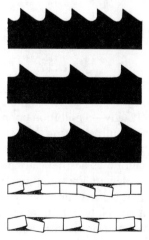

Figure 2.16 Jagged, coarse lines define the shapes and functions of these efficient bandsaw blades.

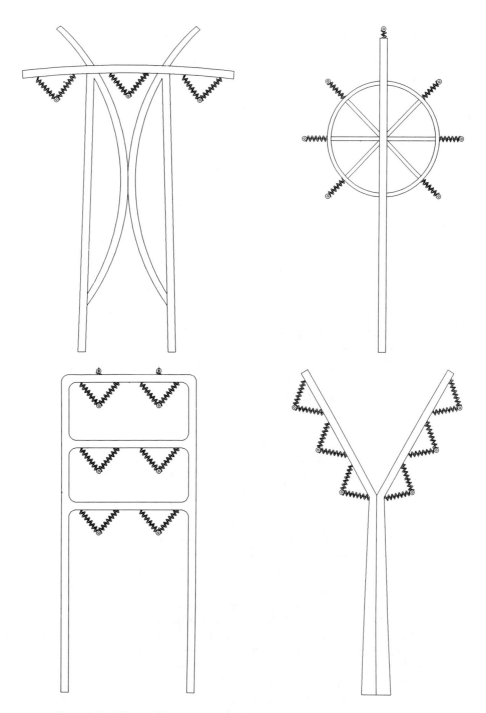

Figure 2.17 These striking power transmission poles are a dramatic departure from the less-attractive styles so familiar to both urban and rural dwellers—proof that attention to the aesthetics of the most mundane artifacts pays off. Courtesy of Edison Electric Institute.

Some configurations are dictated by prevailing standards. Tatami mats are made of of straw covered with bound reeds. They measure about 3 feet by 6 feet and determine the size and shape of the rooms of Japanese homes. Instead of square feet, such a home is described as a 16 tatami structure. Similarly, 4 feet by 8 feet sheets of plywood dictate the modular size of U.S. homes. Designers must work within the limitations of these and many other standards in their work, but must not permit this to unduly restrict creativity.

Forms are the three-dimensional constructions comprised of combinations of lines and planes, and are limitless in variety. They can be as simply geometrical as the office furniture in Figure 2.18, or more contrived in contour as the elegant plastic concept lawnmower in Figure 2.19. The furniture sketches depict a layout that would neatly fit into the work spaces of a functional office. The lawnmower is sculpted to contain the cutting and power mechanisms safely, and to create an efficient form that will repel any accumulation of dirt and grass clippings. The specifications for machine housings, for sporting goods, or for railroad cars can to a certain extent dictate the utilitarian forms for these diverse products. While such considerations can obviously influence structure, it must be remembered that the options for form are limitless. Only an imaginative attention to the potentials of lines, planes, and solids, to direct the senses of both art and propriety, can lead to visual satisfaction.

The faces of planes and solids can be enhanced, embellished, or modified by coloring and texturing. *Surface quality* then becomes the fourth design element. Such treatments or characteristics can add interest or emphasis to a design and thereby generally contribute to appearance. A side chair of pale bentwood and patterned upholstery (see Figure 2.20) exemplifies this treatment with woven and plain surfaces in contrast to each other. This serves to produce a very striking appearance that is obtained through varied yet compatible elements. The reason for this positive human reaction to surface quality is that product surfaces serve as light reflectors and absorbers. (see Figure 2.21.) The ability of a surface to reflect the light striking it is called *value*. White surfaces reflect all the light and lie at the top of a *value scale*. Black, with theoretically no light-reflecting ability is at the bottom of the scale, with all colors and tones falling between. (See Figure 2.22.) Color becomes a significant part of all well-designed products.

Individual reactions to color are frequently based on past associations. By frequent identification with some idea, faith, or individual experience, color becomes symbolic. This explains why purple is so often associated with royalty, greenish-yellow with sickness and disease, blue with atmosphere or despair, green with freshness and youth, white with purity, and yellow with sunshine and happiness. Furthermore, associations with basic foods have so conditioned people that they could not enjoy a meal of purple bread, green milk, or black potatoes. Similarly, most Americans associate color and taste in some learned, cultural way. Green jelly beans should have a lime taste, yellow is lemon, black is licorice, orange is orange, red is cherry, white is peppermint, and purple is grape. Research also has led to the realization that color plays a role in the workplace, as suggested in Chapter 5.

In addition to its aesthetic appeal, color has a definite effect on other design elements. Certain combinations of color can change the relative size relationships of

Creating space-efficient, elegant reception
areas is made easier with Support Cabinets.
They allow visibility while providing enclosure, storage, and work space.

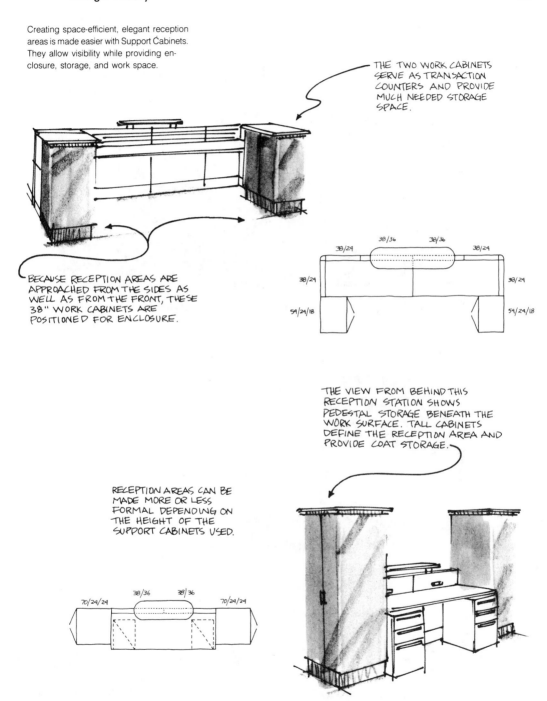

THE TWO WORK CABINETS
SERVE AS TRANSACTION
COUNTERS AND PROVIDE
MUCH NEEDED STORAGE
SPACE.

BECAUSE RECEPTION AREAS ARE
APPROACHED FROM THE SIDES AS
WELL AS FROM THE FRONT, THESE
38" WORK CABINETS ARE
POSITIONED FOR ENCLOSURE.

THE VIEW FROM BEHIND THIS
RECEPTION STATION SHOWS
PEDESTAL STORAGE BENEATH THE
WORK SURFACE. TALL CABINETS
DEFINE THE RECEPTION AREA AND
PROVIDE COAT STORAGE.

RECEPTION AREAS CAN BE
MADE MORE OR LESS
FORMAL DEPENDING ON
THE HEIGHT OF THE
SUPPORT CABINETS USED.

Figure 2.18   Rectilinear structures are ideal for office furniture because they fit the shape of
room layouts and provide functional work surfaces. Courtesy of Herman Miller Inc.

Figure 2.19 Plastic materials were used in this functional and sensitive conceptual lawn mower. Courtesy of The Dow Chemical Company.

Figure 2.20 Textured and patterned upholstery material provide a gentle contrast to the smooth bentwood structure of this side chair. Courtesy of Kinnarps AB, Stockholm.

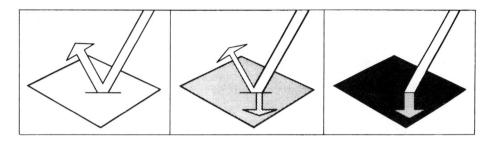

Figure 2.21 Relative absorption and reflection of light by white, gray, and black surfaces. Color and texture also affect the amount of reflected light.

adjoining areas or masses. Though shapes may be exactly the same size, one can appear larger than another because of its color. Another type of illusion produced by color is an advancing and receding effect. When yellow, a warm advancing color, is used with violet, a cold receding color, a three-dimensional effect is produced. And so, as suggested in the discussion of lines, colors, too, can produce certain emotional feelings and can, therefore be used in conjunction with line to produce a desired effect upon the beholder.

Specifically how is color experienced? First of all, the sun radiates energy in the form of wavelengths. Sunlight contains all the colors of the spectrum, and each is expressed by wavelengths within a certain range. When light sources such as the sun or artificial lamps reach an object, some light rays are absorbed and some reflected, depending on the colors in the light source and color of the object. Only the reflected light is seen by the eye. A painter mixes colors that will selectively absorb or reflect the specific hues to impart a desired color to an object. The eye defines the shape and color of the object by transforming the radiant energy into chemical energy, energizing nerve endings, and sending impulses to the optic nerve. The optic nerve registers the message and sends it to the sight center of the brain, at which point the individual becomes aware of the color of the object. This is a rather cursory explanation of what happens in color perception, but the point is that humans have succeeded in identifying and separating these spectrum colors, and reproducing them with great accuracy in order to colors objects.

To guide this activity, a number of different color systems have evolved, one of

Figure 2.22 This scale represents 10 different values ranging from white, limited by the integrity of the paper, to black, a characteristic of the ink. The grays in between are mixtures of black and white carefully controlled to vary the amount of reflected light.

the more usable of which is the Prang color wheel in Figure 2.23. Note that three primary and three secondary hues are indicated. Pairs of adjacent colors on this wheel are known as *harmonious* hues. Those opposite each other are called *complementaries*. Adding white to one of the primaries creates *tints*, and *shades* are produced by mixing in black. The art of controlling surface color must be carefully studied by the designer as a means of increasing both contrast and interest in a product design. Color must be rated as a design element second only to line and form.

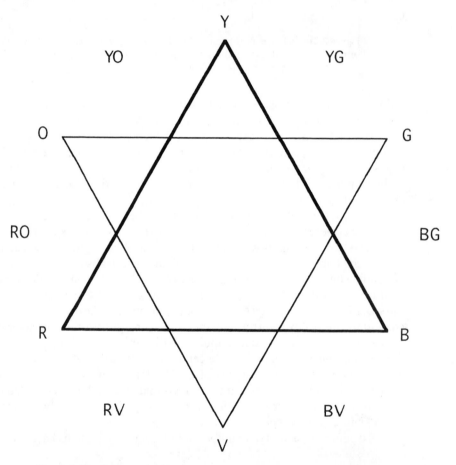

Figure 2.23 As shown in this chart, the Prang color system is comprised of three groups of hues. The *primaries* are connected with a heavy line, and the *secondaries* with a light line. The *intermediates* lie between the primaries and secondaries. Primaries are those from which all other hues are made.

A second reflective quality of a surface is obtained with the property of *texture*. Texture is actually a pattern of contrast in light reflection that identifies the surface rather than the form of an object. Natural textures are characteristic of many materials, such as the smoothness of an egg or the roughness of burlap cloth, and it is understandable that people often identify texture with a sense of touch. Every material has a personal texture that can be described as fine or as coarse, and that perhaps is more indicative of the feel of the surface than of the visual impact. Not all textures are inherent or naturally occurring. Some result from a processing technique, such as embossing, which imparts visual as well as functional qualities. For instance, automobile tires have tread patterns cast into them during manufacture to provide a positive gripping action on wet or snowy highways. (See Figure 2.24.)

Figure 2.24   The tread pattern on this automobile tire is both functional and visually pleasing. The designer used repetition of form to provide for good surface gripping and safety. Courtesy of The Goodyear Tire & Rubber Company.

Textures provide for better hand purchase, as with a knurled metal or rubber tool handle, or a grooved control knob on a television set. A roughened wooden rod (or a light coat of mild adhesive) on a trouser hanger will prevent the garment from slipping to the closet floor. Textured patches applied to the floor of a shower stall will prevent falls while bathing. An embossed metal or plastic sheet on a gasoline dispensing pump will not readily show scratches. A soft fabric covering on an automobile instrument panel top will prevent sun glare in the eyes of the driver. In addition to the matter of visual enrichment, texture also has an important functional role.

Design principles.   It has been said that a beautiful painting is to be looked at but not talked about—an adage that reflects the difficulty of putting into words the emotions and feelings that exist in visual artistry. The professional designer is equally hard-pressed to describe the superior qualities of a well-designed industrial product. How can one remove a part from the whole and evaluate it as a separate entity? Although it is virtually impossible to thus break a design into components, or speak of the method of organization that gives it unity, one can generalize about some of the attributes common to well-designed things. Such characteristics are called *design principles* and they relate to the arrangement of the four elements of design into meaningful wholes. Authors and designers identify constituent principles according to what makes most sense to them. This author has selected *unity and variety*, and *balance and proportion* as the most representative of these principles.

Unity and variety.   Unity and balance are closely related, for within a design there must be a sense of belonging or similarity (unity) among the component parts to achieve order and wholeness. Concurrently, there must be sufficient diversity or contrast (variety) among the parts to display interest and to relieve monotony. This is well illustrated in the 1937 Elgin Skylark bicycle pictured on the cover of this book. It is a study in style with the teardrop pedals and sweeping fenders, and was marketed by Sears Roebuck as "the prettiest thing on wheels." The graceful arcs of the grilled skirt and front frame members contrast nicely with the linear struts to create a striking rhythmic pattern. It is an elegant machine. Unity and variety contribute to the means by which the overall effect of a design is analyzed. Unity is harmony, similarity, agreement, repetition, affirmation, and continuity. If a design has unity, it means that everything in it is woven together according to some well-developed scheme. Variety provides the aesthetic contrast.

Unity and variety relate to all art forms. For example, they are recognized as the essence of music, unity being achieved by the repetition of a basic theme, and variety by contrasting patterns on this theme. A unifying motif runs through the course of Beethoven's *Fifth Symphony*, which causes this musical piece to present a sensation of totality or wholeness; yet the monotony is relieved by contrasting musical structures. Compare this scheme with Figure 2.25. Note how the monotony is relieved in a brick structure by the random introduction of bricks of contrasting shades. The wall is an example of perfect unity achieved by the predictable pattern of bricks of identical

Figure 2.25   Brick walls exemplify unity through repetition of similar shapes. This wall also displays variety by the random placement of contrasting-color bricks for emphasis and a nice visual effect.

size and shape as well as spacing. Variety comes about with the inclusion of the shaded bricks to provide contrast and emphasis. Unfortunately, some architects specify distressed and malformed bricks to emphasize this contrast, as in Figure 2.26. The result is visual discordance and emotional chaos. This same concern holds true in poetry, where the countermotion of phrases provides a variation or contrast. Variety, then, implies the use of contrasting elements, so controlled and placed as to hold and retain attention. Variety means interest—the opposite of monotony.

Rhythm is the flow or movement of the viewing eye as controlled by the repetition of either similar or varying elements, and is therefore closely related to unity and variety. Familiar examples of rhythm are found in music as the listener senses the regular recurrence of beats that establish a definite pattern. The listener learns to anticipate what is to come from what has occurred. Painting, too, concerns itself with a rhythmic pattern of identical forms, and rhythmic patterns can also be observed in architecture. Such terms have meaning, to be sure, in all artforms, and they have a similar meaning to the visual expression that lies in a product. For

Figure 2.26 Contrast in this wall is overdone by using rather grotesque distressed bricks with little thought. The wall is contrived and visually unpleasant.

example, the Finnish hunting knife, or *puukko* (see Figure 2.27), is a vivid expression of unity and variety. Note the clean controlling line that gives shape to this piece. It can be experienced by tracing the outline of the knife with the finger. Another unifying line enters midway through the sculptured handle and is carried into the blade and continues to the point. Variety is achieved with two contrasting materials: the smooth, black nylon handle and the brushed finish on the stainless steel blade. This is a nice harmonious blend of design elements and a very usable product.

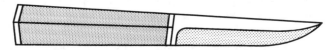

Figure 2.27 This hunting and carving knife is a good illustration of unity and variety in product design. Notice how the outline is unbroken by any linear irregularities.

   Rhythm also has some interesting product safety and convenience relationships, as evidenced in the regular pattern of the steps in a stairway. A person
descending a stairway exhibits a learned expectation that the step tread will have a
depth of about 10 inches and the riser a height of about 7 inches. This is especially
important in a poorlylighted stairway area. Random tread and riser sizes would be
confusing and dangerous, and could result in a serious accident. (See Figure 2.28.)
Similarly, the placement of light switches at a standard height of 42 inches above the
floor is both a convenience and a visual feature. Such natural mappings (see Chapter
6) are examples of applied rhythm and are quite important design considerations.

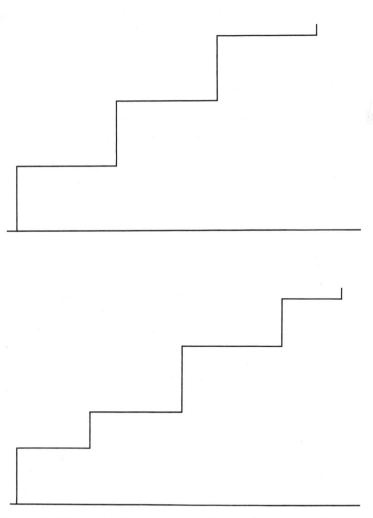

Figure 2.28   The top stairway displays a functional rhythmic pattern. The lower
example is erratic and confusing, and difficult to use.

Balance and proportion.   Balance and proportion are two other familiar principles. The features of the wheelskate in Figure 2.29 are balanced both visually and physically. The horizontal row of identical wheels counterbalances the vertical structure of the shell and blade. Proportion, as a principle, deals with the relationship of the size of one part to the whole and is directly related to the concept of balance. In the case of the wheelskate, the relationships of element sizes work visually. Balance is the quality of equilibrium achieved and sustained through the proper proportioning of the parts of any whole. Ratios of approximately 1:3 or 2:3 are generally considered visually good.

Figure 2.29   An improved rollerblade, called the Metroblade, incorporates removable shoes to permit users to skate freely without excess baggage, and to walk safely when they reach their destinations. It is also a functionally attractive device, well proportioned and balanced. Photo courtesy of Rollerblade, Inc.

Geometric designs can be based on proportions derived from the *golden section*, a system is attributed to the Greeks who devised it as a means of securing the pleasing proportions of their magnificent buildings, sculptures, and artifacts. The diagram is generated by starting with a square, and a radius is scribed about point (A) which bisects a side of the square. The sides of the newly formed rectangle have proportions of 1.618 to 1. By proceeding according to the drawing in Figure 2.30, the system can be repeated to secure a series of rectangles all having well-defined size relationships. The arrangement of these forms can result in pieces that reflect the pleasant space divisions shown in Figure 2.31. Such geometric spacings can become the bases for graphic layouts, wall cabinets, or tabletops. In design, proportion is one of the most effective means of creating unity among the various components. The use of proportionate elements—whether of lines, dimensions, areas, colors, or textures—helps to establish a feeling of fullness and unity, binding all elements together so tightly that removing or altering a single element would disturb the whole design. Proportion also is an element of shape, as suggested earlier in the example of the tatami mats.

Balance relates to proportion in that it is not only a biological necessity in people's makeup but also something we look for in all visual objects. What's more, people seem to be able to recognize this property easily. Balance that can be seen in

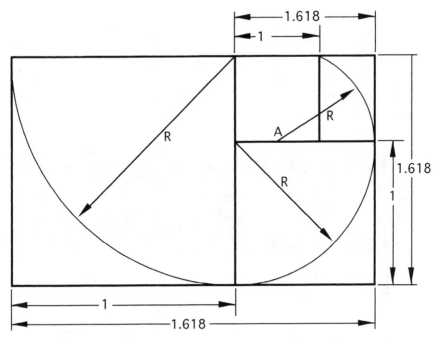

Figure 2.30 The Golden Section is one method of developing sensitive proportions in space division problems.

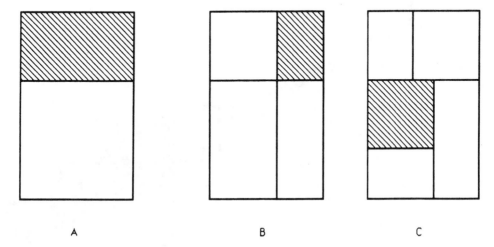

A   B   C

Figure 2.31 These geometric shapes are variations of the Golden Section, and are applicable to graphic design and storage shelving problems, for example.

an object is known as *optical balance*. When the two halves of an object are exactly alike on either side of an axis, the relation is known as *formal symmetry*. A design can also be symmetrical when organized radially around a center point. Wheel discs in automobiles are good examples of radial symmetry. (See Figure 2.32.) However, as illustrated, balance need not be a strictly formal arrangement, for there is such a thing as *informal symmetry* in which balance is perceived just as surely as in the formal scheme.

Figure 2.32 Automobile wheel disc designs are typical creative applications of radial symmetry. Courtesy of Ford Motor Company.

The screwdriver in Figure 2.33 is an uncommonly attractive tool, the handle of which is composed of a linear body with half-circle ends capped with concentric circles, one a utilitarian hole for tool hanging, the other a shallow depression serving as a finger purchase to aid in turning the integral driver shaft and point. The rhythmic pattern of sturdy ridges presents a textured gripping surface, functional as well visually satisfying. A satin black durable finish ends the composition.

Sensitive, well-designed objects are never achieved by memorizing a long list of

Figure 2.33 This common screwdriver is uncommonly attractive. Good use was made of the design elements and principles, and it is usable as well.

design principles and a second list of rules or generalizations regarding their proper application. Designers do not set out to consciously create well-proportioned, well-balanced lamps, or unified tables or chairs to exhibit variety. Instead, they embark on the design mission with open, creative minds, searching for form, experimenting with space combinations, sketching possible contours, and seeking solutions that reflect a good organization of elements. Experience, practice, study, intuition, and reflection will lead to the ability to discriminate among sensitive and awkward forms. When this feeling of "rightness" about an object is present, the parts of a lamp will be in proportion, the table will look as though all the parts belong together, and the chair will display an interesting structural variety. The absence of this feeling of rightness causes visual tension, or the sense of strained, pulling forces in a composition. The constructions in Figure 2.34 provide a summary of preferred spatial relationships.

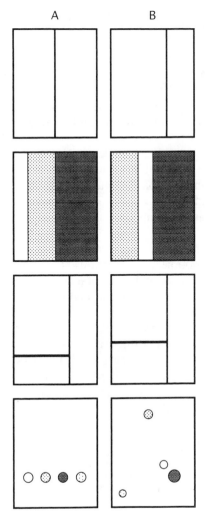

Figure 2.34 Preferred spatial relationship examples are shown in Column B.

INDUSTRIAL DESIGN

Industrial design is a term coined by Americans in 1913 as a synonym for "art in industry," which was to provide society with usable and visually pleasing artifacts. According to the Industrial Designers Society of America, it is the professional service of creating and developing concepts and specifications that optimize the function, value, and appearance of products and systems for the mutual benefit of both user and manufacturer. Industrial designers translate consumer needs into consumer products, ranging from chairs to toys to home appliances. Some develop functional and attractive product shells and machine coverings, such as automobile bodies, drill press housings, and typewriter cases. Others undertake the total design of an artifact, such as a coffee pot or a medical device. Manufacturers who are properly concerned with product quality will engage teams of engineering and industrial design people to create artifacts that work, look nice, and are easy to produce. This collaborative effort is of course the essence of the concept of design for manufacture.

Industrial designers are, by training and inclination, especially capable of working with the visual aspects of a design problem. They can examine the engineering specifications and details of the workings of an automatic washer, and provide for its cover and its ergonomics. An interaction with the technical specialists will ensure that it is producible, safe, easily maintained, and economical. Some examples and analyses of their work follows.

The attractive tracklight in Figure 2.35 is a pleasing example of the work of quality industrial designers. The light is usable and convenient, provides illumination where needed, is strong and durable, and features contours that will fit into any interior setting. One designer's narrative (which follows) provides insights into the industrial design process as it reveals the thinking and experimenting involved in creating a chair.

The Eli chair (named for the designer's son) was planned to provide comfort for dining and other short-period seating situations. In developing the piece, Bruce Sienkowski highlighted the beauty of the manufacturing processes used to transform the materials from their raw state to the finished product.

By its very nature, a chair is in contact with its user at all times during seating. When creating Eli, the designer injected elements of form and material that add pleasantly to this use. An analysis of the chair reveals that its components contain some human-like characteristics. (See Figure 2.36.) From a straight frontal view, the taper of the chair legs relates to a trouser leg joined to the "belt" (front stretcher), creating a pleat at the joint. At the center of the belt is the buckle or "navel." The arms flair out from the seat as though the "hands" were on the "waist," and the "elbows" are protruding strongly. The "back," attached at the waist, terminates at the top corners, representing the strength of the "shoulders." The holes make reference to buttons of a collar while offering an integrated hand hold. At the base of the leg is the finishing touch—the "shoes."

During the development of the chair, a number of alternatives were considered,

Figure 2.35 An attractive and usable tracklight. A good example of the work of an industrial designer. Courtesy of Lumiance, Haarlem, Holland.

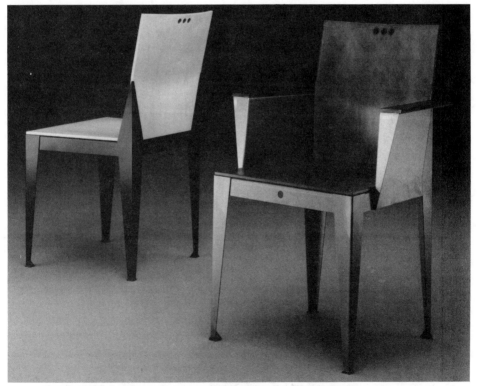

Figure 2.36 The Eli chair displays a good mix of wood and metal materials. Courtesy of Charlotte, Inc.

particularly in the glide (or "shoe"). The initial concept was much more literally a shoe. Although unique and in keeping with the theme, this concept was impractical in application. Each of the succeeding ideas became more refined and universal, as the same glide was used on all four legs, as shown in Figure 2.37. The familiar pediment or pad nature of the glide was maintained throughout the design process. Alternative arm arrangements were also studied, and each evaluated for the interplay of form and function. (See Figure 2.38.)

The juxtaposition of wood and steel calls attention to each of the materials separately while at the same time highlighting their integration of form. The steel provides a rigid hard edge of strength, characteristic of a strong base, while the wood, molded with soft curves, provides support in all areas of user contact. The wood provides "warmth," both physically and mentally, essential to the comfort of the user. A working sketch of one of the chair impressions appears in Figure 2.39.

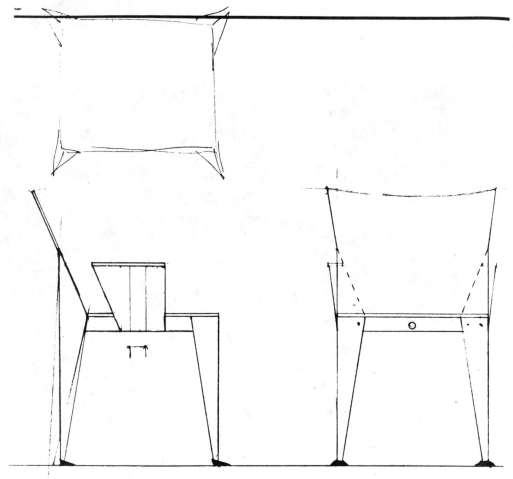

Figure 2.37   Foot pad design for the Eli chair. Courtesy of Charlotte, Inc.

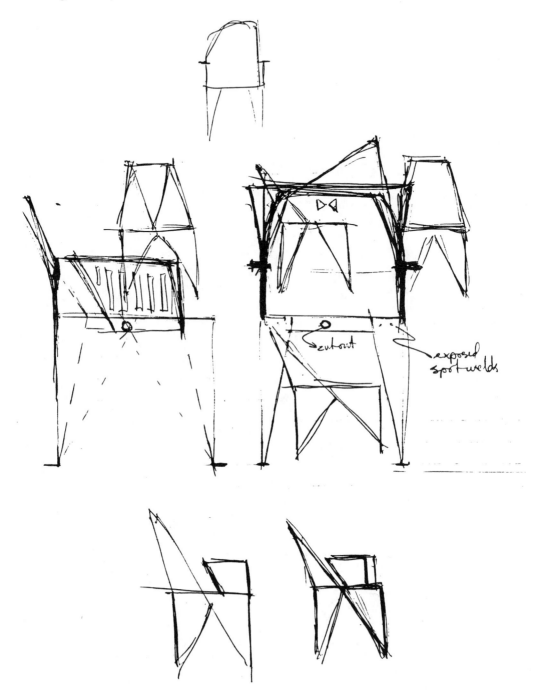

Figure 2.38 Designers typically study a number of arrangements for chair elements, such as these arm structures. Courtesy of Charlotte, Inc.

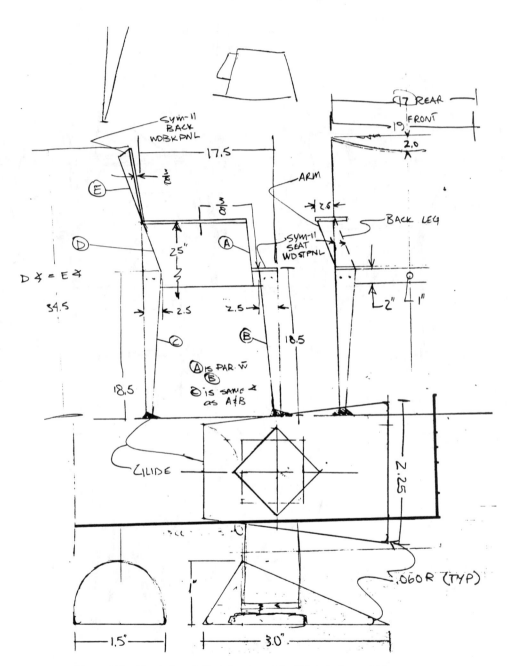

Figure 2.39 The designer's final sketches of the Eli chair, from which a prototype was made for analysis and modification. Courtesy of Charlotte, Inc.

The available chair finishes accentuate the materials used in its construction. The tough protective clear coat on the steel base allows the inherent visual excitement of the manufacturing process to show through. The base is also available in opaque paint finishes, allowing an opportunity to be playful with color. The molded plywood seat, back, and arms are available in wood tones, maintaining the natural character of the wood. As his work progressed, the thoughts of Sienkowski reveal how ideas were generated, modified, rejected, and refined, leading to the ultimate Eli chair.

Industrial designers also play an important role in creating sensitive housings or shells for production equipment. The design of the industrial sanding machine in Figure 2.40 was cited as "a successful response to a complex design problem that was very well thought out and executed." The team of industrial and engineering designers created a new model of an old machine, which displays greatly improved ergonomics, safety, and visual features, and consumes less energy in its operation as well.

Figure 2.40   A team comprised of industrial and engineering designers guided the planning for this efficient and dynamic industrial sanding machine. Courtesy of Timesavers, Inc.

ENGINEERING DESIGN

Engineering design may be defined as the process by which a need is transformed into an actuality. It should be noted that achieving this actuality may involve any or all of the disciplines of engineering, and that the need itself may be very specific or general. The activities of inventing, drafting, analyzing, synthesizing and shaping all may be contributors to the design process individually or in groups. It seems that through the design process a need is met by creating something real. The something created may be an electronic circuit, the apparatus for lowering a wheelchair from van, a machine, a chemical, or any of a myriad of other things generally defined as useful. The designer's attention is somehow drawn to a need that requires satisfaction. Engineering design requires that engineering principles involving analysis be involved in the process. The air-powered angle drill in Figure 2.41 illustrates this. The turbine and power mechanism design had to be based on principles of pneumatics and gearing to

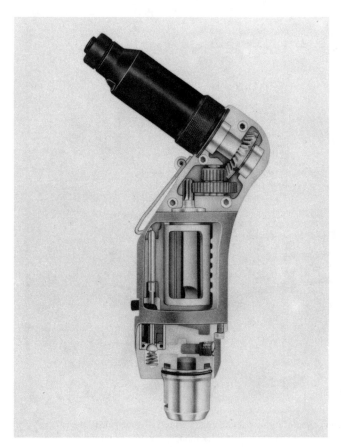

Figure 2.41 Mechanisms for products such as this angle drill are good examples of engineering design tasks. Courtesy of Sioux Tools, Inc.

match the product specifications. An industrial designer-ergonomist would provide input regarding the requirements of a tool such as this used in manual parts assembly.

Design engineers are responsible for large vehicles such as powerful commercial trucks. These engineers create cleverly functional automatic production machinery, the bottle filler and capper in Figure 2.42 being a classic example. They also invent special apparatus for lifting a wheelchair into van to simplify in some small measure the life of a person who has a disability. They design engines, bridges, airplanes, space vehicles, telephones, and off-road construction equipment. Material scientists were responsible for the new extended mobility automotive tire that can

Figure 2.42 Equipment for bottle filling and capping is complex and requires a broad knowledge of machine design. The functional and material requirements must be carefully analyzed to assure product success. Courtesy of Cozzoli Machine Company.

perform at zero pressure. (See Figure 2.43.) In short, engineering designers concern themselves primarily with technical devices and systems, many but not all of which also require the services of industrial designers.

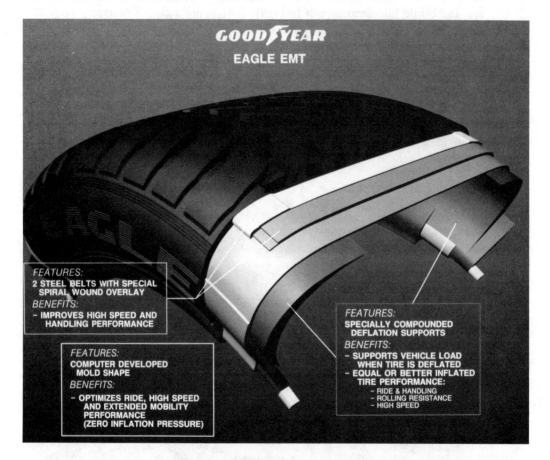

GOODYEAR
EAGLE EMT

FEATURES:
2 STEEL BELTS WITH SPECIAL
SPIRAL WOUND OVERLAY
BENEFITS:
- IMPROVES HIGH SPEED AND
HANDLING PERFORMANCE

FEATURES:
COMPUTER DEVELOPED
MOLD SHAPE
BENEFITS:
- OPTIMIZES RIDE, HIGH SPEED
AND EXTENDED MOBILITY
PERFORMANCE
(ZERO INFLATION PRESSURE)

FEATURES:
SPECIALLY COMPOUNDED
DEFLATION SUPPORTS
BENEFITS:
- SUPPORTS VEHICLE LOAD
WHEN TIRE IS DEFLATED
- EQUAL OR BETTER INFLATED
TIRE PERFORMANCE:
- RIDE & HANDLING
- ROLLING RESISTANCE
- HIGH SPEED

Figure 2.43 Materials scientists and tire performance specialists collaborated on the design of this extended mobility tire, which will support a vehicle load when deflated. Courtesy of The Goodyear Tire & Rubber Company.

According to another definition, engineering design is an activity where scientific techniques are employed to make decisions regarding the selection and use of materials to create a system or device that satisfies a set of specifications. Furthermore, engineering design by both definition and implication concerns itself with solutions that diminish the significance of aesthetics or visual order in the problem solution. Function is supreme, as indicated in the engineering design problem narrative that follows.

Every engineering designer has some personal design problem format that seems to work. In general, the steps include defining the need clearly; examining strategies used by others; defining constraints such as weight, size, and cost; weighing alternatives; and then considering how to synthesize available technology to meet the product needs. If the specifics of the problem are not fully understood, it is unlikely that the attempts to meet them will be successful. By examining the best works of other manufacturers and by identifying common product faults, the designer can avoid having to retrace old steps or repeating old mistakes. This does not mean that new mistakes will not be discovered, such as errors that are not always the result of poor judgment or analysis. It is almost axiomatic that new designs require some degree of testing. When new or different arrangements of elements are advanced, unforeseen or unpredictable behavior may result. Safety, reliability, or function may be compromised because constraints or needs have been met. Testing is usually the only way to approach this problem.

Many design project proposals originate from criticisms identified by users of other products. The salespeople of a manufacturing company are in a particularly good position to refer product ideas from their customers. Of course, not every such idea will become a marketable product, nor does every design proposal come through the sales department. The example described next did not quite follow this path, but it serves to illustrate the initiation process. The following was a problem used by a university professor in his engineering design class, and is typical of such academic activities.

Dentists and other health-care providers are properly concerned with HIV and other infectious diseases, and so it is important that they maintain their equipment in a sterile condition. Most of the tools and appliances found in a dental office are easily sterilized in an autoclave, which is a high-pressure steam chamber. The major exception is the dental handpiece, or drill. Generically, a handpiece consists of an air turbine that rotates at over 200,000 revolutions per minute. It has two or three small tubes entering the handle, at least one of which is for air and one for a water coolant. The turbine is about 6 or 7 mm in diameter, with notches around the periphery to produce the turbine effect. At either end of a hollow turbine shaft there are two tiny ball-bearing assemblies. (See Figure 2.44.) These bearings are a problem source during sterilization, for heat and moisture in the autoclave tend to damage them, or at least make disassembly and relubrication necessary. Because of the small size and complexity of the bearing assembly, the task is time consuming and costly.

One concerned dentist decided that a disposable handpiece would solve the problem, and figured that the drill would be cost effective if it could be sold for under $15. It was also determined that the drill would have to work with a common office air and water supply, and would have to survive at least 30 minutes of use. Here, then, was a tentative product description along with some general specifications, gleaned from studies of handpieces and test data that had appeared in dental journals. The dentist then presented the problem to two student research teams in an engineering design class.

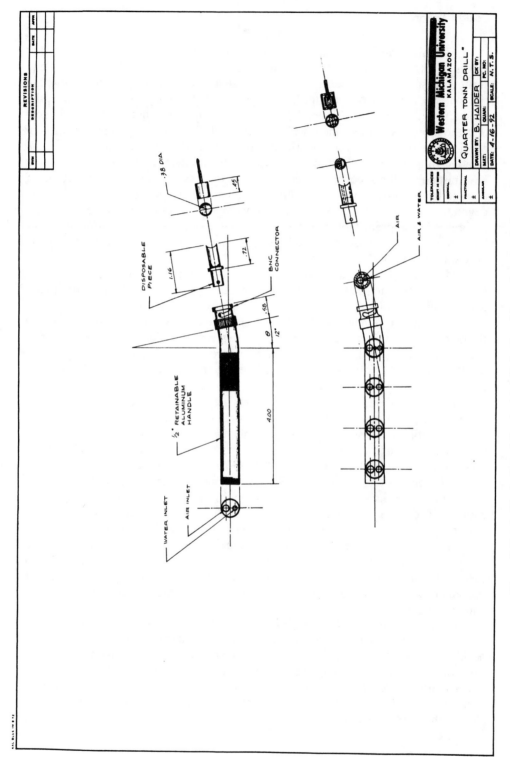

Figure 2.44 An illustration of the dental drill developed by engineering design students. Courtesy of Dr. Jerry Hemmye.

Following the discussion, the teams now knew what the product was, what performance was required, and what the cost limitations were. This design project did not involve inventing something that had never existed before, nor did it require the use of exotic materials. A number of similar disposable dental drills already were on the market. Aside from the high-speed operation and the small size of the components, the project did not seem to involve anything unconventional. A search of the literature and examination of a number of handpieces made by different manufacturers led the teams to believe that it would be possible to meet the tool requirements.

Further review led the students to the conclusion that there were only a few technologies involved, and that air turbines used in such drills appeared to be rather unsophisticated. They considered that a significant improvement in performance might be possible by doing a fluid dynamics study of the turbine. However, they felt that such a study would be time consuming and inflate the product cost, so they agreed to postpone it unless it became a critical matter. The groups also learned that the expense of the miniature ball bearings was a major factor in the overall handpiece cost; they elected to investigate the use of sleeve (journal) type bearings using available food grade plastics; and they decided that a molded plastic handle would reduce product cost and make it more environmentally sound.

In this design problem, a number of crucial design decisions were made early, most of them driven by cost. From this point on, there were two design schemes followed. Had there been more teams, there would have been more schemes. One team chose to design a handpiece with hose attachments in the conventional fashion, with the complete assembly disposable. The other team chose to make the handle of stainless steel with just the turbine end detachable and disposable. Both teams worked on making turbines and bearings in the most economical manner.

The two teams succeeded in making turbines and bearings that met the specifications, but one turbine was measured at 420,000 revolutions per minute. Because of time constraints, neither team produced a finished handpiece, although all the elements necessary to its production were completed. It is interesting to note that handpiece manufacturers have not ignored the original problem. At least one supplier has redesigned the handpiece to allow for easy cleaning and relubrication of the bearings at no great increase in product cost. At the moment, it appears that this tool is the market leader. If the disposable tool manufacturer had a product with a lower profit initially, perhaps the market would have been captured. Because of the conservative nature of the usual dentist, it is not likely that disposable handpieces will account for any sizable part of the market.

There are countless examples of such successful technical products. A unique engineering device has made available to the product design community a spring with many applications. (See Figure 2.45.) The spring is an integral unit consisting of an inner steel shaft surrounded by a molded rubber cylinder with outer half-shells of high-strength metal alloy. The cylinder is bonded to both shaft and shells to form a durable, compact, lightweight part. (See Figure 2.46.) The spring is designed to absorb impact through the rubber cylinder as it is twisted by the movement of the inner shaft. The unit replaces the conventional metal leaf or air spring for suspension

Figure 2.45 This rubber torsion spring was designed to create a durable, reliable
spring with many applications. Courtesy of BFGoodrich Company, Engineered
Polymer Products Division.

applications in busses, heavy equipment, and military vehicles. Smaller more compact configurations are available for the seating market. (See Figure 2.47.)

A good illustration of the proper mix of a technical product's functional,
material, and visual requirements may be seen in the award-winning Crown FC
Series electric forklift truck. (See Figure 2.48.) One's first impression is with the
clean flowing lines and shapes organized to create a visually satisfying structure. The
smooth wheels present a nice collage of color, texture, and form, and the clever
placement of the wheel assembly hexagonal bolt-heads add pleasing contrast points.
This material-handling vehicle looks functional and it is, and the attention to
ergonomics results in good operator efficiency, safety, and comfort. There is a
sensible selection of appropriate metals and plastics for strength, durability, and
finish. This machine is a striking example of the collaborative efforts of engineering
and industrial designers.

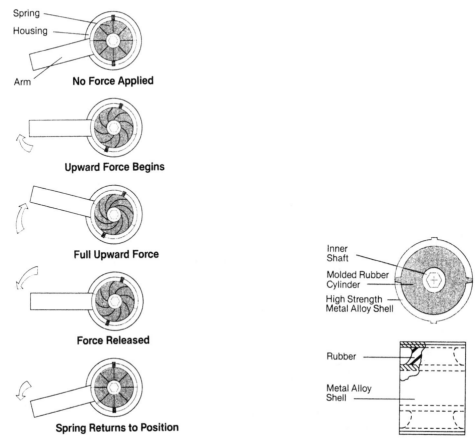

Figure 2.46 A diagram showing the operation of the torsion spring. Courtesy of BFGoodrich Company, Engineered Polymer Products Division.

CRAFT DESIGN

The tremendous range of sensitive craft pieces available today attests to both the skill and inventiveness of the designer-crafter. Pottery, glassware, jewelry, carved wood, furniture, and metal pieces all result from the basic premise that for some creations, little attention should be given to mass production. These are essentially hand-crafted, one-of-a-kind artifacts with a primary emphasis on blending the skills of the artist-crafter. Such products transcend the anonymity of the mass-manufactured article, fulfilling a basic human need of both maker and buyer. The delicate crystal glaze on a porcelain pot, the graceful form of a raised sterling bowl, or the delicate details of a jewelry piece exemplify the efforts born of reflective experimentation and

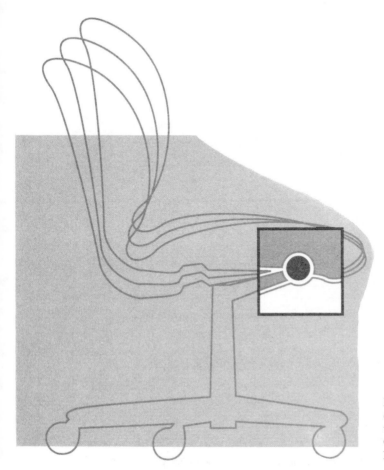

Figure 2.47 Office chairs such as this employ the torsion spring to self-adjust to body loads. Courtesy of BFGoodrich Company, Engineered Polymer Products Division.

a sensitive hand and eye. The inventor-maker satisfies a personal desire to create a unique article, such as the candlestick described earlier in Figure 2.15, where the subtleties of the craft are not compromised. The consumer experiences some vicarious thrill of the joy of creation by possessing an object that satisfies the human need for aesthetic pleasure. The bookend in Figure 2.49, for example, was created by a crafter with considerable metal machining skills. He experimented with a metal shaper to produce on a metal surface a series of gradually spaced intersecting lines, resulting in a pleasing visual effect.

It is difficult to differentiate sharply between the work of the designer-crafter and the industrial designer, for their paths cross, join, and then part at some point, as dictated by the demands of their respective tasks. It might be said that crafters generally possess the ability to invent unique forms and then to render these ideas into skillfully executed pieces. Designers for industry most frequently function as

Figure 2.48 This award-winning forklift truck is handsome and functional, and the result of a good collaborative effort by engineering and industrial designers. Courtesy of Crown Equipment Corporation.

members of research and development teams, where they contribute their visual expertise. Essentially, their tasks are directed toward the design of quality goods for mass production. The fire tools in Figure 2.50 are a functional composition of teak wood and black conversion-finished mild steel structural members. The combination of round and rectangular shapes, harmonious materials, and nicely spaced handle screws provides for unity and variety. This craft product could be easily replicated, and many such craft examples end up being mass-produced. Before this can happen, the fire tools would have to be analyzed by manufacturing engineers to determine their producibility and to recommend necessary modifications. The aim here would be to produce a marketable item without sacrificing the integrity of the original, unique artifact.

The attractive wood-carving against a flat-black background is a nice blend of textures. (See Figure 2.51.) This pleasant marine motif exemplifies the work of the

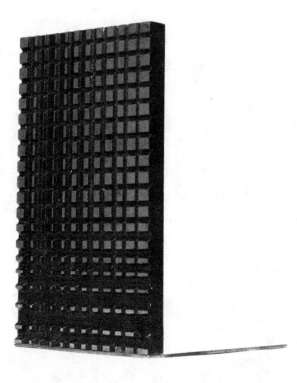

Figure 2.49 Craftwork is exemplified by its uniqueness. This bookend of 0.5-inch mild steel was created with a metal shaper. The problem involved a study of the visual effects possible using this machine.

skilled designer-crafter, a sensitive creation depicting fish suspended amongst sinuous weeds. This carver had an intimate understanding of the subject and executed it with graceful forms.

THE DESIGN PROCESS

Designing is properly considered to be a process because in order to achieve optimum problem solutions, several factors and subtasks must be undertaken in an orderly fashion. Different authors and practitioners may use different process formats, but the aim always is to provide an effective design system. A typical diagram of this process is shown in Figure 2.52, followed by brief descriptions of its constituent elements.

Phase 1: Identify Problem. A design task may emerge as a perceived need for an existing product, such as a better safety switch for a chainsaw, or as a new product, such as a high-performance engine on a concept automobile. Conversely, the task may be totally innovative and with greater visual requirements, as exemplified by a new line of furniture or the refitting of an aircraft interior. Whatever the

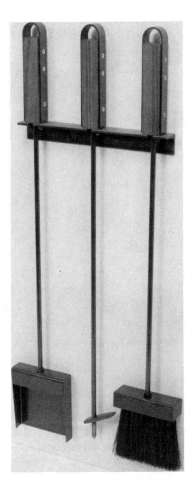

Figure 2.50 Fire tools lend themselves to any number of material and form combinations. This teakwood and blackened steel set is a good example. Design by the author.

source, such tasks must be carefully analyzed to ascertain their appropriateness for a specific company, and for their potential as successful products. Design is costly and time consuming. As a new product is conceived or an existing one improved, the primary objective is to present a commodity that will meet a need or render a service in a manner superior to that of any former or similar product. This can result only from a judicious problem definition.

Phase 2: Collect Data. This is the research phase of product design. Once the problem has been defined and given focus, pertinent data must be gathered, information retrieved, benchmarks examined, market assessments made, and cost analyses undertaken. Any factors that bear on the problem must be studied, leading to the presentation of tentative product specifications. An evaluation of existing

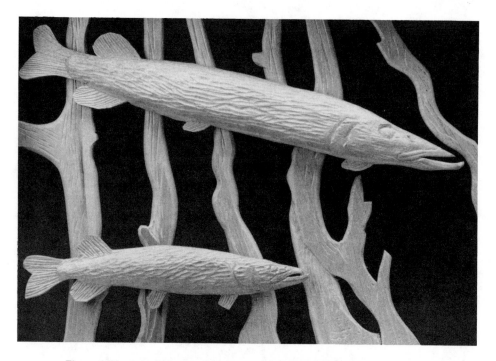

Figure 2.51 A striking example of a carved wall-hanging—graceful and well executed. Courtesy of Craig Spink.

facilities may lead to the conclusion that such an undertaking is not cost effective because of possible extensive refitting or equipment acquisition. Data collection is equally valid for its negative as well as positive conclusions.

Phase 3: Hypothesize. This activity is the concept-development stage, where intuition and technical experience merge to produce a range of possible problem solutions. Pondering the task, considering alternatives, taking creative leaps, sketching possibilities, studying the visual requirements, weighing them against purpose and materials, modifying, and discussing are all necessary to this phase. Continual reference must be made to the original problem and its analyses. Hypothesis is the heart of the design process, the stage at which potential configurations emerge and are evaluated.

Phase 4: Experiment. At this point, the possible solutions are refined and tested, and the prototypes are built, evaluated, modified, and compared to earlier data and analyses. Is the product strong enough? Does it work? Is it visually correct? Is it safe, easily maintained, producible, vandal-proof, and marketable? Experimentation is precisely what the term implies—the opportunity to critically examine the various problem solutions and to detect and correct errors before the article is put into production.

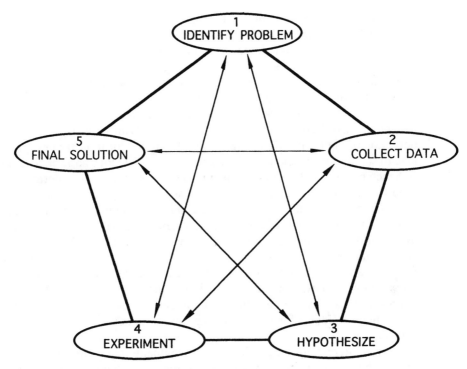

Figure 2.52 This chart depicting the stages of the product design process serves as a guide to creative design activities.

Phase 5: Final Solution. The logical result of the previous four phases of work is the ultimate problem solution. At this stage of the process, considerable attention must be given to those refinements needed to satisfy special methods of manufacture, to simplify the product so that it can be easily assembled, to modify it so that standard components may be used, or to improve its appearance, among others. Obviously, this is a final opportunity for all members of the design-for-manufacture group—the industrial designer, the engineering designer, the production engineer, the safety engineer, the product maintenance specialist, the ergonomist, and others—to meet and congratulate each other on the elegant product they have created.

Benchmarking

Special mention should be given to an important source of data to be used in the design process. This is an innovative technique called *benchmarking*. According to the dictionary, a benchmark is a land surveyor symbol cut into a bronze plate to serve as a reference point or standard by which something can be measured or judged.

What this term implies in product design is an organized process of measuring a company's operations and products against those leading performers and competitors. The ultimate goal is to know the industry's best practices, and incorporate these into company operations in order to strengthen the market position. Although benchmarking is a relatively new practice (first surfacing in the early 1980s), its use is growing because of its obvious benefits.

In his very fine book on the subject (the book itself being something of a benchmark), Robert C. Camp (see Chapter 11: Camp, 1989) provides a working definition of the term: "Benchmarking is the search for industry best practices that lead to superior performance." Furthermore, he offers four basic philosophical steps of benchmarking as being fundamental to success: know your operation; know the industry leaders or competitors; incorporate the best; and gain superiority.

The process itself is straightforward and simple. A manager will generally begin by deciding which aspect of corporate management is in need of benchmarking, such as operations, product development, automation, service, or the like. Next, specialists in that area determine which competitive company is the very best at that function, and then collect data to exchange and discuss with that company. This is followed by an analysis of the data, leading to the incorporation of the most effective approaches used by the benchmarked company. Product design and methodology are prime candidates for benchmarking, and designers would do well to adopt its principles.

It is interesting to speculate how a company might proceed when a specific product has been for benchmarking, for example, a refrigerator. A team assigned this problem would most likely identify the three best refrigerators on the market and analyze them to learn why they are the industry leaders. They could study such factors as energy efficiency, user-friendliness, silent operation, simple but effective door closing, size and shape, durability of finish, storage convenience, and the like. Assume that one negative feature is present in all three target models—that of automatic door closure. One person may then concentrate on determining the causes of door failure. Improper seal material may wear in time so that the door would not close completely. Is it a material fault, a door seal design error, a poor hinge mechanism, or a manufacturing error? Continued analysis would pinpoint the fault, which could then be corrected and incorporated into a new design concept. All such features could be systematically studied, leading to an optimum appliance design. The next step would be to perfect the manufacture of the product, so that it could be made better and at less cost and in turn emerge as the industry leader. This super-refrigerator would later be benchmarked by the competition.

QUESTION AND ACTIVITIES

1. Contrast and compare Aristotle's four causes, Scott's four causes, and the three design requirements described early in the chapter. Present your opinion as to their relevance as approaches to modern design.

2. Study the material on the definition of design and write one of your own.

3. In your own words, describe what is meant by the functional, material, and visual requirements and how they guide the design process.

4. Select three products and describe how they do or do not meet the aims of the three design requirements.

5. Cite some instances where you feel that materials simulation might be justified (e.g., when plastic laminate is finished to simulate wood or marble).

6. Do you feel that the author's comments on the nutcrackers in Figure 2.7 is justified? Explain your reasons.

7. Sketch some product examples that illustrate the meanings of the terms *unity*, *variety*, *proportion*, and *balance*.

8. Refer to Figure 2.17 and sketch some other impressions of telephone poles.

9. Define the terms *value, hue, shade*, and *tint*.

10. Give some examples of functional textures in product design.

11. Describe the differences between industrial design and engineering design by selecting and analyzing a product example of each.

12. Select a few examples of good crafts design and analyze them from the standpoint of producibility. How many design changes were necessary to improve their producibility? Did these changes affect the integrity on the original craft piece?

13. Write a short paragraph on how the five-step design process can aid in creating quality products.

14. Describe the benchmarking process, and search the literature on the subject to identify some examples of it.

CHAPTER 3 | Design Materials

Materials are the stuff from which products are made, and designers must select from the array of metals, plastics, woods, ceramics, and composites available to them. The material requirement for product design was introduced in Chapter 2; the additional information presented here is directed at aiding designers in making intelligent decisions regarding material choice. The selection of materials for the screwdrivers in Figure 3.1 is a good illustration. The tough steel interchangeable tips surpass the minimum torque standards and will not twist or bend, and therefore fail to perform the function of turning screws. Note that the chrome-plated shaft emerges from a cluster of tips for convenient storage and accessibility. The patterned polyvinyl chloride (PVC) plastic handle is of sufficient size to be comfortable and to assure a firm grip for more power and less muscle strain. The tool is functionally, materially, and ergonomically correct, and also very attractive.

MATERIAL PROPERTIES

A number of factors or qualities influence materials selection. *Physical* properties describe a material's melting point, density, moisture content, porosity, and surface texture. *Chemical* properties relate to resistance to corrosion and dissolution. *Thermal* properties are measures of the effects of temperature on materials. *Electrical* properties determine material conductivity and resistance to electrical charges. *Acoustical* properties indicate reactions to sound, and *optical* properties reactions to light.

104

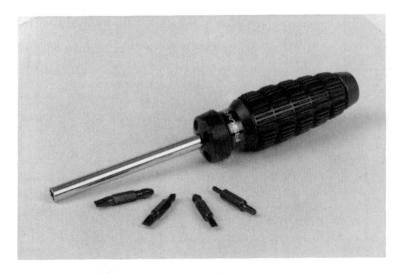

Figure 3.1 A good selection of materials for this screwdriver design: A tough steel shaft and point and a comfortable, efficient plastic handle. Courtesy of Fiskars Inc.

Mechanical properties are especially important because they are indicators of strength, producibility, and durability. For example, *tension* is a force that tends to stretch a material; *compression* is a force that applies squeeze pressure; *torsion* is a twisting or torquing force; and *shear* involves two opposing forces tending to fracture a material, as in shearing a piece of paper or metal. The drawings in Figure 3.2 illustrate the actions of these mechanical forces on wood samples. Similar results would be obtained with other materials. A knowledge of such forces, and the ways in which materials react to them, is valuable in determining which material to use in a specific application. These forces provide measures of material hardness, toughness, abrasion-resistance, brittleness, and ductility, among others.

Finally, the designer must consider those indefinable characteristics such as appearance, odor, feel, and general impression that result from special uses and combinations of materials for esthetic purposes. The boldly designed chair shown in Figure 3.3 features an elegant brushed aluminum frame and a molded plywood seat and back finished naturally. The piece is very comfortable due to the ergonomics of the backrest design. Other striking combinations include wood and steel frames and upholstered seats and backs. This is an excellent example of a pleasant contrast of material and color, and a consonance of form.

Product design materials, as stated earlier, fall into the general categories of metals, plastics, woods, ceramics, and composites. Designers are forced into situations where suitable choices must be made from databases of available materials, and there are many published rosters of engineering properties of these substances. The notations presented here are restricted to general descriptions of material character-

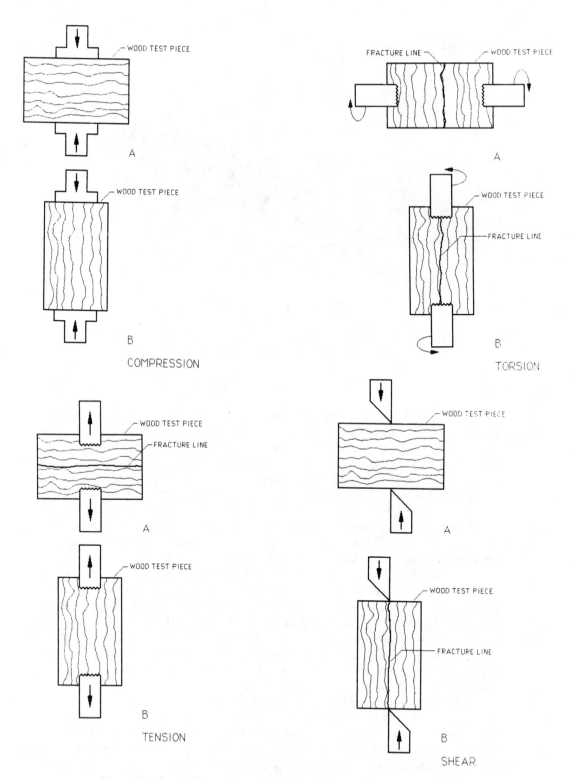

Figure 3.2 Graphic illustrations of the mechanical properties tests applied to wood samples. The importance of wood grain direction is shown as "A" and "B" in the samples.

Figure 3.3 The prize-winning Paletti chair has a frame of tough aluminum, satin-brushed to provide a pleasant contrast with the molded birch plywood back and seat. Courtesy of Design Forum Finland.

istics and their applications. The initial design specifications must be followed by discussions with materials specialists and suppliers regarding properties, characteristics, process technology, availability, and costs. This information will aid in the material selection decisions.

METALS

There are two basic classifications of metals: *ferrous* metals are those comprised primarily of iron; *nonferrous* metals contain little or no iron. The many kinds of steel belong to the first group, whereas copper, nickel, and aluminum are among those in the nonferrous category. Metals are seldom used in their pure states. Instead, base metals are combined with other metals or chemicals to form *alloys* to secure the requisite properties for a specified use.

Iron

Iron is a basic ferrous material and, prior to 1850, saw wide use in both the wrought and cast forms. Since that date, its use has been largely supplanted by that of steel. Wrought iron is essentially pure iron containing small quantities of slag and carbon.

Though not as strong as steel, it is soft, ductile, and easily worked into a variety of shapes such as decorative ironwork. Iron has good corrosion resistance and it is used in the manufacture of water pipes, grating, marine sheathing, and heavy hardware.

There are several kinds of cast iron, which is essentially a brittle, hard material whose characteristics can be altered by adding carbon and other alloying ingredients. It is this alloying that makes it possible to acquire a strong, machinable, and ductile casting metal. It finds wide use as hardware, machine tool bases, manhole covers, engine blocks, and numerous other large and small cast shapes. The common fire hydrant is a cast-iron product that is both reliable and durable. It is tough enough to withstand considerable use and abuse, and yet will function when a fire emergency arises. The large, strong threads on the hose stubs can be wrench-loosened, and seldom fail. This product also is practically vandal-proof and reflects a good use of materials. (See Figure 3.4.)

Steel

Carbon and other alloying elements are added to iron in specific amounts in order to make steel. This provides the engineer and the designer with a range of ferrous materials that are strong, hard, non-brittle, malleable, ductile, and corrosion-resistant, in varying combinations of these factors. Steel is an extremely versatile material and it is the most widely used metal. It has many applications, from huge

Figure 3.4 Fire hydrants are functional, durable, and reliable cast-iron products. Although not generally designed with aesthetics in mind, the model in the foreground is especially nice.

Figure 3.5 This stove of mild steel displays an innovative use of forms to create functional and attractive structure. Courtesy of Peter Haythornthwaite Design Limited, New Zealand.

structural steel beams to high-carbon tools and razor blades, and sheet steel housings for electrical appliances, automobiles, and railway cars. For example, the wood stove in Figure 3.5 is cut from 5 mm mild steel plate, with a welded and bolted assembly, and finished with heat-protective aluminum paint. The clever combination of rectangular and round elements results in a most visually pleasing and functional product.

Stainless steels are expensive but almost indestructible, and are available as bars, castings, forgings, or sheets for applications requiring high strength and anticorrosion resistance. These steels can accept high-quality, durable, hard-polished, or matte finishes. They are used extensively for marine hardware, boiler tanks, mechanical fasteners, kitchen and laboratory utensils, medical apparatus, tools, and woven mesh screening. The attractive and durable camp ax is a good example. (See Figure 3.6.)

Steel is a material of choice in applications such as athletic lockers for clothing storage. This equipment is subject to considerable normal wear and tear, as well as vandalistic abuse. It must be strong, have a tough and durable finish, have ventilation features, and be fitted with secure closures, locks, and hinges. Steel sheet of about 1.20 mm thickness (18 gauge) would be appropriate for this product. The only drawback is that these lockers are impossibly noisy.

Copper and Its Alloys

Copper is probably the oldest metallic material known and it continues to be used in a variety of industrial and product applications. It is a reddish-brown, very soft material that can be worked by most metal processes. Because of its high thermal and electrical conductivity, it is a primary metal in modern communications systems. Copper is alloyed with zinc to form brass, and alloyed with tin to form bronze. (By alloying tin and copper in inverse proportions to that of bronze, pewter is produced). Brass is a popular material for bathroom fittings, appliance hardware, and household fixtures. In these situations, it is generally plated with nickel or chromium because of

Figure 3.6 This unique camping ax is comprised of a stainless steel blade and a tough plastic handle. Note the clever plastic case for safe storage and portability. Courtesy of Design Forum Finland.

its tendency to tarnish easily. It also is used in such marine applications as hardware, propellers, and shafts. Bronze is stronger than brass, is quite resistant to both wear and corrosion, and is a metal of choice for bearing and bushing surface components for moving machinery shafts.

Aluminum

Aluminum is bluish-grey, rather soft, and light in weight. Because of its versatility and durability, it is challenging steel as the most important single product metal. This remarkable material can be alloyed to meet almost any specification for metal products, and it has a high workability in that it responds easily to virtually every metalworking process. Its high thermal and electrical conductivity are characteristics that make it desirable in electrical and electronic applications. Product applications range from aluminum foils to air-frame coverings for aircraft, pots and pans, tools, extruded structural and decorative trim members, and many components on modern automobiles. Its light weight makes it an ideal material for beverage cans.

Another attribute of aluminum is its acceptance of a spectrum of finishes and finishing techniques. It can be polished bright or given a soft, satin luster, and it may be textured by embossing. The gentle form of the torch shown in Figure 3.7 is

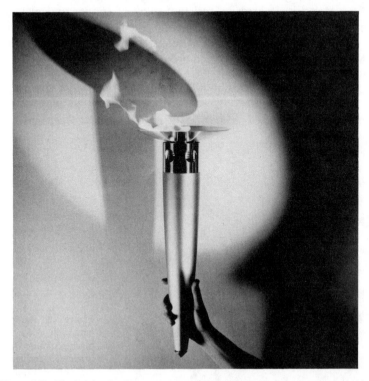

Figure 3.7 The bright aluminum contrasts nicely with the black carbon fiber on this Olympic Torch. Courtesy of Peter Haythornthwaite Design Limited, New Zealand.

enhanced by the bright machined aluminum handle and the contrasting black carbon fiber combustion unit. Aluminum also can be anodized, a process by which the surface of the material is altered chemically to resist corrosion and staining, as well as to prepare it to accept coloring dyes.

Other Metals

Zinc is a soft, corrosion-resistant metal, useful as a plating material to provide a galvanized coating to steel sheets and structural shapes. Zinc alloys are commonly used to produce accurate and intricate die-cast product shapes, such as the robust industrial caster. (See Figure 3.8.)

Magnesium is expensive, lighter than aluminum, and essential to the aircraft and space industries. It is tough, durable, machinable, and used in die-casting. Other applications include storage tanks, industrial tools, and parts for printing and textile machinery.

Titanium is a tough, ductile, corrosion-resistant material equal in strength to steel, but only half as heavy. It finds use in components for aircraft, missiles, fuel containers, and jet engines. Additionally, it is used increasingly for pumps, propeller shafts, bearings, reactor vessels, and marine rigging.

Monel is a corrosion-resistant alloy of copper and nickel, employed in the manufacture of products for the marine, chemical, food-service, and power industries.

Refractory metals—such as tungsten, molybdenum, and tantalum—have very high melting temperatures and are used in the manufacture of rocket motors and turbine engines.

Precious metals include gold, silver, and platinum. These costly materials are

Figure 3.8 An industrial caster made with a die cast zinc body and a white plastic coating. The grey rubber tire provides a cushioned, nonmarking rolling surface. Courtesy of Jilson Casters, Incorporated.

used in jewelry, coins, special electronic contacts and wiring, and the spinerets (dies) used to produce plastic and glass fibers.

PLASTICS

So many adjectives and superlatives have been applied to the field of plastics that it is presumptuous to attempt to give proper treatment to this remarkable material in a few paragraphs. However, the designer may find some value in general information relative to the types of plastics and their characteristics and uses.

Plastics are synthetic materials of relatively recent discovery, the first being celluloid, which was developed as a substitute for ivory billiard balls in 1868. Indeed, and unfortunately, plastics were for many years relegated to the position of being merely substitutes for other more premier materials. This is no longer the case. They are made from raw materials such as coal, tar, petroleum, and milk and wood derivatives, and are available in rod, bar, sheet, film, and special shapes. Plastics can be machined, cast, molded, bent, blow-formed, and worked in many, other ways. Attractive and colorful artifacts can be thus produced, such as the clever index card flip file. (See Figure 3.9.) This useful product features a simple acetyl spring for the indexing movement and a thumb wheel instead of a twist knob.

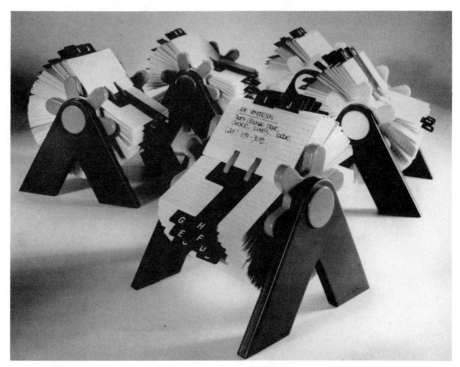

Figure 3.9 An attractive index card file of molded plastic. Courtesy of Peter Haythornthwaite Design Limited, New Zealand.

One of the best reasons to specify plastic is its light weight. A common peanut butter jar made of plastic weighs 50 grams, whereas a glass jar weighs 300 grams. Both containers hold 500 grams of peanut butter. Dozens of other plastic bottle and package designs are in use because the material is tough and can be formulated to be tasteless, and therefore will not taint the edible contents. A stainless steel surge tank for a diesel truck requires costly forming and welding operations and weighs 10 pounds. The glass-reinforced nylon tank to replace it was blow-molded and weighs only 4 pounds. (See Figure 3.10.)

The problem with this topical area is that it is difficult for the practitioner to keep abreast of new developments. Ten years ago, there were but a few dozen types available, today there are hundreds and the inventory continues to grow. What makes the material so desirable as a product medium is this very fact of type proliferation. No matter what product application presents itself, one is sure to find a plastic to meet design specifications. Plastics are either heat-resistant or they can be easily melted. They are available as structural shapes approaching the strength of steels or as tough, pliable films. They are also excellent adhesives, lubricants, and finishes, and are waterproof, corrosion-resistant, odorless, and colorful, as well as convenient to use. An ordinary water bucket becomes a colorful, useful, lightweight

Figure 3.10 A lighter, less expensive, more producible fuel surge tank of plastic (left) replaced the heavier stainless steel on the right. Courtesy of DuPont Company.

container when made of plastic. (See Figure 3.11.) Plastics attributes also make them ideal design materials for products such as baby bottles. One flexible, shatterproof, heat-resistant, tasteless clear bottle reduces bubbles and trapped air, and its unique angled neck promotes feeding babies in an upright position, which improves swallowing control. (See Figure 3.12.)

There are two basic groups in the family of plastics. *Thermoplastic* materials, or thermoplasts, become soft when exposed to sufficient heat and harden when cooled, no matter how often the process is repeated. They are comprised primarily of chain-like linear molecules, and consequently undergo a physical, not a chemical, change during the forming process. *Thermosetting* materials, or thermosets, are set into permanent shapes when heat and pressure are applied to them. Reheating will not soften them. They are comprised primarily of a cross-link molecular structure where the individual chain segments are chemically attached to one another. Regardless of forms or properties, all plastics fall into one of these two groups. There are hundreds

Figure 3.11 A durable and colorful plastic material was used to produce this handsome bucket. Note the addition of a shaped mass at the top of the handle for comfortable carrying. Courtesy of Reisenthel-Programm, Munchen, Germany.

Figure 3.12   The unique design of this angled-neck baby bottle makes it a convenient and safe plastic product. Courtesy of Wynnovations, Inc.

of specific kinds of plastics; the categories charted in Figure 3.13 indicate some of the more common ones.

WOODS

Perhaps no other substance has enjoyed the popularity of wood as a design material. For years it has been used in craft, construction, and furniture applications. Because of increased utilization and technological improvements in the field, wood continues to enjoy a prominence. It also has limited use in the building of boats and marine apparatus, but wood as a medium for the product designer is primary in the construction of furniture. The material always has been desirable in this application, for it has warmth, character, and presentation. The oak structure shown in Figure 3.14 is comfortable and durable, and will withstand many years of use as restaurant seating. The table surface is covered with tough, colorful plastic laminate for easy cleaning.

Softwood and *hardwood* are terms used to distinguish between trees that have needle-like or scale-like leaves and those that have broad, flat leaves. Generally

TYPE	CHARACTERISTICS	USES
ABS Acrylonitrile Butadiene Styrene	Thermoplast. High impact, tensile and flex strength. Very good temperature and chemical resistance. Overall electrical qualities and dimensional stability good.	Battery and radio cases, tool and utensil handles. Football and safety helmets, automotive trim, and water pipes.
Acrylic Polymethyl methacrylate	Thermoplast. Strong, hard to break, withstands weathering and cold temperatures. Excellent insulator. Excellent optical clarity, pipes light. Broad color range. Odorless, tasteless, and nontoxic.	Signs, aircraft windows, light fixtures, radio, TV and refrigerator parts, lenses, lamps and trays. Examples: Plexiglas® and Lucite®.
Cellulosic	Thermoplast. Strong, tough, and scratch-resistant. Clear or full range of colors. Good electric insulator. Tasteless and non toxic. Can be used outdoors.	Toys, lampshades, packages, tool handles, camera parts, glass frames, ping pong balls, heel covers, and piping.
Fluoroplastic	Thermoplast. Extreme chemical, temperature and moisture resistance. Excellent anti-stick qualities. Excellent electric insulator. Weather resistant.	Chemical gaskets and pump packing, antenna insulators, high-temperature insulation. Cooking utensil coatings. (Teflon®)
Polyurethane	Thermoplast and thermoset. High resistance to chemicals and solvents. Water- and rot-proof. Can be flame resistant. Unusual tear and abrasion resistance.	Flexible and rigid foams used for mattresses, rug underlays, and insulation. Solid articles include automotive parts and airplane structures. Widely used as a coating.
Vinyl	Thermoplast. Available as films, sheets, coatings and filaments. Strong; hard to break, tear or abrade. Resists chemicals and water. Full range of colors, transparent to opaque. Good electric insulator. Can be odorless, tasteless, and nontoxic.	Wire insulation, floor and wall tile, hose, signs, packaging, lamp shades, table mats, fabrics, luggage, raincoats, toys, records, baby pants, and appliance parts.
Nylon Polyamide	Thermoplast. Resists extreme temperatures. Strong, high impact and tensile strength. Odorless, tasteless, and nontoxic. Affords long frictionless wear. Good electric insulator. Resists oils, common chemicals and stains.	Gears and bearings in machines; fishing lines, reels; door hinges, roller parts in stoves and refrigerators; slide fasteners, tumblers, funnels, brush bristles, and fabrics. Examples: Nomex® and Kevlar®.
Polyethylene	Thermoplast. Most popular plastic. Can be a film or coating, semirigid or rigid. Unbreakable, strong, and flexible. Permits passage of oxygen. Good electric insulator. Odorless, tasteless, and nontoxic. High resistance to chemicals, heat, cold and water. Full range of waxy colors.	Squeeze bottles, bags, toys, mixing bowls, storage boxes, ice trays, juice containers, tumblers, table cloths, chair covers, and shrink-wrap packaging.

(continued)

Figure 3.13 A plastic materials chart, showing characteristics and uses.

TYPE	CHARACTERISTICS	USES
Polystyrene	Thermoplast. Second most popular plastic. Full range of bright colors, translucent, transparent or opaque. Not affected by freezing temperatures. Good wearing surface. Good electric insulator. Odorless, tasteless, and nontoxic. Resists common chemicals and water.	Wall tile, storage boxes, tumblers, picnic ware, toys, costume jewelry, shelves, radio and air conditioner housing, and battery cases. Foams used for insulation and packaging (Styrofoam$^{®}$).
Polypropylene	Thermoplast. Strong, flexible, excellent chemical resistance.	Fibers, filaments, films, containers, caps, and machine parts, and automotive batteries.
PVC Polyvinyl chloride	Thermoplast. Good chemical resistance, flexible or rigid, excellent electrical properties, good resistance to weather and moisture.	Floor covering, rigid home exterior siding, gutters, windows, piping, and upholstery material.
Polyester	Thermoplast. Easily made into large, laminated products in combination with reinforcing materials. Strong, good surface hardness. Withstands weathering, water, chemicals and heat. Full range of bright and pastel colors.	Air conditioner housings, chairs, light panels, bath tubs, skylights, luggage, boats, sleds, fishing rods, sports car bodies, patio roofing, archery equipment, and aircraft parts. (Fiberglas$^{®}$). Fabrics. (Dacron$^{®}$).
Phenolic Phenol formaldehyde	Thermoset. Hard, brittle, and scratch-resistant. Good heat and electric insulator. High resistance to moisture and chemicals. Colors on dark, opaque side. Heat-resistant.	Handles, appliance parts. radios, cabinets, telephones, light plugs and switches, cameras, furniture parts, and novelties. Automobile ignition system parts. (Bakelite$^{®}$).
Melamine	Thermoset. Hard to break or scratch. Shock-resistant. Full range of colors. Good electric insulators. Odorless, tasteless, nontoxic. Resists oils, common chemicals, moisture, and heat.	Dinnerware, mixing bowls, counter tops, electrical equipment, auto parts, light fixtures, shaver housings, and buttons. Used to make plastic laminate sheets. Examples: Formica$^{®}$ and Panelyte$^{®}$.
Polycarbonate	Thermoplast. Rigid, high impact strength, good chemical restance.	Sporting goods, helmets, clear sheets for doors and windows, and lenses.

Figure 3.13

speaking, the hardwoods are stronger and harder than the softwoods, though this is not always true; thus, the prefixes *hard-* and *soft-* have little literal significance. *Softwoods* are often referred to as coniferous, needle-bearing trees, and almost all of them are evergreen. The tougher, leaf-shedding *hardwoods* are generally more desirable in the construction of furniture because their grain patterns are most interesting and their toughness results in more durable finishes and structures.

Wood is available in many forms in addition to the traditional lumber products. Plywoods are commonly used for furniture in situations where strong panels are required in construction and where dimensional stability is a necessity. Frequently, a

Figure 3.14   Tough oak was used in this durable and attractive seating and eating structure. Courtesy of Plymold Booths.

thin veneer of a more expensive hardwood is used to make durable and attractive plywood panels that have many applications in furniture and cabinet construction. Additionally, there are particle and pressed composition boards, honey-comb veneer panels, and pressed fiberwood shapes in the catalog of wood materials available to the designer. The attractive and usable paper trimmer is one example of the use of these particulate materials. (See Figure 3.15.) Each of these wood products has a number of important properties and characteristics that should be considered before the designer proceeds to material selection. For example, particle board does not hold screws as well as solid wood or plywood, and may require special gluing. This can be an important factor when specifying materials for cabinet doors and furniture.

Many exterior windows and doors are fashioned from milled wooden pieces and painted to match the color of the structure. One of the subsequent problems is that these painted units will eventually weather and chip and peel, and will need repainting, despite the advances in exterior paint technology. An excellent solution to this problem is to clad window parts with extruded vinyl plastic sheaths. (See Figure 3.16.) This provides the homeowner with a prefinished, durable, low-maintenance structure.

Figure 3.15 This very usable paper trimmer is made of durable wood particle board. Courtesy of Hunt Manufacturing Co.

The density of wood is affected by the amount of water present. It is notable that the density of hardwoods is generally much higher than that of softwoods—a factor that contributes to their strength and durability. The thermal expansion of wood, and its toughness, is different along and across the grain of the material— factors that must be accounted for in furniture construction. What this means is that wood shrinks and swells with humidity changes in a room or out of doors; thus, when designing drawers and doors in furniture cabinets, proper finishing techniques must be used, proper clearances provided, and wood materials possessing low thermal expansion characteristics must be selected. The more common mechanical properties of wood vary according to the specific material; these are elasticity, bending strength, and hardness. Descriptions of some of the more commons woods follow.

Hardwoods

Oak   Oak is hard, tough, heavy, and durable, and it is available as red and white groupings. This close-grained material splits with difficulty and cracks if not very carefully seasoned. It takes a high polish and resists marring and denting, making it desirable for interior finishing, cabinet work, and commercial furniture.

Figure 3.16 An example of an exterior window featuring vinyl-clad wooden parts. Courtesy of Andersen Corporation.

The pores of white oak are tighter than red oak, and it is therefore more suited to tight cooperage and finer furniture.

Hickory   The toughest of all North American hardwoods, hickory is shock-resistant, heavy, very hard, and commonly employed for tool handles, furniture, dowels, gymnasium apparatus, and agricultural implements.

Maple   A beautifully patterned, heavy, strong, tough, close-grained material, maple takes a good polish and is light in color. It is used as flooring material, interior finishing, veneers, cabinetwork, toys, and furniture.

Ash   Ash is heavy, hard, and resilient, and employed for interior finishing and cabinetwork, baseball bats, handles, and oars. It also finds limited use in furniture.

Birch   Strong, hard, heavy, shock-resistant, and close grained, birch warps badly and is rather difficult to work. Products include veneered plywood, woodenware, cabinets, and occasional furniture.

Black walnut   This is one of the most popular materials for architectural woodwork and furniture construction because it is hard, durable, rich, and takes a fine finish. Especially desirable are furniture panels made from patterned walnut, in which a beautiful gradation of color is present because of sapwood running through the solid grain. This material is available, as are most hardwoods, as either solid stock or veneers and plywoods.

Mahogany   Grown in several parts of the world, the African and Honduran varieties of mahogany are somewhat lighter in color than the Philippine variety, which is characterized by a reddish appearance. These woods are widely used in cabinetwork and furniture, and as veneered panels. It is also a very important wood in boat construction.

Teak   Teak, which comes from Thailand, has emerged as a material almost symbolic of contemporary furniture and attractive woodcraft items. It is used all over the world and is characterized by a golden grain appearance, with occasional dark sap streaks present, as evident in the lovely turned bowl in Figure 3.17. Because of its oily nature, it is most generally finished with oils, or oil and wax in combination, in order to enhance the appearance of the final product. It is strong and easy to work, but it wears edge tools rapidly because of the solidified grains of oil in its structure.

Figure 3.17   This attractive lathe-turned candy bowl reveals the beauty of the teak-grain pattern.

Rosewood   A brittle, very hard material, rosewood is difficult to work. Its use results in some of the most expensive pieces of furniture and assumes a beautiful finish either as a satin luster or a high gloss. Veneers and panels are common forms for this material.

Ebony   This wood is extremely difficult to work, hard, brittle, and checks badly if not dried properly. It is seldom used for anything other than furniture appointments such as drawer pulls and knobs, and some veneers utilized as accent paneling or overlays.

Softwoods

White pine   This is a light, very soft, straight-grained, easy-to-work material with a clear beige color. It is not a very strong wood; consequently, it is used primarily for interior finish work, toys, and pattern making. The knotty variety is popular for occasional paneling and cabinetwork.

Douglas fir   Hard, strong, and durable, fir splits easily and is rather difficult to work. It is used in all kinds of heavy construction, occasional finish work, and some modern furniture.

Cedar   A very fine grain characterizes this wood, which is extremely durable and light in weight. Red cedar is an aromatic variety and has some applications in clothing chests and closets. White cedar is used extensively for interior finish work, and because of its durability, exterior fencing, posts, shingles, and boat construction.

CERAMICS

Ceramic products are made from nonmetallic, inorganic materials, generally involving high temperatures in their manufacture. This definition includes the classic articles and most of the newer ceramic products. In general, ceramic materials possess a very desirable combination of properties: good tensile and high compressive strength; resistance to chemical attack and weathering; excellent electrical resistance; and great stability when exposed to high temperatures. Ceramics are brittle and difficult to work after the clay has been fired or the raw material hardens. The knowledgeable designer will specify openings and special shapes while the product is in a plastic state.

Ceramic products are encountered daily in the home and are popular in many industries. The field includes such categories as abrasives, cements, enamels, glass, refractories, structural clay products, and the whiteware dinner table pieces in Figure 3.18. Modern homes are built on concrete foundations with brick exteriors and plaster walls inside. Glass windows are necessary, and window walls play an increas-

Figure 3.18   Ceramic dinnerware of simple form and restrained embellishment.
Courtesy of Danish Design Centre.

ingly prominent role in the contemporary home. Ceramic household products include dinnerware and serviceware, enameled cooking utensils, and the easy-to-maintain enameled surfaces of the stove, dishwasher, and other appliances. The bathroom features wall and floor tile in addition to the enameled or porcelain fixtures. The numerous industrial applications include ceramics for acid-resistant tanks, pipes, and valves. Ceramics are essential not only for products of daily living and modern industry but also for equipment with high temperature and strength requirements, such as missiles, rocket motors, jet engines, and nuclear reactors.

The most common material used in the manufacture of ceramic products is clay. This material is mixed to specifications, formed while in its plastic state, and fired at temperatures of around 1000°C (1832°F) to fuse it into a hard, solid mass. A glaze—which is finely ground, glass-like clay—is frequently fused to the surface of the article through an additional firing process. A porous, reddish-brown, unglazed, and rather coarse ware, called *terra-cotta*, is perhaps the oldest type of pottery. (See Figure 3.19.) The following categoric treatments identify some of the more important ceramic materials.

Brickware

Made primarily from clay or shale, the principal products in this category are building, flooring, paving and decorative brick, sewer pipes and fittings, drain tile, wall and floor tile, and other special shapes. These may be considered as relatively

Figure 3.19 These pleasant hurricane lamps of terra cotta and glass feature simple, functional forms. Courtesy of Crate and Barrel.

inexpensive products, the general character being one of permanence and requiring no paint, preservatives or termite-proofing.

Whiteware

The term *ceramic whiteware* generally refers to a glazed or unglazed ceramic body that is usually white and of fine texture. Whiteware includes earthenware, a glazed or unglazed nonvitreous whiteware having greater than 10 percent water absorption; fixtures made from glazed vitreous bodies (vitreous signifies less than 0.5 percent

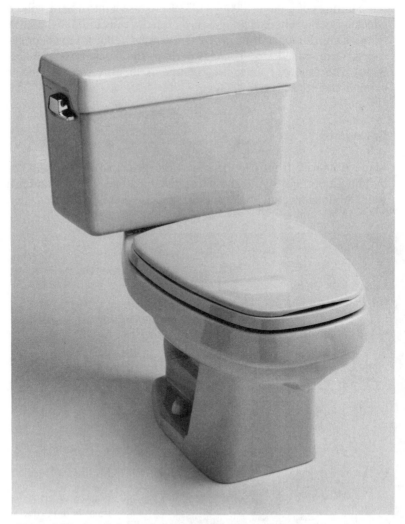

Figure 3.20 A well-designed glazed ceramic toilet, nicely formed and proportioned. Courtesy of Eljer.

water absorption); sanitary plumbing fixtures. (See Figure 3.20); and tiles made in a range of colors and textures that present a hard, permanent, easy-to-clean surface. Other products include various dinnerwares, which are basically hard porcelains consisting of kaolin, clay, feldspar, and quartz. This strong, vitreous, translucent material is commonly called *china* and is often glazed to provide durable, decorative surfaces. Much ceramic artware falls within the whiteware category. (See Figure 3.21.)

Glass

Any material that has cooled from a liquid state to a rigid condition without crystallizing is properly called a *glass*, regardless of its chemical composition. Most ceramic materials are formulated so that they will be at least partly, and sometimes almost wholly, crystalline in structure, whereas glasses are noncrystalline. Even though this distinction exists, both natural and human-made glasses are ceramic products.

Glass possesses certain characteristics that make it a desirable production material. It is resistant to corrosion, wear, and electricity, and it is hard and durable. A distinguishing feature is its transparency, and accordingly it is made into windows and optical devices. The material lacks ductility and has poor resistance to thermal shock. Glass can be strengthened by tempering, and modern processing techniques continue to widen the range of application in product design. In addition to fenestration and optical uses, its employment includes attractive, functional glassware (see Figure 3.22) and glass filament cloths and batts.

Figure 3.21 A lovely, glazed-ceramic, covered baking dish. In Honor of Marc Hansen.

Figure 3.22 This pitcher and tumbler were handblown from partially recycled crude glass, and wrapped with a rod of cobalt blue glass. Courtesy of Crate and Barrel.

Refractories

There are many nonmetallic materials that are temperature resistant, but all processes involving extreme temperatures are dependent on ceramic refractories. The word *refractory* refers to the property of being resistant to softening or deformation at high temperatures. As such, these materials are used as furnace linings, crucibles, fire brick, and research apparatus for high-temperature experimentation.

Porcelain Enamel

Porcelain enamels provide high-quality, durable, protective, and decorative coatings for metal surfaces. They possess physical and mechanical properties such as hardness and excellent chemical resistance, but are given to chipping. Porcelain enamels are commonly defined as glossy inorganic coatings fused to metal at temperatures above

427°C (800°F). They are often designated by the metal to which they are applied (e.g., aluminum enamel, cast-iron enamel, copper enamel, sheet steel enamel, etc.).

Enamel coatings are applied to dryers, refrigerators, washing machines, electric appliances, plumbing fixtures, and kitchen utensils, such as the striking teakettle in Figure 3.23. The enamels are available in many beautiful, durable colors. With modern techniques, the products are less subject to chipping and peeling.

Cement and Concrete

Ceramic cements are inorganic materials that, when mixed with water, will set and harden without the application of heat. A variety of cementitious materials are used in building construction, paving, cast products, and mold making. The common

Figure 3.23   Metal products frequently are finished with a fused coat of colorful and durable porcelain enamel, such as these striking tea kettles. Courtesy of Umbra USA, Inc.

characteristic of these materials is the formation of durable, noncrystalline products. Although there are many types of cements used in construction, portland cement is by far the most common and important. This material is essentially a blend of clay and limestone, sintered and crushed to specification.

The terms *cement* and *concrete* are often used interchangeably. This practice, however, is incorrect. More accurately, portland cement is an adhesive paste that coats the surfaces of aggregate particles, and binds this mixture into a rock-like mass as the cement hardens through chemical action with water. Thus, cement is the binding material and concrete is the resultant monolithic mass.

COMPOSITES

When two or more materials are combined and retain their identities by not dissolving or merging completely into each other, the result is a *composite* material. Such a material has a different composition (hence the name) and has properties different from any of the individual materials that were made. The substances combined to form a composite are called *constituents*. Since the constituents retain their individual identities, the properties of the composite are influenced by the properties of the constituents. Thus, a composite can have the strength of one constituent, the density of another, and the thermal properties of a third. Any number of property combinations are therefore possible. Depending on the form of the constituents, composites can be classified as fibrous, laminar, or particulate.

Fibrous Composites

Fibrous composites consist of at least two constituents: the fibers and the matrix. The fibers may be natural or synthetic, and metallic or nonmetallic. Matrices are mostly synthetic resins, but other materials also may be used. All materials are stronger as fibers than in bulk form. Glass fibers, for example, are over 400 times as strong as ordinary glass.

The properties of fibrous composites are influenced by the following:

1. The types of fibers and their arrangement in the composite
2. The type of matrix
3. The process used to produce the composite

The common fiber arrangements are random mat, woven fabrics, and continuous or wound filaments. The simple *random mat* system is used to produce fiberglass-reinforced plastic films, synthetic leathers, and insulating materials. A characteristic of such composites is that they have the same strength in all directions, because the fibers have been randomly distributed throughout the mat. However, composites produced with fibers *woven* into a fabric have greater strength in the direction of the weave. This method is used for fiberglass-reinforced boat hulls, bathtubs, shower enclosures, automobile body parts, and storage containers, among others. The bathtub in Figure 3.24 is a composite of thermo-formed acrylic with glass fiber reinforcement. This tub is not only strong and durable but it also has a nubby slip-resistant surface for safety, is ergonomically correct, and is attractive.

An innovative application of advanced materials technology occurs with the molded carbon fiber Velocity chair in Figure 3.25. First used in the aerospace industry because of its exceptional strength and light weight, the molded fiber has been incorporated into the design of aircraft, bicycle frames, boat hulls, sporting goods, and safety apparatus. For example, the steel air tanks worn by firefighters as part of their emergency breathing system weigh 10 kilograms (22 pounds). Many are now replaced by carbon filament tanks weighing 3 kilograms (6.6 pounds). Unfortunately, it is a technology virtually unexplored in the design of furniture and household products. There were five stages to the development of the Velocity chair, according to the design team's research report.

1. *Design.* One begins by studying composite fiber characteristics: high strength-to-weight ratio, moldability into complex compound shapes, and flexibility. One then determines that the technology is perfect for exploring the concept of the structural shell. (At the same time, one is studying the Italian Futurist movement, the furniture of Carlo Mollino, and the sculpture of Brancusi.) Drawings and study models are made, then several full-scale foam shapes are carved to test ergonomics.

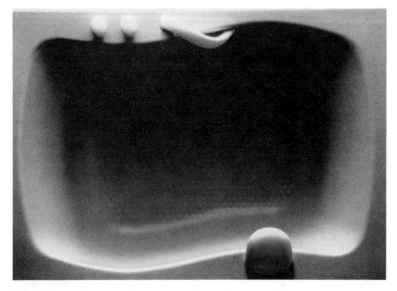

Figure 3.24 A safe, usable, and nicely formed bathtub of plastic composite materials. Courtesy of American Standard.

Figure 3.25 The innovative, striking, and functional molded carbon fiber Velocity chair. Courtesy of Vance Trimble.

2. *Plug.* Blocks of foam are carved to final shapes. These are covered with fiberglass for strength, then painted and sanded smooth. There are plugs for five parts: back, seat, connector, front legs, and back legs.

3. *Mold.* The plug is waxed, painted with mold release, then painted with a layer of epoxy resin mixed with carbon powder. This gives the inside of the mold a tough surface. Layers of fiberglass are applied, the whole assembly is vacuum bagged to force out air bubbles and excess resin, and the piece is placed in a hot box to cure. When cured, the mold is cut away from the plug.

4. *Layup.* Inside the mold is waxed, painted with a layer of epoxy resin, and then layers of carbon fiber are laid into mold. This is vacuum bagged and left to cure in a hot box.

5. *Finish.* The parts are released from the mold and the edges are trimmed. The surface is natural carbon fiber and requires no further finish work. The parts are drilled using special jigs, and then the chair is assembled using aircraft bolts.

A study of the preceding account can provide the reader with a valuable insight into the technical process of chair design, and the thoughts that influenced the product aesthetics.

If maximum strength in one direction is required, it can be achieved by making the composite with *continuous* filaments. The filaments are unidirectionally arranged parallel to each other, like the wires in a cable, to impart their maximum strength in only one direction. This method has been used to produce metallic composites of extremely high strength and high temperature resistance. Continuous filaments can also be wound over a symmetrical mandrel at intersecting angles to produce strong composites. Products include rocket boosters, aircraft components, and tubular structural members requiring strength and stiffness but light weight. The matrix material can be applied to the fibers by spraying, brushing, dipping, or casting. Some of the advantages of using fibrous composites in aircraft component manufacture are strength, stiffness, lightness, dimensional stability, and resistance to corrosion and heat. (See Figure 3.26.)

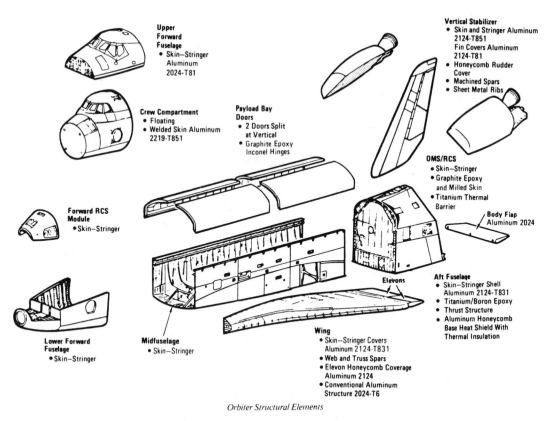

Orbiter Structural Elements

Figure 3.26   Note the many plastics and composites used in the manufacture of this space vehicle. The requirements for such transport are demanding, and must be matched by appropriate materials. Courtesy of NASA.

Laminar Composite Materials

When two or more sheets of similar or dissimilar materials are layered, the resulting material is a *laminar composite*. These are among the oldest and best known of fabricated composites, having been used for over 3,000 years for such applications as swords and armor, where metal sheets were joined by hammer-forging. Examples of more current laminar composites are plywood, laminated lumber, countertops, covers of hardbound books, and laminated plastics. The layers (lamina) involved in laminar composites may be of the same thickness throughout, as in some plywood, or of different thicknesses, as in counter tops and veneered wood panels. In most laminar composites, the layers are combined by adhesion. There is an almost infinite combination of materials that can be laminated, such as metal to metal, metals to plastics, plastics to glass, leather to wood, and paper to gypsum. The plastic laminate countertops, available under many tradenames, are made from layers of heavy kraft paper and thin plastic color or wood pattern sheets, adhered with interspersed coats of melamine resin, and cured under heat and pressure. (See Figure 3.27.) The table in Figure 3.14 has this kind of top.

Laminar composites can provide decoration and protection; strength; toughness and stiffness; wear-, corrosion-, and heat-resistance; thermal, acoustical, and electrical insulation or conductivity; color and light transmission; and control of distortion. Such composites can increase the strength-to-weight ratios for airplanes and submerged vessels by combining strong and light materials. They are produced to increase resistance to wear, corrosion, and heat, as in *clad* metals, by which the metal to be protected is overlaid with a thin layer of some more resistant metal.

Laminar composites are also produced for safety purposes. For instance, in safety glass, plastic sheet and plate glass are combined to increase the impact strength and to minimize shatter. Bimetallic laminar composites with different coefficients of expansion are used in thermostats to regulate temperatures in refrigeration and heating units. Composites in the form of "structural sandwiches" are used for increasing strength, stiffness, and heat or sound insulation as in carton boxes made of different papers and honeycomb structures made of aluminum for airplanes and space vehicles. They also are used for decoration in car interiors, where synthetic and natural leathers or wood veneers are laminated to metal or plastic surfaces.

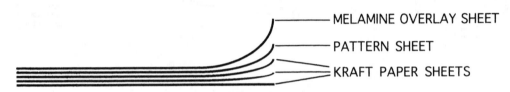

MELAMINE OVERLAY SHEET

PATTERN SHEET

KRAFT PAPER SHEETS

Figure 3.27 The familiar plastic laminates are composites of plastic resin and sheets, and paper sheets. They are often used in fabricating desk and countertops.

Particulate Composite Materials

When various-sized particles of a material are bonded together by a matrix, the resulting material is called a *particulate composite*. A common example is ordinary concrete, discussed earlier in this chapter. Concrete, as you will recall, consists of aggregate particles mixed in a portland cement matrix and hardened by chemical reaction with water. Materials difficult to combine by any other known process can be made into particulate composites having unique properties and applications. Successful combinations include metals in metals, metals in plastics, ceramics in metals, organic materials in inorganics, as well as combinations of these. The particles in these composites can be bonded together in a matrix, as in concrete, or they can be sintered (heated without melting), as in construction bricks and powder metallurgy products.

In addition to portland cement concrete, several other concretes are produced by bonding particles in a matrix of a different material. Among the best known are the asphalts, in which particles of sand or other aggregate are bonded in a petroleum base to form asphalt concrete used for road pavements. Metal powders are bonded in plastic to produce cold solder and various types of bearing materials. Ceramic particles are mixed in metal matrices to produce *cermets*, which are used for high-speed metal-cutting tools, high-temperature valves, and hot-drawing dies, among other applications.

The reason for this rather extensive treatment of composite materials is because of their great potential for use in the design of consumer products. A criticism commonly directed toward industrial designers is that too often they restrict their designs to the more traditional materials, neglecting the possibilities of the state-of-the-art materials. It is important for these designers to become familiar with this array of engineering materials in order to expand their inventories of available design substances. A working knowledge of these can lead to many innovative product creations.

FIBERS

A *fiber* is a particle of inorganic or organic matter whose length is at least 100 times its diameter. Many fibers exist naturally, but only those that are pliable, strong, and at least 5 mm (0.02 inch) long are usable commercially. Some fibers, such as cotton and wool, are short and must be spun together to form yarns. These are called *staple fibers*. Fibers of an indefinite length (such as silk) and manufactured fibers (such as nylon) are called *filaments*. Fibers are ultimately made into yarns that are woven, knitted, or otherwise made into fabrics for use in apparel, home furnishings, or household textiles. But fibers are also useful in industry when mixed with binders to make fiber construction panels and brake shoes, when twisted into rope, and when used to produce paint or cleaning brushes. Strands of metal or plastic wire are woven into screen cloth, strainers, and filter baskets for a broad range of industrial applications. (See Figure 3.28.) Wire cloth fabricators employ several different weave patterns, such as those shown in Figure 3.29, to create this range of products. Some fibers, such as linen and cotton, are natural plant products, whereas others, such as silk and wool, are of animal origin. Fibers emerging from plant and animal sources are governed by their respective biological growth processes and possess different properties and characteristics.

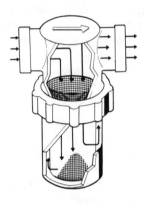

Figure 3.28 A T-type, molded nylon, in-line strainer unit. The metal filter screen is clearly shown in the disassembly. Courtesy of RON-VIK, Inc., Minneapolis, MN.

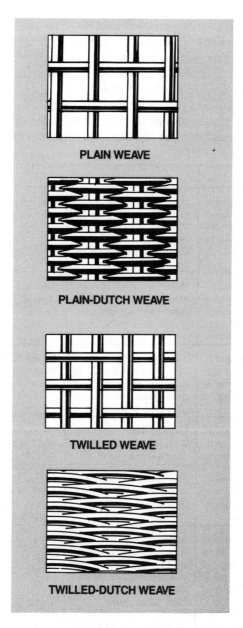

PLAIN WEAVE

PLAIN-DUTCH WEAVE

TWILLED WEAVE

TWILLED-DUTCH WEAVE

Figure 3.29 Typical wire cloth weaving patterns. The dutch weave styles are generally employed for filter cloth applications. Courtesy of RON-VIK Inc., Minneapolis, MN.

With few exceptions, manufactured fibers are polymeric materials, characterized by molecules of great length. Examples include nylon (Antron), polyester (Dacron), and acrylics (Orlon). There are also inorganic fibers such as asbestos and glass for use in composite materials.

Organic fibrous materials consist of large molecules in which covalently bonded atoms of carbon, hydrogen, and oxygen predominate. The individual polymer molecules may also be cross-linked to each other through covalent bonds. This results in stronger and more rigid fibers, some of which are described in Figure 3.30.

Synthetic fibers can be engineered to possess specific characteristics to fulfill a particular need. Such fibers are classified as *cellulosic* and *noncellulosic*. Rayon, acetate, and triacetate are classified as cellulosic fibers. Noncellulosics include nylon, polyester, and acrylic, and are chemical derivatives of petroleum. Synthetic fibers are made from polymers, which must be either dissolved in a solvent or in a molten form.

Fiber	Description	Characteristics	Uses
Linen	Bast or inner bark fibers of flax stems; seed pods crushed to extract linseed oil; fibers up to one meter in length.	Most expensive cellulose fiber; has great strength and high rate of water absorption; crushes easily; inelastic.	Household textiles and clothing.
Cotton	Seed fiber of cotton plant; fibers separated from seed by ginning; average length, 50 mm; most widely used fiber.	Moderate strength and elasticity; easy to dye; good wet strength.	Clothing, household textiles, industrial fibers.
Jute	Stem fiber of jute plant; second only to cotton in amount used; fibers 2 to 3 meters in length.	Jute fibers (or gunny) are inexpensive, of medium strength, and deteriorate quickly in presence of moisture.	Burlap bags and sacks, rope and twine.
Sisal	Leaf fiber of sisal (heneguen) plant; fibers about one meter long.	Deteriorates in salt water; inexpensive; stronger than jute.	Floor mats, light cord, and binder twine.
Hemp	Bast or stem fibers of hemp plant, up to 2 meters in length; similar to flax in processing.	Has great strength and water resistance; coarser fibers than linen.	Rope, cordage, sailcloth; waste fibers (oakum) are used for caulking.
Wool	Hair fiber of sheep, staple length from 25 to 250 mm; most important of the hair fibers; lanolin oil is a byproduct of fleece.	Moderately strong, highly elastic, resilient, crease-resistant, easy to dye, extremely absorbent, pliant because of high oil content.	Clothing, blankets, carpets.
Silk	Filament fiber extruded from the spinning gland of the silkworm and hardened by exposure to air; filaments 3000 to 5000 meters long.	Stronger but less elastic than wool; lustrous filament which readily accepts dyes.	Fine clothing, textiles, fabrics and scarves; largely replaced by synthetic filaments such as nylon.

Figure 3.30 Common fibers and their uses.

The liquid polymer material is then extruded through the tiny holes of a spinneret to form continuous filaments. These filaments solidify as they pass through either air or a chemical bath. The long filaments may be processed into yarns or cut into staple lengths and blended with natural fibers. Blends are important in the production of textiles because they capitalize on the best qualities of each fiber, combining the look and feel of the natural fiber with the easy-care properties of the synthetics.

Fibers are the basis of fabric manufacturing. Any fabric made from fibers by any method can be broadly defined as a *textile*. Textiles include nonwoven fabrics, woven fabrics, and knitted fabrics. Fibers used to make nonwoven materials, such as felt, are arranged and processed directly into a fabric by moisture, heat, and pressure. Bonded fabrics are formed from a web or sheet of fibers held together by adhesives or treated with heat or solvents, which soften the fibers and bond them into position. Most textile fibers, however, are formed into long strands called *yarns* before being made into fabrics. Regardless of the type of fiber, natural or manufactured, the three primary processes employed to fashion them into usable products are spinning, knitting, and weaving, as exemplified in the woven hammock in Figure 3.31.

Figure 3.31   This comfortable and attractive hammock is a loose weave of soft-spun cotton strands. Courtesy of Crate and Barrel.

Fabrics are finished in order to change or improve their characteristics, appearance, or feel before being made into consumer products. The term *greige goods* is used to refer to fabrics that have not received any finishing process. Depending on the fiber content, certain finishes are routinely applied in preparing a fabric for end use. Cotton, for example, would typically be singed, desized, scoured, bleached, mercerized, tentered calendered, and inspected. A different sequence of procedures would be utilized for other fabrics. Additional finishes might be applied for specific purposes. Some common finishes include singeing, mercerization, compressive shrinkage, and water-resistance.

Singeing eliminates projecting fiber ends that cause roughness, pilling, and dullness if left on the fabric surface. Fiber ends are brushed up and subjected to open flames or heated plates to remove the unwanted fiber ends. Fabrics of natural and blended fibers may be singed.

Mercerization imparts a sheen to fabrics such as cotton and linen. The material is treated with a cold caustic chemical solution that swells the fibers, allowing them to reflect light and thus achieve a lustrous, silky appearance. Additionally, mercerization increases strength, improves dyeability, and partially reduces residual shrinkage of the fabric.

Compressive shrinkage produces cotton and linen fabrics that will have less than 1 or 2 percent shrinkage during laundering. Slightly damp fabric placed over a thick blanket is fed over a roller and against a heated cylinder causing the fabric to retract, which compresses or shortens the fabric. A fabric with the trade name *Sanforized* is an example of compressive shrinkage.

Water-resistant finishes utilize agents such as silicones, fluorochemicals, and ammonium compounds to coat individual fibers, causing them to resist wetting. Water-resistant finishes do not alter the appearance of the fabric, are comfortable to wear, and repel stains.

Some finishes are applied to fabrics with mechanical equipment, while others utilize chemical solutions to produce the desired result. Finishes are considered durable if they can withstand "normal" wear; others come off during wear or cleaning and can be reapplied. Whatever the requirement or specific characteristic, finishes offer the consumer many wear, care, and comfort benefits.

QUESTIONS AND ACTIVITIES

1. Prepare a report on the use of wood, fiberglass, and graphite, among others, as tennis racquet materials. List some of the advertised differences in the relative performance of these various racquets.
2. A reference was made in this chapter to the use of heavy steel sheet as a material for athletic lockers. They are strong and reliable but terribly noisy. How could steel doors be made less noisy? What other door materials might be used that may be quieter and yet would not sacrifice the factors of strength and durability?

3. Design a heavy, durable park bench employing a support structure made of cast iron or concrete, with seating planks made of wood or recycled plastic lumber.

4. List some typical uses of hardwoods and softwoods in product design.

5. Define the terms *ferrous, nonferrous, hardwood, softwood, thermoset, thermoplast, whiteware*, and *glass*.

6. Prepare a report of consumer products made of carbon fiber materials. List some of the special attributes of such products.

7. Select a simple product, such as a tool or a household utensil, and analyze it as to the appropriateness of the materials used in its construction. List any faults and suggest some design or material specifications changes.

8. List some of the strengths and weaknesses of plastic as a material for automobile bodies.

9. Design a chair made entirely of woven metal wires.

10. Select a product you have used that has failed because of an improper construction material. Redesign it to prevent a recurrence of the failure.

Manufacturing Processes

Manufacturing is the act of transforming or changing raw materials into usable products. Manufacturing *processes* are the special operations by which these material transformations are achieved. The first impression one gets when studying these processes is that there are so many of them. How does a designer or a process engineer select the best cutting process, for example, to make a specific part? The purpose of this chapter is to provide an overview of these processes to aid those who design products for mass manufacture.

PROCESS THEORY

For convenience, manufacturing processes can be classified as cutting, forming, assembling, and finishing operations. Essentially, these are the only things one can do to a material to change it into something useful. Additionally, it can be said that these imposed changes produce modifications in the mass or bulk of the workpiece, and cause a subsequent change in its geometry or shape. These modifications can be described as *mass reducing* (cutting), *mass conserving* (forming), *mass increasing* (assembling), and *mass conditioning* (finishing). Stated another way, an object can be given a particular shape by removing unwanted material (cutting), by redistributing the material (forming), or by joining pieces of material together (assembling). Finally, the surface of the material can be enhanced or conditioned (finishing). By these methods, products such as the simple but effective ski exerciser (see Figure 4.1) are created.

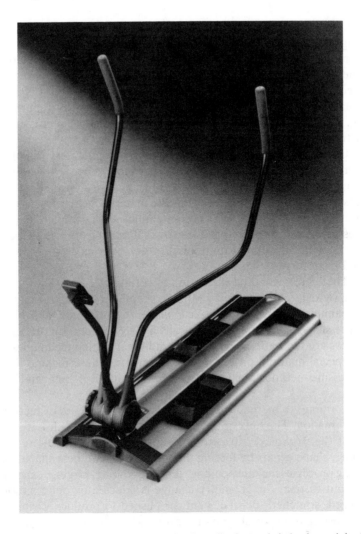

Figure 4.1 This ski exerciser is made of anodized extruded aluminum, injection-molded glass-filled nylon thermoplastics, finished steel stampings, and bent tubing. Many manufacturing processes were required to create this simple, usable, elegant exercising machine. Courtesy of Fitness Master.

GENERAL PROCESS DESCRIPTIONS

In *cutting* , shape change is achieved by *removing* material particles until the work-piece has the desired geometry. Cutting also is a *separation* process, where a part size is diminished by sawing or shearing to a specified dimension. Thus, cutting differs from hot or cold forming, by which bulk deformation is employed to create a new shape, but where no gross section of the workpiece is removed. An advantage of cutting is the production of intricate shapes having extremely accurate dimensions. Generally, material removal can be more precisely controlled than can plastic flow. Furthermore, the problems of shrinkage in hot-formed parts and springback in cold-formed parts are eliminated. Because cutting does not rely on plastic flow to modify the shape of a workpiece, a cut part can have the same properties as the starting piece. For example, machined metal parts can retain the benefits of prior cold work or heat treatment.

Cutting can be more economical than forming for very complex parts, or when only a few pieces are to be produced. For most shapes, however, cutting is more expensive and must be used judiciously. In comparison with metal forming, machining is slower, results in considerable waste material, generates heat, and may require greater crafting skill. Cutting is often employed to complete parts made by other processes. Some pieces are produced by first making a rough shape by forming, and then completing the part by machining. Automotive axles, for example, are generally turned from forged blanks.

There are several processes in the cutting category. Sawing and shearing are cutting by *separation* methods. Milling, shaping, planing, turning, drilling, boring, and reaming commonly are called *machining* processes and involve *removal* to produce shapes. *Abrading* is technically a material-removal process, although abrasive saws are separation devices. Nontraditonal cutting techniques are employed to meet special material processing requirements. These include thermal, chemical, and pressure methods, such as lasers, photochemical etching, and hydro-abrasive jet cutting.

Forming is a method where shape change is achieved by material manipulation or deformation. The bulk of the end product is equal, or almost equal, to the initial bulk of the workpiece. In cutting, a shape is produced by material removal. In forming, shape occurs by forcing the material to assume a shape. The fundamental forming processes include bending, drawing, rolling, forging, extruding, and casting. In all but casting, the geometry change is effected by deformation. Metal can be squeezed, stretched, twisted, bent, or pushed to produce various shapes.

Casting is a special type of forming where a fixed amount of soft material is forced into a mold cavity. The material is conserved; only the amount needed goes into the end product. Little is cut away to achieve the final part geometry. Forming is a most significant method to produce pieces to near final shape and size at great economy, and with certain structural attributes.

Assembling involves joining together several components to create a final product. As such, it is a bulk-increasing process because the total mass of the final product is the sum total of the masses of the individual parts. Parts can be assembled with mechanical fasteners such as nuts and bolts, by fusing them together as weldments, or by adhering them with glues and solders. Some assemblies are meant to be permanent, while others are meant to be disassembled as required.

Finishing involves workpiece surface modification for the purposes of protection or appearance. A fine piece of furniture is lacquered both to protect its surfaces and to make it more attractive. It is a mass conditioning process, inasmuch as the bulk of the end product remains essentially unchanged. A board may be sanded or a metal part surface-ground, methods whereby modest amounts of material are removed. Wood may be painted or a metal part chrome-plated, and in the process minute additive surface changes will occur. However, in both sets of examples, the bulk change is generally insignificant. Examples here include polishing, buffing, plating, painting, staining, anodizing, and heat treating. With the last three examples, no surface additions or subtractions occur. The surface structure or color is the sole resultant change.

It is significant to note that any discussion of material processing leads to the ultimate conclusion that process involves selection and option. For example, holes may be *generated* in a piece of metal in a variety of ways. They may be drilled, punched, milled, laser-cut, chiselled, chemically etched, pierced, abrasive-cut, liquid jet-cut, or thermally cut with a oxyfuel torch. How does one cut the hole? The decision is based on factors such as the quality and accuracy of the hole, the reason for having a hole in the first place, the kind of material being cut, the available equipment, the shape of the hole, the matter of automated systems, numbers of parts, and speed or economy. The many people involved in the design and manufacture of the part must make the final decision and select the most appropriate process. The same considerations must be made for the forming, assembling, and finishing processes. The following charts, illustrations, and discussions will provide additional manufacturing process information.

CUTTING PROCESSES

The chart in Figure 4.2 has been prepared as a means of organizing this set of processes for better comprehension and comparison. More detailed explanatory information can be found in the individual process paragraphs.

Sawing

This method of cutting typically is used to prepare workpieces for subsequent operations. For example, round bars of metal or sticks of wood are sawed to length for use as workpieces in lathe turning processes. This preparatory cutting generally is done with an automatic band, reciprocating, or circular saw. (See Figure 4.3.) Band saws also are valued tools to cut wood, metal, and plastic to irregular shapes, which may then be completed with additional machining operations. All sawing operations are chip-producing, leave rough surfaces, and generally require finishing operations.

Shearing

Although certain wooden parts and metal bars can be sheared, the process generally is used to cut sheet materials. It is an effective and economical method of preparing

CUTTING

The process of removing or separating pieces of material from a workpiece, thereby reducing the mass of the workpiece

TYPE	DESCRIPTION	EXAMPLES
SAWING TOOL WORKPIECE	Separating a workpiece using a tool with pointed teeth, equally-spaced in a straight line along a metal band, or about the circumference of a metal disk; chip-producing	Reciprocating hacksaws for metal and plastic; bandsaws and circular saws for wood, metal, and plastic; manual, power, or automatic equipment used
SHEARING TOOL WORKPIECE FIXED TOOL	Separating a workpiece by placing it between two sharpened blade tools which slide across one another; non-chip producing	Manual, power, or automatic tools used to separate wood, metal, plastic, and soft materials, by nibbling, punching, blanking, perforating, notching, trimming, slotting, slitting, lancing, and similar force operations
MILLING TOOL WORKPIECE	Removing material from a metal workpiece moving into a rotating, multi-toothed cylindrical tool; chip-producing	Operations on vertical and horizontal, power or automatic milling machines; precision contour-milling using computer-controlled equipment

Figure 4.2 Cutting process chart. Note that some cutting operations involve material removal, while others involve material separation.

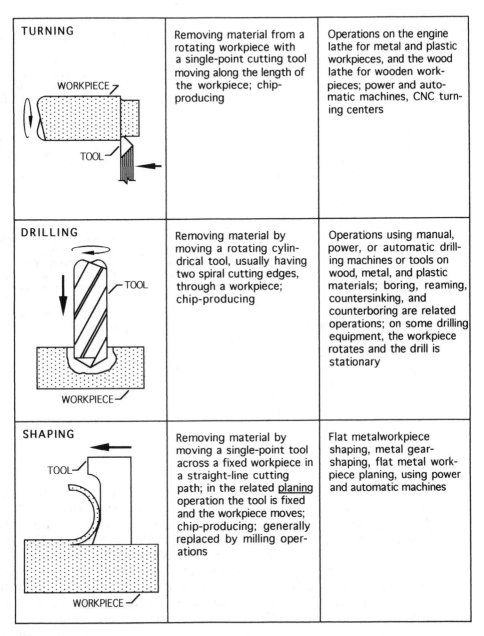

TURNING	Removing material from a rotating workpiece with a single-point cutting tool moving along the length of the workpiece; chip-producing	Operations on the engine lathe for metal and plastic workpieces, and the wood lathe for wooden work-pieces; power and auto-matic machines, CNC turn-ing centers
DRILLING	Removing material by moving a rotating cylin-drical tool, usually having two spiral cutting edges, through a workpiece; chip-producing	Operations using manual, power, or automatic drill-ing machines or tools on wood, metal, and plastic materials; boring, reaming, countersinking, and counterboring are related operations; on some drilling equipment, the workpiece rotates and the drill is stationary
SHAPING	Removing material by moving a single-point tool across a fixed workpiece in a straight-line cutting path; in the related planing operation the tool is fixed and the workpiece moves; chip-producing; generally replaced by milling oper-ations	Flat metalworkpiece shaping, metal gear-shaping, flat metal work-piece planing, using power and automatic machines

Figure 4.2

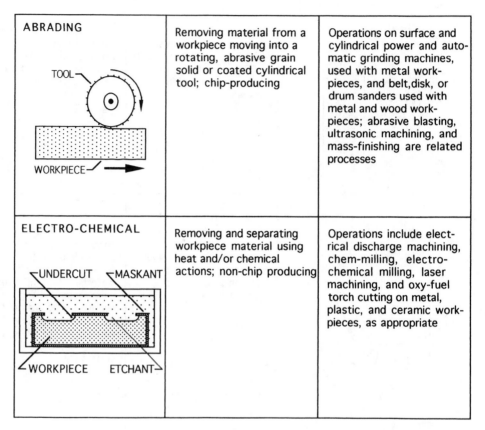

| ABRADING | Removing material from a workpiece moving into a rotating, abrasive grain solid or coated cylindrical tool; chip-producing | Operations on surface and cylindrical power and automatic grinding machines, used with metal workpieces, and belt,disk, or drum sanders used with metal and wood workpieces; abrasive blasting, ultrasonic machining, and mass-finishing are related processes |
| ELECTRO-CHEMICAL | Removing and separating workpiece material using heat and/or chemical actions; non-chip producing | Operations include electrical discharge machining, chem-milling, electrochemical milling, laser machining, and oxy-fuel torch cutting on metal, plastic, and ceramic workpieces, as appropriate |

Figure 4.2

sheetmetal blanks (or workpieces) and for subsequently giving an outline shape or inner detail to such blanks. The basic shearing processes are shown in Figure 4.4, and they have many applications in manufacturing. For example, the stretchout pattern for a galvanized sheetmetal electric box is produced by blanking, punching, slotting, lancing, and notching. (See Figure 4.5.) The side panels for the box are similarly produced. Subsequent operations include tapping or threading some of the punched holes, shallow-drawing depressions, twisting and bending tabs, and bending the box to shape. Final operations include attaching the side panels with machine screws to complete the box assembly. The cutting and forming processes used to manufacture such a box are performed using sets of progressive upper and lower dies positioned on a single machine for greater productivity. These are *pressworking* or *stamping* processes and include cutting and forming.

Shearing equipment is of two basic types. There are large *squaring shears*, which cut metal sheets and plates up to two meters (six feet) long, similar to using a paper cutter. Other equipment requires the use of *upper and lower dies* (or a punch

Figure 4.3 Using a circular saw to cut metal parts is generally called *cold sawing*. A similar machine is used to cut wood or plastic. Courtesy of Wagner Machinery Corporation.

and die), to effect such cutting as shown in Figure 4.6. Computer-controlled *flexible manufacturing systems* (FMS) are especially adaptable to fabricating metal sheet products. (See Chapter 8.)

Milling

This is one of the most versatile and widely used methods of removing material from a workpiece to create flat or contoured shapes. Milling differs from turning in that in milling the workpiece generally is stationary, and the tool moves to effect the cutting action. Although there are many different milling machine configurations, there are but three basic types, as shown in Figure 4.7.

The *vertical* mill has a rotating spindle that holds a cutting tool in a plane perpendicular to both the table and the secured workpiece, similar to a drill press. The tools resemble drilling tools, but shaped cutters also are used for slotting and dovetailing. On some machines, the head can be turned to different angles to position tools for special cuts.

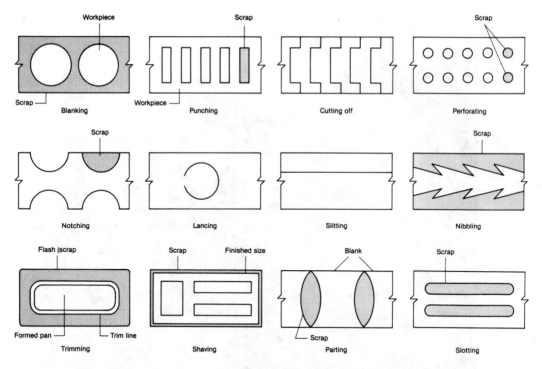

Figure 4.4 These basic shearing operations are used to produce parts from metal sheets. Plastic, paper, fabric, and wood are also sheared.

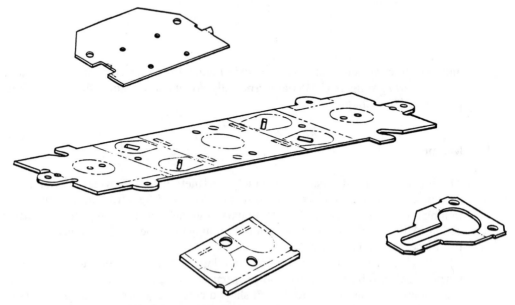

Figure 4.5 This sheet metal blank was processed by several shearing and forming operations to produce the workpieces for an electrical outlet box.

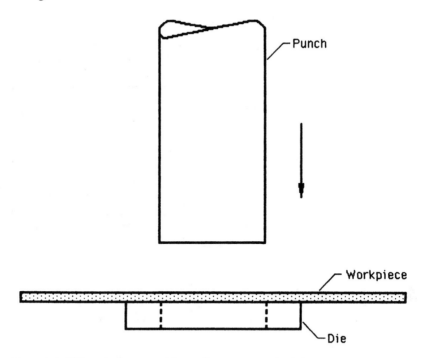

Figure 4.6 This typical upper and lower die arrangement is similar to punching a hole in paper. These tools are commonly used on automated equipment.

The *horizontal* mill has an arbor that holds a cylindrical cutter. The arbor fits into the spindle, which supplies the power to rotate the arbor and cutting tool. The table moves in x, y, and z axes to position the workpiece for cutting. A variety of cutting tools are used with this machine, to make both flat and shaped cuts.

Contour mills are sophisticated, computer-controlled machines used to produce parts requiring accurate, complex shapes. These machines are similar to a vertical mill with a movable head, except that they have a far greater range of tool movement. All three types are commonly employed in computer-controlled machining centers. (See Figure 4.8.)

Turning

This is a common method of fashioning cylindrical shapes of plastic, wood, and metal. The basic lathe is designed to hold workpieces between supporting centers located in the headstock, which is the source of turning power, and in the tailstock. Workpieces also can be chuck-mounted to permit the removal of material from the face of the workpiece or to bore holes. Single-point cutting tools are used. (See Figure 4.9.)

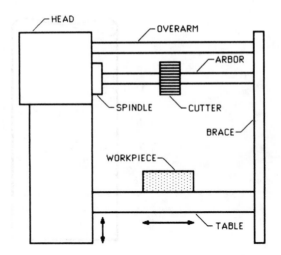

HORIZONTAL MILL
SIDE VIEW

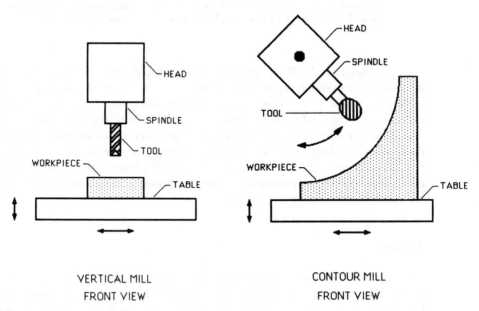

VERTICAL MILL
FRONT VIEW

CONTOUR MILL
FRONT VIEW

Figure 4.7 The three basic milling machines used to produce part geometries.

Figure 4.8 This modern computer-controlled copy milling center provides accurate duplication of two- and three-dimensional production dies and molds. Courtesy of Cincinnati Milacron.

Production machines used for turning operations are of several types, the most common of which are the *indexed tool* lathes. These machines feature a turret in the tailstock position, which contains a series of cutting tools. The turret can be indexed or rotated to present the required cutting tool. Some machines also have rear and front cross-slides, which hold additional tools to enhance the cutting capabilities. The availability of several cutting tools on one machine makes it possible to perform a series of operations on a single lathe, making it unnecessary to move the workpiece from machine to machine. For example, a chucked workpiece can be turned, grooved, drilled, countersunk, tapped, and chamfered in progressive operations, and under computer control. (See Figure 4.10.)

Drilling

Drilling is the most common method of machining round holes in workpieces. There are many types of drilling tools available, each designed for a specific hole-producing

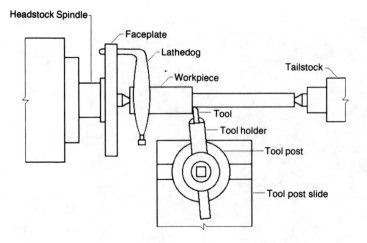

Turning between centers

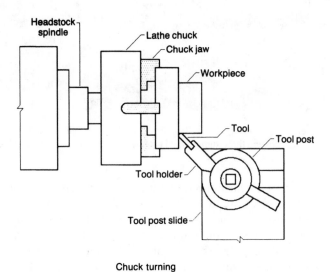

Chuck turning

Figure 4.9 Lathe workpiece holding systems.

operation. The twist drill is the common, general-purpose tool, applicable for use in most materials. Deep, accurate holes are made with long gun drills; there are others for making holes of various sizes in wood without tearing or slivering; and still others are used on concrete or ceramic materials or for truing cored-holes in metal castings. Additional drilling-type tools are used for reaming, countersinking, boring, and other similar operations. (See Figure 4.11.)

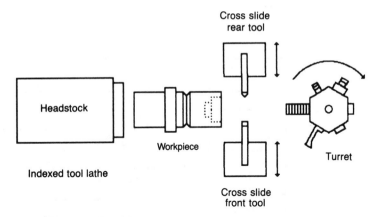

Figure 4.10 Indexed tool lathes are typical of the turning centers used in high-efficiency part production.

The machines used in these hole-producing operations range from hand drills, to drill presses, to computer-controlled, multiple-spindle machines that drill several holes in a workpiece simultaneously and automatically. The rate of production and the design of the part dictate the type of equipment to be used. (See Figure 4.12.)

Shaping and Planing

Aside from the highly complex, computer-controlled gear-shaping operations, shaping and planing as metal machining processes are not commonly used in mass production work. They find some use in toolmaking and machine repair, but such processes have generally been replaced by the faster, more efficient milling operations. The special wood shaping and wood planing processes are described later.

Abrading

Grinding is the most common form of abrading or abrasive machining, and is used to remove measurable quantities of material from a metal workpiece. Grinding wheels are made of compacted, bonded, abrasive grains, and are available in varying degrees of coarseness and in many shapes. The primary attribute of grinding is that it can be used to cut very hard materials, which are difficult or impossibile to cut by other machining methods. Grinding is used to produce precision flat or contoured cuts, and for interior or exterior cuts on cylindrical workpieces. (See Figure 4.13.)

Polishing is generally done with coated abrasive belts, drums, disks, and flat sheets. In these instances, the abrasive grains are adhered to a tough cloth or fiber substrate, to create the familiar "sandpaper." A variety of abrasive grains such as

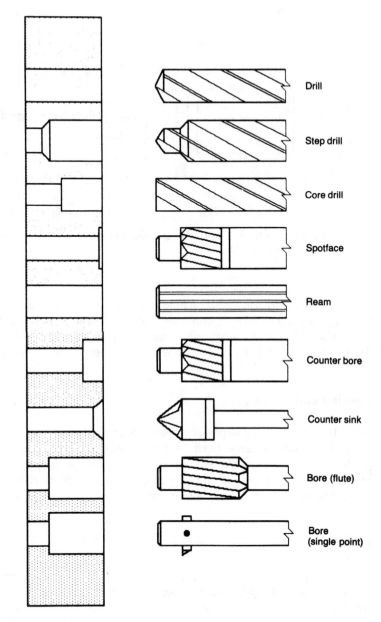

Figure 4.11 Common types of tools used to produce shaped holes with drilling machines.

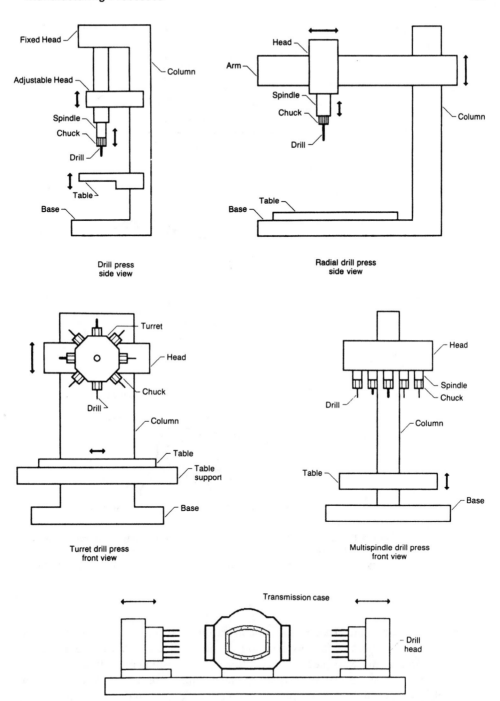

Figure 4.12   Common types of drilling machines, each having a special production function.

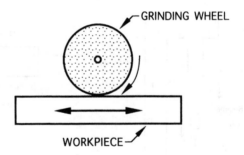

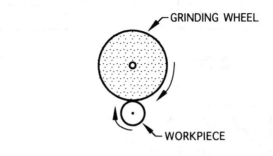

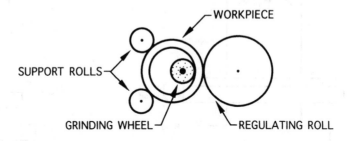

Figure 4.13 Basic grinding systems: (Top) Surface grinding, workpiece held on magnetic table. (Middle) External cylindrical grinding, workpiece held between centers. (Bottom) Internal cylindrical grinding, workpiece held by support rolls.

garnet, aluminum oxide, and silicon carbide are used in industrial applications. Polishing generally follows grinding in many abrasive machining applications, to produce a finer workpiece surface. *Buffing* is a finishing operation used to produce a high luster on objects such as silver-plated tableware, and is described later in this chapter. A number of equipment types are used in coated abrasive machining, such as single- and multiple-head polishing (or sanding) belt and drum systems.

Electrochemical

This category deals with nontraditional systems, all of which involve the use of electricity, heat, or chemicals, separately or in some combination. The methods are employed in situations where the cutting either cannot be done by any usual process, or where the dictates of economy, efficiency, or high accuracy come into play. Some of the more common methods are described here.

Electrical discharge machining (EDM) is a process by which metal is removed from a workpiece by the erosive action of a controlled electric spark to produce intricate holes and surface cavities. *Wire EDM* is a related process employing a moving brass or copper wire as the electrode. The cutting action is similar to that of a bandsaw but with considerably more precision, and it produces complex solid shapes. (See Figure 4.14.) Chemicals can be combined with electricity for related applications. *Electrochemical machining (ECM)* removes metal from a workpiece by chemical dissolution aided by an electric current. A noncontact tool controls the cutting, and material is removed in the form of atom-sized particles instead of chips. Both the tool and the workpiece are immersed in the electrolyte in these cutting processes. In *chemical milling (CHM)* operations, metal is removed by the chemical attack of acid on the workpiece. (See Figure 4.2.) No tool or electricity is used, and areas of the workpiece coated with a resistant will not be subject to the acidic action. *Photo chemical milling (PCM)* is similar to CHM, but employs a photographic process to prepare the workpiece for especially fine work.

Ultrasonic machining (USM) processes employ the action of an abrasive grain and water slurry rapidly oscillating between a tool and a workpiece. USM is especially applicable to hard, brittle materials. *Water jet cutting* involves the use of a narrow, high-velocity stream of water directed at a workpiece. The process finds increasingly wide use in separating nonmetallics such as plastic, cardboard, wood, cloth, and food products. It is a noncontact cutting technique that is dustless, clean, and nonheat generating. With the introduction of abrasive particles in the jet stream, steel plate up to 75 mm (3 inches) thick can be cut.

Several familiar welding techniques can be used in *thermal cutting*. Oxyfuel and plasma arc processes are typically employed. Plasma arc cutting provides a narrow kerf and heat affect zone (HAZ), resulting in low metal workpiece damage and distortion. Also, this method is especially adaptable to computer-controlled systems. Oxyfuel systems are used to cut heavy steel plate. They have a wide HAZ and a resultant rough cut requiring further treatment.

Laser technology is used industrially to cut, weld, drill, scribe, heat-treat, and measure a variety of workpiece materials and sizes. (See Figure 4.15.) The system employs a collimated beam of light focused on a workpiece to cut it, and has a similar narrow beam, narrow HAZ as the plasma arc process, but with the added advantage of a safer, more easily controlled heat/light beam. It, too, is readily adapted to computer control. Extreme accuracy is possible with laser technology.

All of these thermochemical systems provide a reservoir of special noncontact cutting methods to meet the demands of modern material processing.

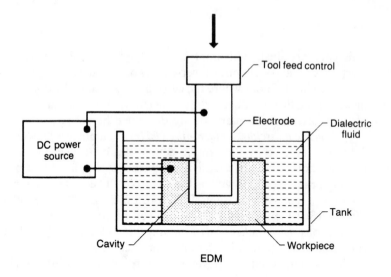

EDM

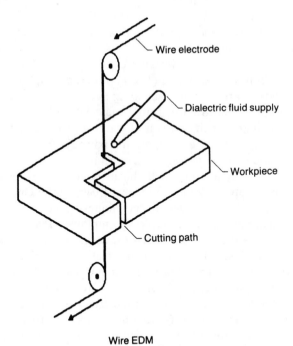

Wire EDM

Figure 4.14 Electrical discharge machining is a nonontact technique used to produce precision parts. Note the difference between EDM and wire EDM.

Figure 4.15 A typical laser-cutting system. Courtesy of PTR-Precision Technologies, Inc.

WOOD-CUTTING OPERATIONS

There are several wood-machining methods related to the common cutting techniques just described. Typically, the multiple-point cutting tools run at speeds considerable higher than those employed in metal machining. These speeds are necessary to remove the fibrous material from wood workpieces. Slower speeds would result in splintered, rough surfaces, and prominent machining ridges.

In manual wood-cutting operations, a rough piece of lumber is sawed to the desired length, and one surface is *jointed* so that it is flat and true. The board is then *planed* (or milled) to the required thickness. One edge is next jointed to "true" it, the board sawed to width, and the sawn edge jointed. The resultant smoothed and sized board is now a workpiece for subsequent product application.

Jointing and planing operations are further described and illustrated in Figure 4.16. In modern production work, jointing, planing, and sawing operations are highly automated to produce great quantities of boards at a reasonable cost. Special equipment can perform all these operations simultaneously and efficiently.

TYPE	DESCRIPTION	EXAMPLES
WOOD JOINTING 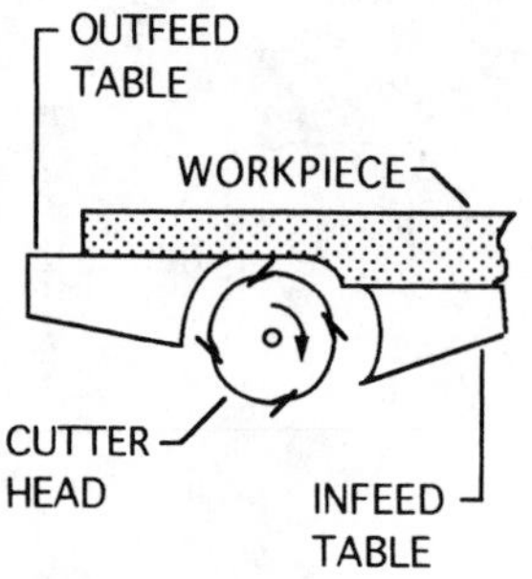 OUTFEED TABLE WORKPIECE CUTTER HEAD INFEED TABLE	Removing material from the bottom surface of a wooden workpiece moving into a multi-blade cylindrical cutting tool; tool (or head) is rotating at high speed; chip-producing	Process used to true surface of workpiece prior to thickness planing; also capable of cutting edges, tapers, bevels, and chamfers; power and automatic jointers
WOOD PLANING 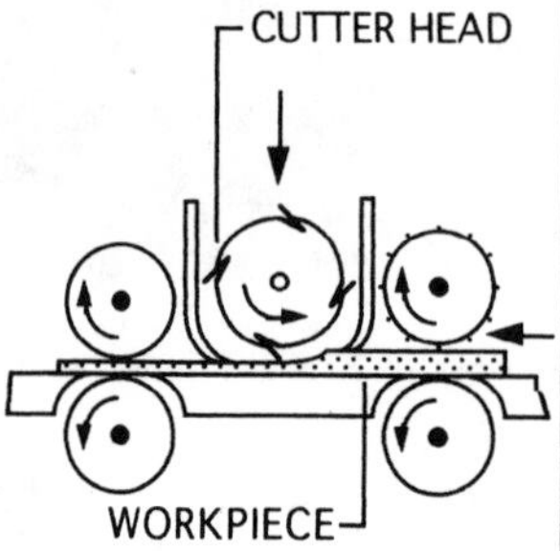 CUTTER HEAD WORKPIECE	Removing material from the top surface of a wooden workpiece moving into a multi-blade cylindrical cutting tool; tool (or head) is rotating at high-speed; chip-producing; also called wood milling, and wood surfacing	Process used to cut boards to thickness after jointing; special, shaped cutting heads used to produce moldings and trim pieces; power and automatic planers; high production machines mill to thickness and width in one simultaneous operation

Figure 4.16 Wood jointing and planing machines smooth and true rough workpieces.

Routing and shaping are techniques used to produce contoured workpieces, to shape edges, to cut out parts, and to remove surface material. (See Figure 4.17.) Computer-controlled equipment is employed to produce a variety of products. Here, a flat wooden workpiece is secured on a vacuum table and is moved under the fixed router head to effect the cutting. Single or multiple parts can be cut from the workpiece.

To conclude this discussion of wood processes, the chair in Figure 4.18 is presented as an example of a modern chair built using totally manual techniques. The author watched its manufacture in New Delhi, India, in 1967, beginning with a teak log being hand sawn and the subsequent boards hand scraped and sanded. The legs were cut to shape with flat saws, then turned between two steel pin centers mounted in large stones. This rudimentary lathe was powered by a worker spinning it with a bow and string, and guiding the cutting chisels with his toes. Holes were drilled using

TYPE	DESCRIPTION	EXAMPLES
WOOD ROUTING 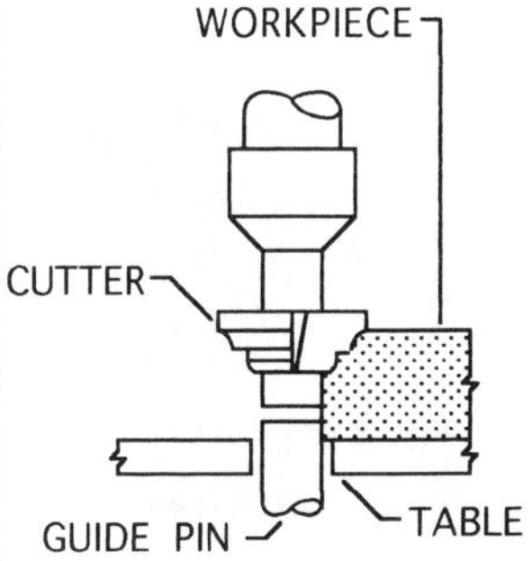 WORKPIECE CUTTER GUIDE PIN — TABLE	Removing material from the top surface or edge of a wooden workpiece moving into a shaped, overhead cutting tool; tool enters the workpiece as drill in a drill press; high-speed operation; chip-producing	Process used to cut recesses, interior and exterior detail, and shaped pieces; computer controlled equipment used for precision surface carving; can function as a mortising, boring, and drilling machine as required; handheld routers used for trimming and edging
WOOD SHAPING 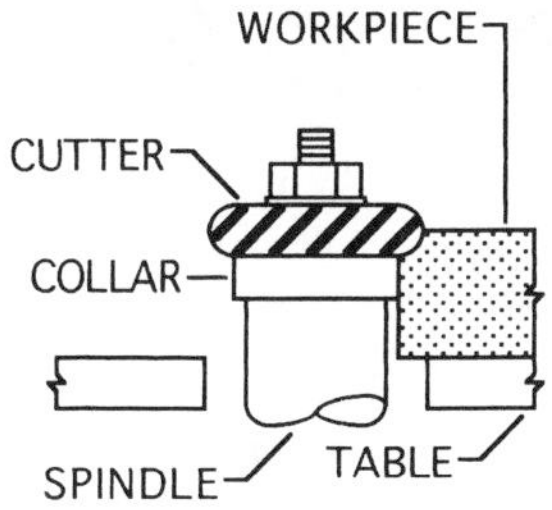 WORKPIECE CUTTER COLLAR SPINDLE TABLE	Removing material from the edges of a wooden workpiece moving into a cutting tool, bottom-mounted at table level; high speed operation; chip-producing	Process primarily used to shape edges of rectangular or round workpieces; power and automatic equipment used

Figure 4.17 Automatic wood routers and shapers are used to produce contoured workpieces.

a bit rotated by another bow and string arrangement. Additional hand shaping and smoothing operations were employed to complete the pieces, which were then assembled with natural glues and secured with an impossible system of ties and clamps. It was a rewarding experience to observe these crafters at work, creating this magnificent piece. The author purchased it for $24.00.

FORMING PROCESSES

As stated earlier, forming imparts shape by means of workpiece deformation. Some common forming methods are shown in Figure 4.19 and elaborated upon in the paragraphs that follow.

Figure 4.18 A fine example of a teak lounge chair made entirely by hand.

Drawing

There are two primary methods used in this forming technique. *Shallow drawing* involves only a moderate movement of material to produce such objects as automotive hub caps, trays, and pans, and generally requires upper and lower dies of opposing shapes. This is a relatively simple operation achieved on a single-action type of press equipment. *Deep drawing* involves the movement (or deformation) of a considerable amount of material when forming objects of greater depth, such as gas cylinders and beverage cans. This process requires the use of double- and even triple-action presses, so that the drawing occurs as series of gradual draws to avoid tearing the workpiece, or *blank*, as such workpieces are usually called. A comparison of these two methods is shown in Figure 4.20.

In deep drawing, the bottom die is called the *drawing die* and the upper die is known as the *punch*. When forming an aluminum soft-drink can, the drawing die is a

FORMING

The process of giving shape to a workpiece without adding material to, or removing material from, the workpiece, thereby conserving the mass of the workpiece.

TYPE	DESCRIPTION	EXAMPLES
DRAWING 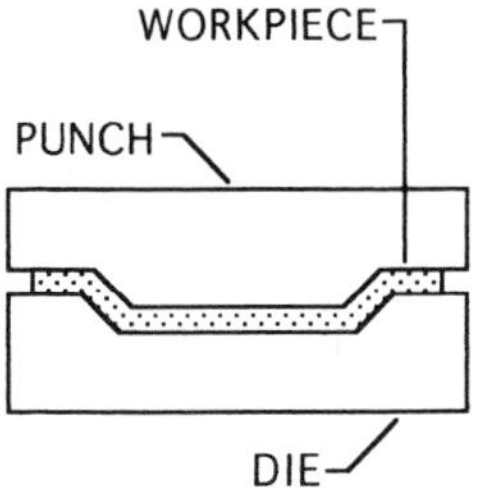	Forcing cold metal sheet materials between two opposing (upper and lower) forming dies	Deep and shallow drawing operations using fixed dies; drawing operations using flexible dies; metal spinning, and embossing, using power and automatic equipment
WIRE DRAWING 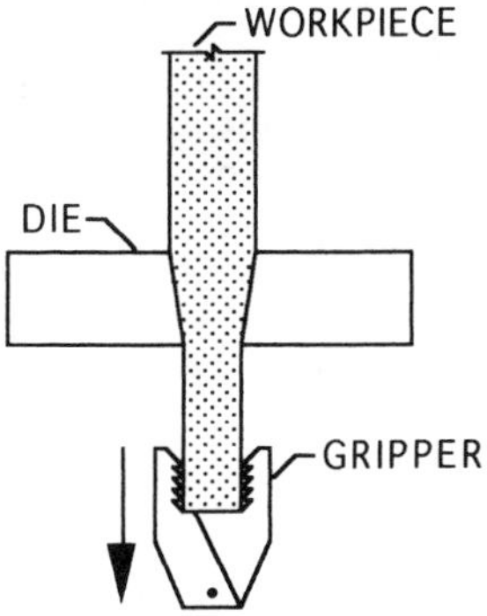	Pulling a cold metal rod through a series of reduction dies to form a wire or a tube	Various tube and wire drawing operations using power and automatic equipment
BENDING 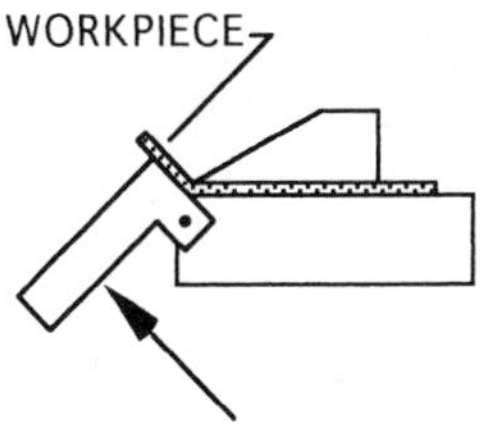	Applying force to a cold workpiece to uniformly strain it around a straight axis	Operations on folding and pressing brakes, bar and tube benders, forming rolls, progressive die roll formers, and edge formers, using manual, power, and automatic equipment

(continued)

Figure 4.19 Forming process chart. These methods generally involve the deformation of blanks to create shapes.

TYPE	DESCRIPTION	EXAMPLES
ROLLING 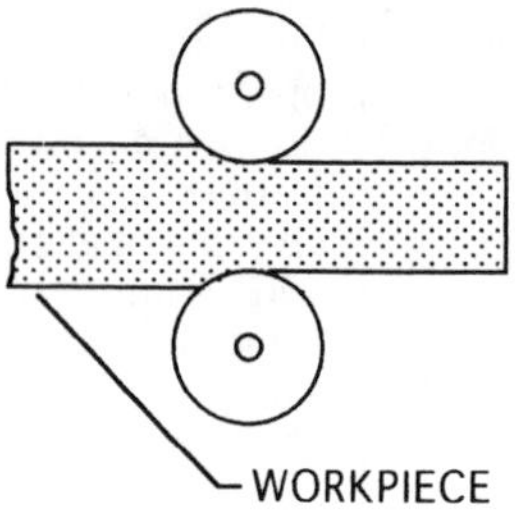 WORKPIECE	Forcing a hot or cold metal workpiece between two pressure-rollers to reduce it to a flat sheet or bar	Various sheet and bar rolling operations using power or automatic equipment; process also used on soft plastic and ceramic materials
EXTRUDING 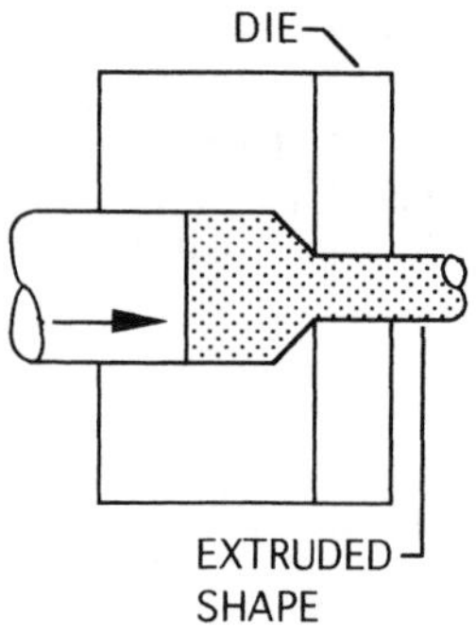 DIE EXTRUDED SHAPE	Forcing a warmed metal workpiece (billet) through a die opening which changes the cross-sectional area of the workpiece; forward, backward, and upset methods also used	Various operations used to produce extruded metal shapes, such as molding, tubing, channels, and trim pieces, using power and automatic equipment; plastic and ceramic shapes also are extruded

Figure 4.19

round hole and the punch a solid cylinder. During forming, the circular blank is pressed in graduated steps through the drawing die, and wraps around the punch to form the can. In subsequent operations, a can lid is shallow drawn and secured to the can after filling. Employing a square-hole die and a square-shaped punch results in a square-shaped container. Obviously, any such shape can be produced with the appropriate dies.

Flexible dies also may be used in both shallow and deep drawing to simplify the tooling needed to produce complex shapes. (See Figure 4.21.) For example, the *Guerin* process employs a tough rubber pad as the concave die mounted on the movable ram of the press. The blank is placed on the punch, and the ram forces the rubber pad against the blank and then down around the punch to form the part. This is an economical method of shallow drawing because the rubber die can be used with any punch shape. Punches are easier to prepare than are concave dies.

TYPE	DESCRIPTION	EXAMPLES
FORGING 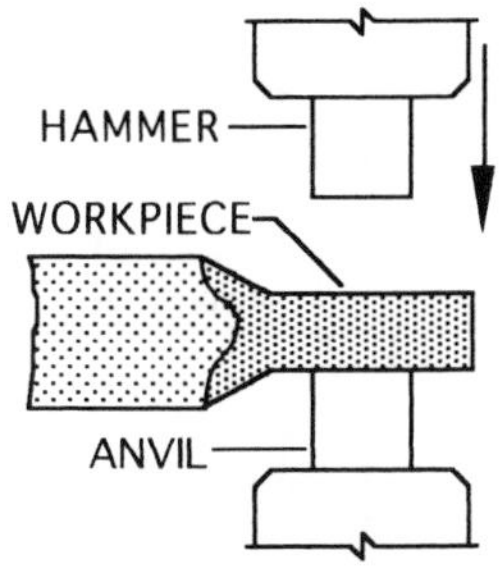 HAMMER — WORKPIECE — ANVIL —	Applying directed blows or pressure to force a hot metal workpiece to take the shape of forging dies; cold operations also used	Various open and closed die forging operations using manual, power, or automatic equipment
CASTING 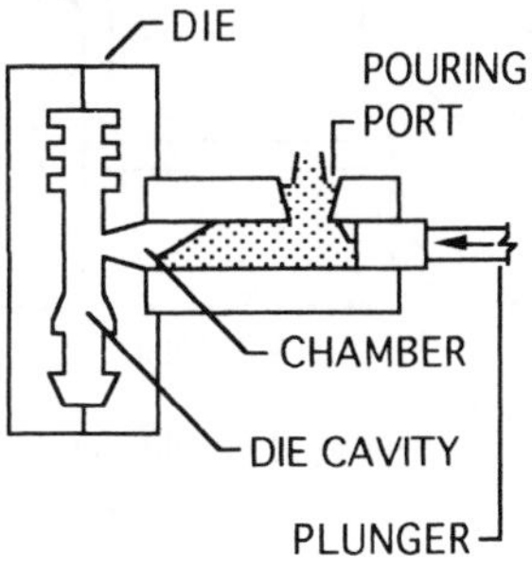 — DIE POURING — PORT — CHAMBER — DIE CAVITY PLUNGER —	Pouring or forcing molten metal, or another semi-liquid into a hollow mold cavity, and allowing it to cool or set	Various metal casting operations such as die, sand, investment, centrifugal, and shell; powder metal operations; and casting plastics and ceramics, using manual, power, or automatic equipment

Figure 4.19

The *Verson-Whealon* process utilizes a special press to apply force to the rubber pad by means of a flexible fluid sack instead of a movable ram. This results in higher working pressures to produce larger, deeper parts. The *Marform* process employs a combination rubber pad and hydraulic piston to produce exceptionally deep (three times the blank diameter) parts. These are among the numerous deep and shallow drawing methods used in product manufacture. *Wire drawing* is a totally different technique used to form wires, tubes, and bars of many configurations and sizes.

Bending

As a forming process, bending contrasts with drawing in that bending is unidirectionally stretching a workpiece along a straight axis, much as one would fold and

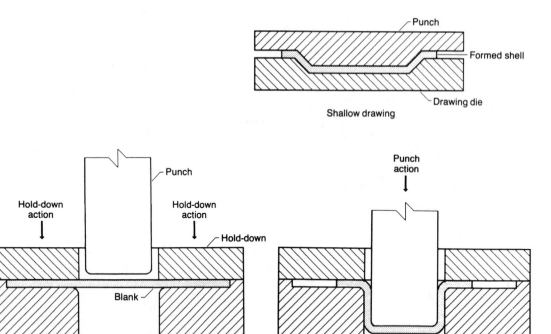

Figure 4.20   Shallow and deep drawing can be done with rigid punches and dies.

crease a piece of paper. It is a common forming operation applicable to sheet, plate, bar, tube, and other structural shapes. A variety of methods and equipment are employed in bending, the simplest of which is *wing brake bending*, shown in the chart in Figure 4.19. *Press brake bending* employs rigid, mated, precision upper and lower dies, or flexible rubber pads (similar to flexible drawing dies) in conjunction with a shaped punch. (See Figure 4.22.)

   Roll bending is a method of producing cylindrical shapes from both sheet and heavy metal plate. Equipment used includes cylindrical rolls for forming cylinders or curves, and wheel-shaped rolls for bending bars and tubes. (See Figure 4.23.) *Roll forming*, different from roll bending, is used to form strips of sheet metal into long, straight lengths of various cross-sections. The material is shaped as it passes through a series of paired concave and convex rollers called *roll sets* . An important use of this method is to produce welded metal tubing in a continuous rolling-welding operation. *Simple rolling*, on the other hand, is squeezing a thick workpiece between rollers to reduce it to a flat sheet or bar. This method is used to produce raw materials for subsequent operations, and can be done on cold or hot metal, plastic, clay, and pizza dough.

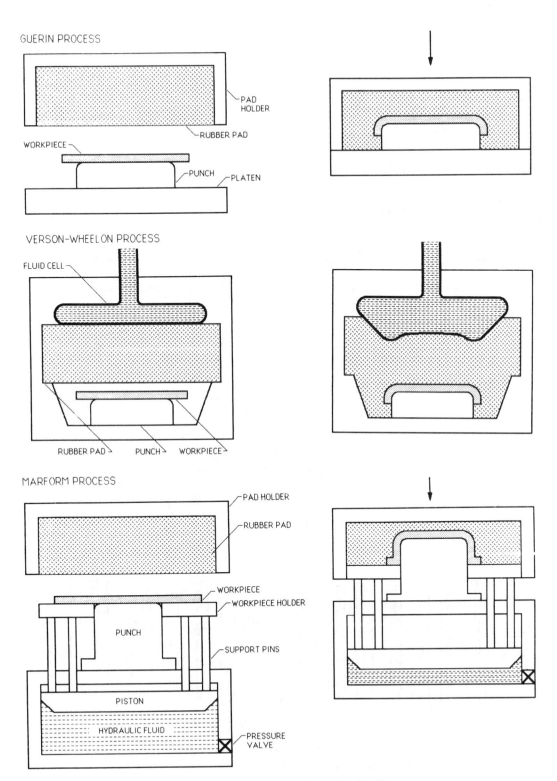

Figure 4.21 Typical systems employing flexible dies.

169

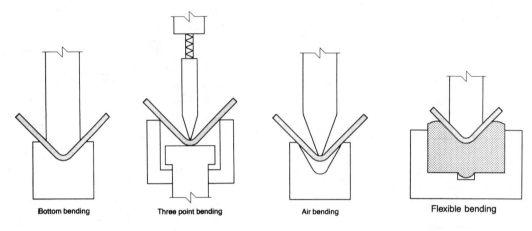

Figure 4.22 Brake bending systems. Note that only the bottom bending method requires a matching punch and die.

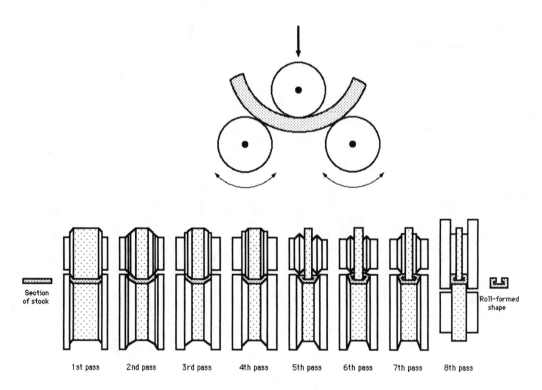

Figure 4.23 Roll bending and roll forming are compared here. Totally different product configurations result from these two roll systems.

Extruding

Technically, extruding is the process of compressing a metal workpiece (or billet) and forcing it through a die opening to produce a part with a cross-section equal to that of the die opening. Plastics, ceramics, as well as other materials can be processed in this way. As a practical, familiar example, toothpaste emerges from its tube opening by extrusion. Materials to be extruded obviously must be in a plastic or semi-liquid form. Metals can be worked cold to good advantage, but often are warmed to facilitate the extrusion. Certain breakfast foods are made in a related extrusion process. (See Chapter 8.)

There are three variations of this process. There is *forward* extrusion if the material is forced in the same direction as the ram (or punch) travels. *Backward* extrusion occurs when the material flow is opposite to the motion of the ram. (See Figure 4.24.) *Upsetting* is a special operation whereby a metal rod is secured and struck on its end to cause a diameter and shape change. It is a process in common use to manufacture "headed" fasteners such as bolts and rivets. It also is called *cold-heading* or *cold-forging*.

Forging

The word *forging* conjures up images of products of great size, strength, and toughness, but in reality most forgings are of modest size. The process is based on the fact that while a cold steel rod 13 mm (0.5 inch) in diameter can be flattened with a large hammer on an anvil, it is much easier to work the metal if it is red hot. Hot workpieces can be shaped by impact or squeeze pressures. (See Figure 4.25.) *Open die forging* is an impact method where a hot workpiece is hammered between flat dies. The dies do not confine the metal, and the shaping of a part occurs by manipulating the workpiece as the flat dies continually hammer it. In *closed die forging*, or impression die forging, the workpiece is shaped between dies that completely enclose the metal and thereby control its shape. This feature makes it unnecessary to manipulate the workpiece during forging. In operation, the workpiece is placed between the dies and squeeze pressure is applied until the hot metal completely fills the die cavity.

A special cold-forging operation is called *swaging*, which is a method of tapering or decreasing the cross-sections of the ends of round metal bars or tubes. This is achieved by the repeated hammer action of rotating dies as they close and open against the workpiece. Swaging is used to produce tapered metal chair legs, hollow needles, bicycle spokes, and precision parts. (See Figure 4.26.)

Casting

Very simply, casting is the process of pouring a liquid material into a cavity where it hardens to form a part. *Molding* is another term used to describe this process, but

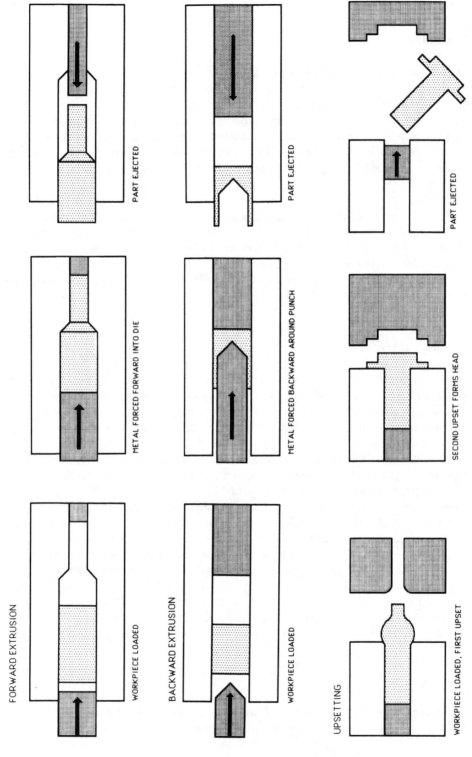

Figure 4.24 Examples of forward and backward extrusion and upsetting. These processes are especially useful in the production of metal fasteners.

172

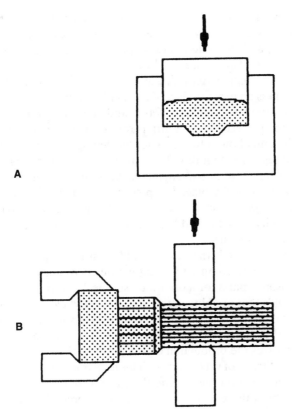

A

B

Figure 4.25 Closed (A) and open (B) forging die systems.

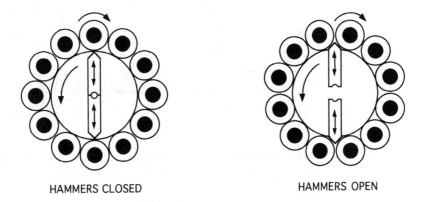

HAMMERS CLOSED   HAMMERS OPEN

Figure 4.26 A typical swaging die, showing the forming hammers open and closed. Two-, three-, four-, and six-die configurations also are used.

actually the mold is the prepared cavity into which the liquid is poured. Metal, plastic, and ceramic materials commonly are used to produce cast parts. Metal casting is an ancient process, and its contemporary applications are numerous and varied. Several important methods are discussed here.

Sand casting is a common casting method. Basically, this involves the preparation of a wooden pattern that is an exact replica of the final cast part. The pattern is placed in a metal containment vessel called a *flask*, and specially prepared sand is rammed tightly around the pattern. The pattern is removed, and molten metal is poured into the mold cavity and allowed to set. After hardening, the sand is broken away to reveal the cast pieces. (See Figure 4.27.) This is an oversimplification. In reality, complex castings such as automotive engine blocks require several patterns, with numerous sand cores strategically placed to provide the coolant passages and cylinder openings. The attributes of such a cast piece are the mass to absorb engine shock and vibration, the ability to withstand high operating temperatures, and the durability to function over years of use.

The problem with sand casting is that a new mold must be prepared for each cast piece. This problem can be overcome by using various *permanent mold* techniques. For example, the bronze parts for a door lock-set mechanism can be cast in specially prepared steel molds that are hinged to swing open and close to provide a mold cavity. This is a typical method for the high production of small parts to close tolerances, and with smoother finishes than obtained with sand casting.

Investment casting involves the production of wax patterns that are attached to sprues and runners to form a "tree" of wax patterns. This matrix is invested in a metal flask of liquid plaster, which sets up rapidly. The flask is baked and the wax melts away to create a mold cavity in the plaster. Molten metal is forced into the cavity under pressure to produce the casting. This process permits the manufacture of

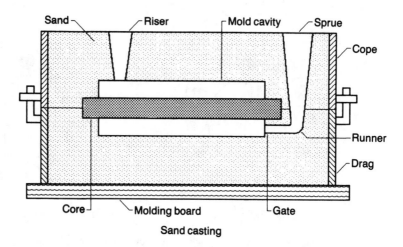

Figure 4.27 Typical flask arrangement for sand casting.

numerous precision parts with complex shapes. A common application of this process is in dentristry, where a wax crown is shaped to fit a tooth, and then cast in gold to produce the familiar gold crown or filling.

Die casting utilizes permanent metal molds to produce numbers of complex parts quickly and economically—parts that seldom require any secondary operations. In this process, molten metal is forced under high pressure into a die cavity, where it cools quickly and is ejected. After ejection, the process is immediately repeated. A die casting unit is shown in the forming process chart. (Refer to Figure 4.19.)

Powder metallurgy (P/M) is a casting-related process used to manufacture complex, accurate, and reliable ferrous, nonferrous, and composite parts. P/M parts, for example, are made by mixing metal alloy powders and compacting this mix in a steel die. The resultant shape is next sintered in a furnace to metallurgically bond, but not melt together, the particles at lower-than-melt temperatures. (See Figure 4.28.) Metallic, plastic, and ceramic powders can be mixed to make parts economically and at production rates. Parts produced in this manner have some unique attributes. Ferrous and nonferrous parts can be oil impregnated to function as self-lubricating bearings, resin impregnated to act as seals, infiltrated with a low-melting temperature metal for additional shock resistance, and heat treated or plated. Composite automotive brake shoes can be made of metals and minerals, so that they are both tough and heat-resistant.

Earlier, reference was made to the casting of plastic materials. A most common method is plastic *injection molding*, which is closely related to metal die casting.

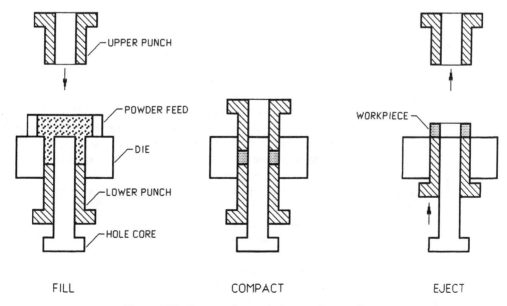

Figure 4.28   A system for producing powder metallurgy parts.

Injection molding involves the introduction of thermoplastic granules into a heating chamber where they are brought to a melt temperature. A screw or a plunger mechanism then forces the hot plastic into metal die cavity where the injection molded part is quickly cooled and ejected. This is a fast method of producing complex and precision shapes. (See Figure 4.29.)

There are several other important plastic-forming methods. *Extrusion molding* (similar to metal extrusion) employs a screw-feed to force heated thermoplastic granules through a die orifice to produce plastic tubing, pipes, and other linear shapes. *Transfer molding* is a special technique for casting thermosetting plastics. Here, a plastic powder charge is preheated and plunged under pressure into a die cavity. Heat-resistant parts, such as automotive distributor caps, are made this way. In *compression molding*, thermoset resin is placed in a heated mold where it is compressed to form a variety of parts. *Vacuum forming* is used to draw heat plastic sheets into a shaped die cavity to form the common plastic containers. *Blow molding* involves the positioning of a limp, hot plastic sack or parison in closed dies, and pressure is applied to form a jar or bottle, some with threads to accept a cap closure.

ASSEMBLY PROCESSES

The components of a typical product are manufactured by the cutting and forming processes described earlier. At some point in the manufacturing cycle, the parts must come together and be assembled to create the final product. The primary categories of the assembling (or joining and fastening) processes are shown in Figure 4.30 and described below.

Mechanical

A wide range of mechanical assembly devices are used in industry to join workpieces permanently or semi-permanently. Examples include nails, staples, rivets, nuts and bolts, screws, sheet metal lock joints, and plastic pins. Some general statements are in order here. Nails and wood screws are meant to join two pieces of wood together permanently; nuts and bolts are not, for they can be taken apart to disassemble a product. Rivets are permanent fasteners, but sheet metal screws are not, so the back panel of a television set is attached with such screws for service convenience. Fasteners are therefore selected on the basis of whether product parts are to be joined permanently or if they must at some time be disassembled.

Another selection factor relates to simplifying product assembly. It is far easier and less costly to join two metal sheets with a self-drilling sheetmetal screw than with bolts and nuts. The former requires one operation, the latter requires several operations, including a tedious placement of the nut on the bolt, and then tightening. Also, consideration must be given to automatic versus manual assembly. Again, the sheetmetal screw wins out in this example.

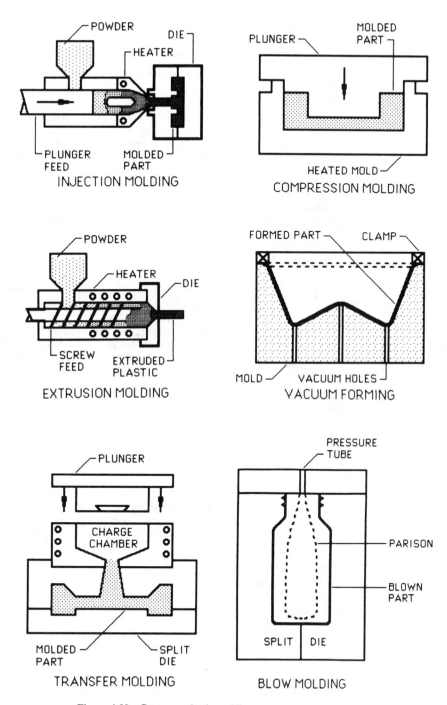

POWDER

DIE

HEATER

PLUNGER
FEED

MOLDED
PART

INJECTION MOLDING

PLUNGER

MOLDED
PART

HEATED MOLD

COMPRESSION MOLDING

POWDER

HEATER

DIE

SCREW
FEED

EXTRUDED
PLASTIC

EXTRUSION MOLDING

FORMED PART

CLAMP

MOLD

VACUUM HOLES

VACUUM FORMING

PLUNGER

CHARGE
CHAMBER

MOLDED
PART

SPLIT
DIE

TRANSFER MOLDING

PRESSURE
TUBE

PARISON

BLOWN
PART

SPLIT   DIE

BLOW MOLDING

Figure 4.29 Common plastic molding processes.

ASSEMBLING

The process of joining or fastening workpieces together, permanently or semi-permanently, thereby <u>increasing</u> the mass of the final workpiece

TYPE	DESCRIPTION	EXAMPLES
MECHANICAL 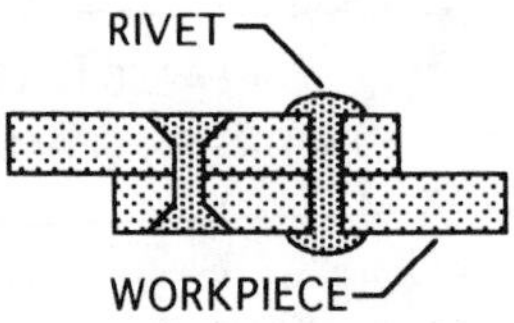 RIVET — WORKPIECE —	Permanent or semi perma-nent joining of like or un-like materials with fast-eners, locking devices, or special joints	Threaded fasteners, rivets, nails, retaining rings, clips, snaps, and sheet metal joints, used to as-semble workpieces of any material, using manual, power, or automatic equip-ment; weaving and knitting methods for fabrics; and metal or plastic screening
ADHESIVE 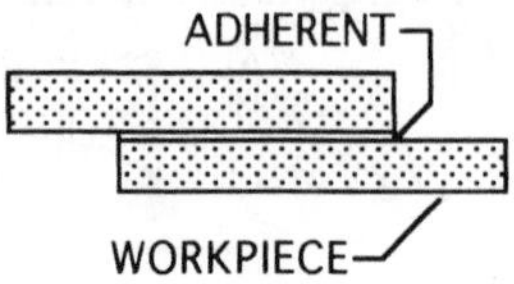 ADHERENT — WORKPIECE —	Permanent joining of like or unlike materials with glues or cements, or metal-lics with molten metals	Solder, brazing rod, and various glues, cements and adhesives, using manual, power, or automatic equip-ment; some adhesives are non-permanent
COHESIVE 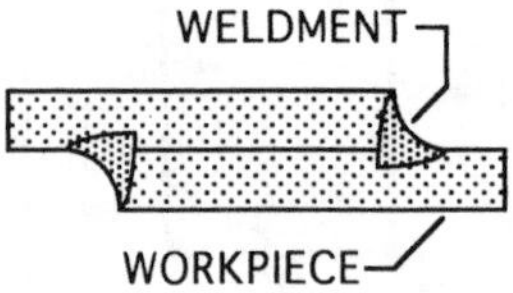 WELDMENT — WORKPIECE —	Permanent joining by fuzing metallics together with molten metal, or heat and pressure	Various forms of welding, using manual or automatic equipment

Figure 4.30 Assembling process chart. These methods involve either permanent or semi-permanent functions.

The strength of the product assembly is yet another factor. The cylinder head of an automotive engine is joined to the block with studs and nuts because this assembly must be strong enough to withstand the tremendous internal combustion forces. Yet it must be taken apart for service and repair, and so must be a semi-permanent type of assembly, with an appropriate mechanical fastening device. (See Figure 4.31.)

Several special mechanical assembly methods are employed in the manufacture woven products, such as cloth and metal screening.(See Figure 4.32.) For example, *weaving* is the process of making fabrics from two sets of threads that interlace at right angles. Those running the length of the fabric are called the *warp*, and those lying across the width of the piece are called the *weft*. There are many weave patterns, such as hopsack, herringbone, tweed, and satin, each designed to produce material that is loose or close, soft or hard, or having different decorative patterns or textures. *Knitting* is used for those fabrics requiring flexiblity and porosity, such as that used for underwear, sweaters, and stockings. Here, the loop is the basic structural unit, with many variations of pattern. These and other methods also are used to make woven plastic and metal screening products. Fabric-finishing methods include washing, dyeing, printing, and drying.

Adhesive

The terms *adhesion, adhesive fastening*, or *bonding* imply that two workpieces are joined permanently by a distinctly different substance. In theory, brazing, soldering and gluing (adhering) fall into this category. However, because brazing and soldering involve the use of molten metal bonders, they will be grouped with welding as part of the section on cohesion. The words *glue, cement,* and *adhesive,* often are used interchangeably and therefore are loosely synonymous. However, the omnibus term *adhesive* generally is used in literature dealing with this technology, and will be used as such here. Adhesives can be *natural*, those having organic origins in animal and vegetable matter, or *synthetic*, which are chemical compounds. Natural adhesives are typically inexpensive, water soluble, easy to apply, and often are for wood-gluing applications.

Synthetic adhesives generally are more expensive and must be applied with strict precision controls for consistent results. There are two classes of synthetics, paralleling those used to define plastic materials. *Thermosets* are cured by chemical reations at specified temperatures, and they cannot be heated and softened after an initial curing. Structural adhesives generally are thermosets. Conversely, *thermoplasts* can be heated and softened repeatedly after setting (and therefore reused). The popular "hot-melts" are a good example of this type.

The use of adhesives in product assembly is common and is continually growing. Several reasons for this include the following:

1. They provide for a uniform distribution of forces over the entire bond area, thereby eliminating stress concentrations caused by rivets, bolts, or spot welds.

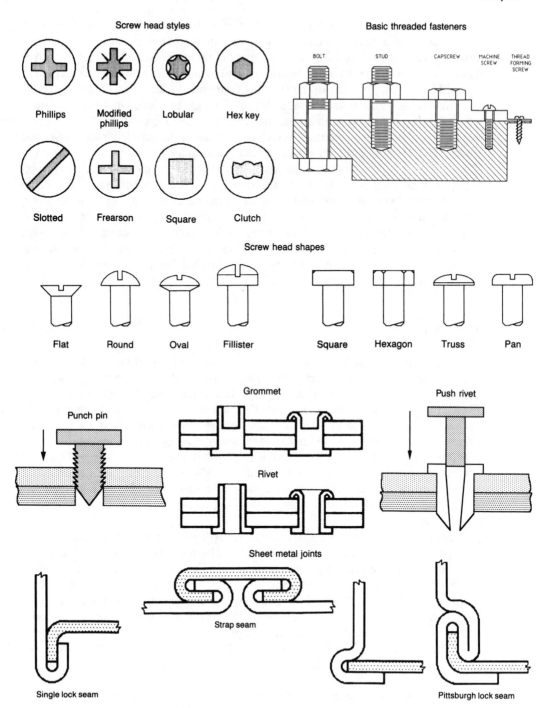

Figure 4.31 Representative mechanical fastenings.

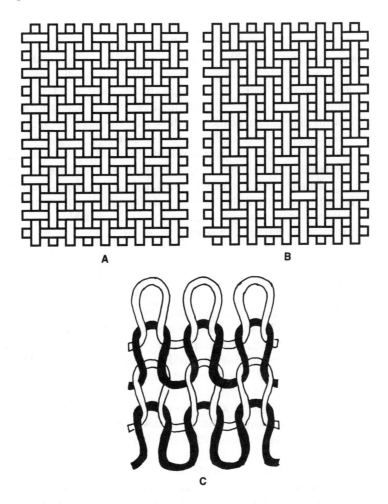

Figure 4.32 Weaving (above) and knitting (below) are special mechanical fastening methods. These examples are representative of many methods.

2. They are effective in joining both similar and dissimilar materials and those that are thin and fragile, or robust.

3. They can serve as environmental seals, heat insulators, and vibration dampers, and can prevent electrolytic corrosion.

4. They result in joints with smooth contours and no surface imperfections.

5. They cure at relatively low temperatures, and thereby prevent part distortion.

6. They are lightweight, generally inexpensive, and readily adaptable to automatic assembly methods.

7. They are applicable to almost any assembly requirements because of the many types of adhesives available.

The disadvantages include relatively low service temperatures, rigid process control and cleanliness, and the difficulty of joint inspection. In reality, adhesives are no more costly or difficult to use, and offer a considerable number of attributes.

As alluded to earlier, adhesives used in product manufacture may be either structural or nonstructural. High-strength *structural* types are those that will join almost any material over extended periods of time and are fundamentally load-bearing. Typical applications include aircraft wings, automotive gas tanks, electronic components, and refrigeration coils. Conversely, *nonstructural* adhesives are employed in no-load bearing applications such as cushions, upholstery, carpeting, automotive trim and molding, bandages, and package labels.

Finally, adhesion is the force (or the adhesive) that acts to hold the adherends (or workpieces) together. The popularity and advantage of this fastening system are in large part due to the numerous adhesives available for use in automated manufacturing, as described in the chart in Figure 4.33.

Cohesive

This type of assembly is exemplified by a series of welding processes where the metal workpieces to be joined are melted together to become one homogeneous part—permanent, strong, and inseparable. Similarly, plastic parts can be melted together with heat or special solvents, and clay pieces can be mixed at the joint surfaces and then fired to produce a strong ceramic product. As shown in the chart in Figure 4.34, there are different welding processes available for distinct applications. Those especially applicable to product manufacture will be described here. It is important to mention that many of the processes used to weld metal parts together can be employed to cut metal workpieces with equal facility. These will be identified as appropriate.

Oxyfuel gas welding (OFW) achieves coalexcence (or melting together) with a flame heat supplied from a mixture of oxygen and other gases such as acetylene. This slow-joining method can cause workpiece distortion and is not generally used in mass manufacturing. The exception is oxyfuel cutting, which is adaptable to automatic controls and is widely used for producing shapes from heavy steel plate.

Arc welding (AW) involves fusing the faying (closely fitted) surfaces of workpieces using heat produced by an electric arc generated between an electrode and a workpiece. The process is fast, lessens the chance of workpiece distortion, and lends itself to automatic process control. *Shielded metal arc welding (SMAW)* is a manual method, using a flux-coated metal stick as a consumable electrode that produces the heating arc with the workpiece. The electrode acts as filler metal for the joint, and the flux shields the material from oxidation.

Another AW process employs argon gas to shield the work area, making fluxes unnecessary. This type is called *gas metal arc welding (GMAW)*, also commonly called *MIG (metal inert gas) welding*. The system has a continuous coil of filler metal wire stored on a reel that is automatically fed to the work area. The advantages of GMAW are clean welds, little part distortion, and automatic controls.

Another gas-shielded system is *gas tungsten arc welding (GTAW)*, also known as *TIG (tungsten inert gas)*. It differs from GMAW in that it features a nonconsumable tungsten electrode to produce the arc, and it can be used with or without a filler metal. It is especially useful in joining pieces of metal without burning or distorting them. Automatic production applications are limited inasmuch as it is essentially a manual process.

Plasma arc welding (PAW), like GTAW, employs a nonconsumable tungsten electrode. The arc, however, is conducted to the workpiece by a controlled stream of plasma, or electrically charged gas. (See Figure 4.35.) This process commonly is used for autogenous welds (those requiring no filler metal) made with sophisticated automatic equipment. The high operating temperatures of about 30,000°C (50,000°F) permit rapid welding, with a nondistorting narrow flame, in thick workpieces, and with great accuracy. Plasma arc *cutting* is commonly used in computer-controlled, high-production rate sheetmetal fabrication operations.

Resistance welding (RW) is fusion welding derived from a combination of heat and pressure. The heat is generated within the workpieces by their resistance to the flow of electricity. The workpieces (generally metal sheets) are held under pressure between electrodes and heated until molten, and these molten surfaces flow together and solidify to form the weld. No filler metal is used. The tight contact held between the workpieces shuts out air and eliminates the need for flux or shielding gas.

The most common type of RW is *resistance spot welding (RSW)*, where lapped metal sheets are joined by local fusion (spots) caused by an electric current flowing between opposing electrodes. Automobile body members are spot welded automatically with welding robots—a method that is economical, relieves human drudgery, and raises product quality.

Ultrasonic welding (USW) is done by applying high-frequency vibrations to workpieces held together under pressure. The combination of clamping force and vibration on the faying surfaces results in a metallic bond being formed. This technique often is used to join thin metal and plastic workpieces.

Electron beam welding (EBW) achieves coalescence from a concentrated beam of high-velocity electrons directed to the weld area. This narrow beam must be generated in a vacuum because the electrons are easily deflected by gas and air molecules that tend to scatter the beam and render it ineffective. Although the EBW equipment is expensive, the method does offer such advantages as speed, quality, automatic control, and very narrow fusion zones, and therefore no product distortion. EBW generates small amounts of radiation, requiring appropriate operator protection.

Laser beam welding (LBW) achieves coalescence of workpieces by directing a narrow, concentrated beam of amplified light to the weld area. The laser resembles the electron beam in that both are concentrated beams of intense energy. The main difference is that the light energy of the laser beam is converted to thermal energy

INDUSTRIAL ADHESIVES

TYPE	CHARACTERISTICS	USES
Acrylic	Thermoplast, quick-setting, tough bond at room temperature; two component; good solvent, chemical and impact resistance; short work life, odorous, ventilation required.	Solar panels, ceramic magnets, fiberglass and steel sandwich bonds, tennis racquets, vehicle body components, metal parts, plastics.
Anaerobic	Thermoset, easy to use, slow curing, bonds at room temperature; curing occurs in absence of air, will not cure where air contacts adherends; easy to use; one component; good cohesive, low adhesive strength; not good on permeable surfaces; can flow into areas and cause unexpected bonds.	Close-fitting machine parts, such as shafts and pulleys, bolts, and nuts, studs, keys, bearings, bushings, core plugs, and pins; (not for use on plated metals; glass, ceramic, and plastic assemblies).
Cyanoacrylate	Thermoplast, quick-setting, tough bond at room temperature; one component; easy to use; colorless; no clamps because of rapid cure; common as "superglue."	Electronic components, gaskets, musical instruments, locking of nuts and bolts, lipstick cases, most plastics, metals, rubber, ceramics, hardwoods, and automotive body trim.
Epoxy	Thermoset, one or two component, tough bond; strongest of the engineering adhesives; versatile, wide variety of formulations to provide different pot life, cure and temperature rates, and bond rigidity; familiar as two-component adhesives; high tensile and low peel strengths; resists moisture and high temperatures; resists most solvents and chemicals; difficult to use, requiring measuring and mixing equipment; one component epoxies cure at elevated temperatures.	Metal, ceramic, and rigid plastic parts, rubber.
Formaldehyde --urea --melamine --phenol --resorcinol	Thermoset, strong wood to wood bonds. *Urea* is inexpensive, versatile, available in powder or liquid forms, must be combined with a catalyst to produce hard, rigid, water-resistant bond at room temperature. *Melamine* is more expensive, cures with heat, bond is waterproof. *Phenol* cures with heat, bond is waterproof. *Resorcinol* forms waterproof bonds at room temperature. *Urea-melamine* and *phenol-resorcinol* combinations can be used with radio frequency heating systems for fast, sure bonds.	All types of wood joint and plywood bonding.

Figure 4.33 Adhesives applications chart.

INDUSTRIAL ADHESIVES (cont'd.)

TYPE	CHARACTERISTICS	USES
Hot Melt	Thermoplast, quick-setting, rigid or flexible bonds; joins permeable and impermeable adherends; easy to apply; familiar hot melt gun use; low heat, and solvent resistance; good moisture resistance; solid, becomes fluid when heated, bonds upon cooling; gap filling; brittle at low temperatures; based on ethylene vinyl acetates, polyolefins, polyamides and polyesters.	Bonds most materials, packaging, bookbinding, textiles, metal can joints, furniture, footwear, carpets, oil and air filters.
Phenolic	Thermoset, oven-cure, strong bond; high tensile and low impact strength; brittle, easy to use; long shelf life; cures by solvent evaporation; one component, resistant to elevated temperatures.	Acoustical padding, cushioning, brake lining, clutch pads, oil and air filter webs, abrasive grain bonding, foundry sand bonding, core bonding, honeycomb structures.
Pressure sensitive	Thermoplast, variable strength bonds, primer anchors adhesive to roll tape backing material, a release agent coated on back of webb permits unwinding of roll; peel, shear and tack strengths, permanent or removable; based on polyacrylate esters and various natural and synthetic rubbers.	Surgical, masking, binding, and electrical tapes, packaging and mailing labels, metal foils, clear acetate film, protective coverings for acrylic and high quality metal sheets, shelf and wall coverings, wood tape, duct tape.
Silicone	Thermoset, slow curing, flexible, bonds at room temperature; high impact and peel strength; rubber-like; one and two component; easy to apply; excellent sealer; long pot life when covered; good chemical and environmental resistance.	Gaskets, sealants.
Urethane	Thermoset, flexible, bond at room or oven cure temperatures; water resistant; difficult to mix and use; good gap-filling qualities.	Fiberglass body parts, rubber, fabrics, and other materials to metal parts.
Water Base --animal --vegetable --polyvinylacetate --natural and synthetic rubbers	Thermoplast, water-soluble and water-dispersed bonds; inexpensive, nontoxic, nonflammable, and set by the evacuation of water. *Animal* bone and hide glues are protein, strong bonds, low moisture resistance; common as "brown" liquid glues. *Vegetable* glues, based on starches, and resins; poor moisture resistance. *Polyvinyl acetate* is a water emulsion, dries quickly, tough bond, low resistance to heat and moisture, common as "white" liquid glues. *Rubbers* are used as contact cements and pressure resistive tapes.	Wood, fabric, paper, and leather workpieces, dry seal envelopes, carpet backing, tire cord bonding, and packaging.

Figure 4.33

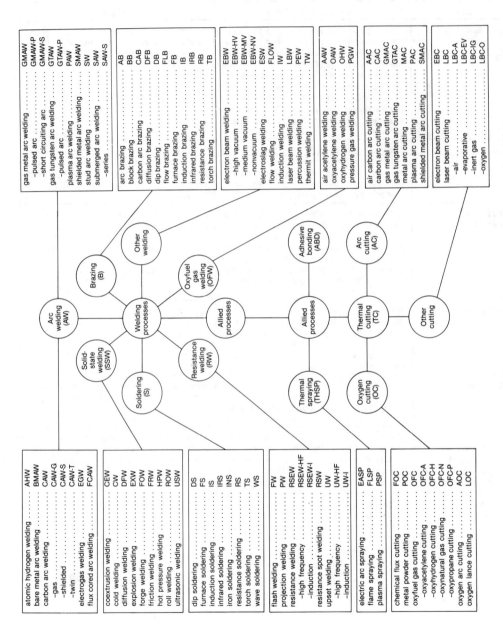

Figure 4.34 Master chart of welding and allied processes. Courtesy of American Welding Society.

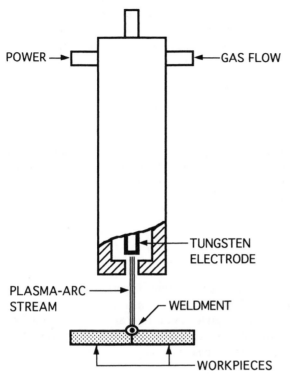

Figure 4.35 Plasma arc welding technique.

when it strikes a nontransparent substance such as metal. This feature makes it possible to weld tiny metal filaments inside a clear glass vacuum tube. Laser welding can be used in sensitive, difficult joining situations, where excessive heat can damage parts.

Brazing and Soldering

Although these two metal-joining methods are often classed as welding processes, they differ in that they involve the formation of metallic bonds by using filler metals with melting points below the melt temperatures of the workpieces. Both are permanent joining systems, but because the workpieces are not melted, the resultant bonds are not as strong as weldments.

Brazing (*B*) is performed by heating the workpieces and then applying a fluxed filler metal that melts to produce the joint. Flux is required. The resultant joint is smooth and neat, and typically is used to join small parts made of steel, brass, or copper. The filler metal generally is bronze, silver, or nickel.

Soldering (*S*) is similar to brazing except that the operating temperatures are lower, and the filler metal is a solder comprised of tin, lead, silver, antimony, and copper. Many lead-free solders are being used for safety reasons. Electronic component assembly generally is achieved with solders.

FINISHING PROCESSES

Finishing is generally the final phase of product manufacture, and its purposes are appearance and protection. According to the chart in Figure 4.36, mechanical, coloring, and coating are the primary systems used to obtain attractive, protective, and durable product finishes.

Mechanical Finishing

This category involves the removal of small amounts of workpiece surface material by polishing, buffing, mass-finishing, and abrasive blasting. Note that these all are related abrasive-cutting operations, used for smoothing and cleaning, and are generally followed by coating or coloring. For example, a walnut chair would be sanded (polished), wiped clean, then stained (colored), and finally varnished (coated) to produce the final chair finish.

Polishing is an intermediate abrading operation that follows grinding and precedes buffing. Its purpose is to remove deep scratches, nicks, discolorations, and other imperfections that may mar the surface of a workpiece, thereby reducing finish quality. Polishing materials typically are the same coated abrasives used in cutting. Polishing equipment includes belt, drum, and disc sanders (or polishers), both manual and automatic, to which workpieces are fed to cause the abrading action. Special contoured polishing heads are used to smooth the edges of workpieces.

Buffing is an operation used to apply a high luster to an object, such as buffing silver-plated tableware. Cloth-buffing wheels, or drums coated with liquid or paste-buffing compounds, are used. The drums rotate above a feed table containing a number of spoons or knives that are moved under the abrasive coated drum to effect the buffing. An important distinction to be made is that while polishing is an abrasive operation used to remove unwanted material, buffing primarily involves the gentle plastic deformation of a workpiece surface. This "flows" material into minute scratch areas to produce a bright or satin appearance.

Mass-finishing, or tumbling, is a process where small metal parts are loaded into a container together with an abrasive medium, such as grains, shells, or pellets. When a vibrating motion is applied to the container, the contents tumble and rub together to clean and enhance the surfaces of the parts. This is a very economical way of conditioning large quantities of small parts.

FINISHING

The process of modifying a workpiece surface for the purposes of protection and/or appearance, thereby <u>conserving</u> the mass of the workpiece

TYPE	DESCRIPTION	EXAMPLES
MECHANICAL 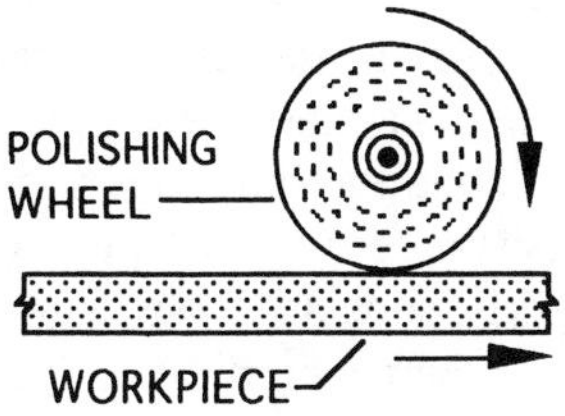 POLISHING WHEEL WORKPIECE	Removing minute amounts of material from a work-piece surface, or displacing surface material, to im-prove appearance, and/or prepare a workpiece for subsequent finishing operations	Abrasive polishing, buffing, blasting, sanding, and tum-bling; chemical etching, hammering, embossing, planishing, and knurling, using manual, power, or automatic equipment on a range of materials
COLORING 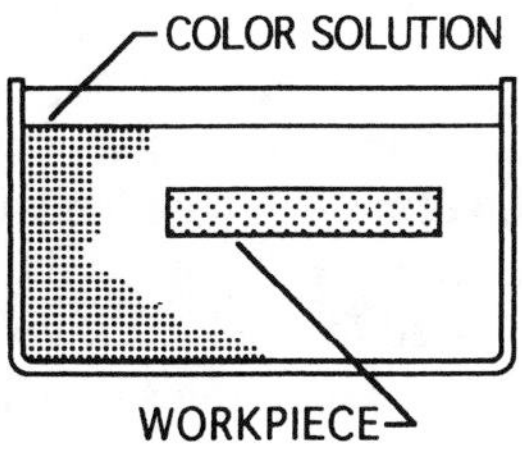 COLOR SOLUTION WORKPIECE	Applying penetrating liquids, chemicals, or heat to a workpiece, to effect a color change for the purposes of appearance and/or protection	Dyes or stains applied to woods or fabrics; surface conversion chemicals such as "gun bluing" to metals; color anodizing aluminum; or heating metals, using manual or automatic equipment
COATING 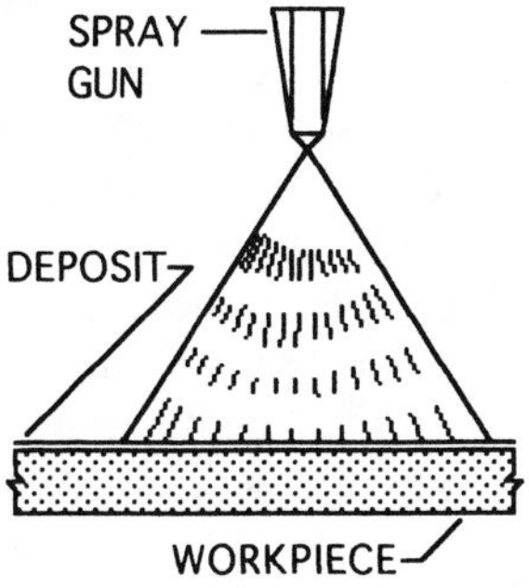 SPRAY GUN DEPOSIT WORKPIECE	Applying a layer of a finishing substance to the surface of a workpiece, for the purpose of appearance, and/or protection	Paints, printing inks, enamels, lacquers, electro-plating materials, and plastic coats, applied manually, or with power and automatic equipment on a range of materials

Figure 4.36   Finishing process chart.

Abrasive blasting is the action of abrasive particles propelled by compressed air on a workpiece. This is a common workpiece-cleaning method, and it also can be used to impart a matte or frosted finish on most materials.

Coloring

Very simply, coloring is staining a wooden part to secure a desired hue or tint. Plastic, cloth, and paper are dyed for the same reasons. Metal parts are colored with heat, or by chemical *conversion* methods. Surface conversions are inorganic films produced by chemical reactions to metal surfaces. *Gun-bluing* is a classic example, as are the less desirable rusting and tarnishing. Conversion coatings (coatings in name only) provide protection against corrosion and wear resistance, improve appearance, and sometimes are followed by organic coating processes, such as lacquering. *Anodizing* aluminum and some other nonferrous metals is an electrochemical conversion process, and colorants may be used.

The many methods of graphic reproduction technology, or "printing," are ways of imparting color or pattern to a range of materials. As such, it is a special form of finishing, and will be described in Chapter 9.

Coating

Coating is exactly what the term implies. Paint, lacquer, or varnish are applied to workpieces by dipping, brushing, spraying, or rolling for protection and appearance. Modern paints are tough, durable, and attractive. (See Figure 4.37.) Porcelain enamel coatings are fine, glassy, vitreous powders fused to steel bathtubs or sinks, or to sheetmetal refrigerator boxes. These enamel coating techniques are expensive but very durable.

Electrolytic coatings are exemplified by objects that have been electroplated, such as a silver tea service or a chrome automobile bumper. This is a method whereby a metal workpiece (the cathode) is coated with another metal (the anode) in an electrolyte solution, as electric current passes between the anode and cathode through this solution. Other electrocoating methods include the use of electrically charged paints or powder particles, which are dispensed in a chamber and are attracted to the workpiece, and followed by curing or baking. (See Figure 4.38.) Some representative finishing methods are shown in Figure 4.39.

The tape measure, Figure 4.40, is a classic example of the range of finishing methods required to produce a quality product. Some finish features include an abrasive-resistant, durable blade coating, and easy-to-read printed blade markings; a high-strength chrome-plated case; embossed areas of the case for gripping; a pressure-sensitive colored graphics overlay for product identification; and colored plastic thumb toggle lock to conveniently hold the blade at a desired measurement. All of these features should be considered early in the design stage, as part of a philosophy of designing for finish quality.

Figure 4.37 A simply styled, attractive, white-painted hardwood nightstand. Courtesy of Crate and Barrel.

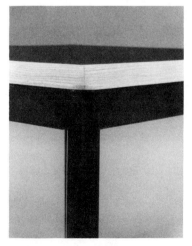

Figure 4.38 Note how powder coating maintains the integrity of the metal ribbing pattern in these attractive leg and rail details. Courtesy of Krueger.

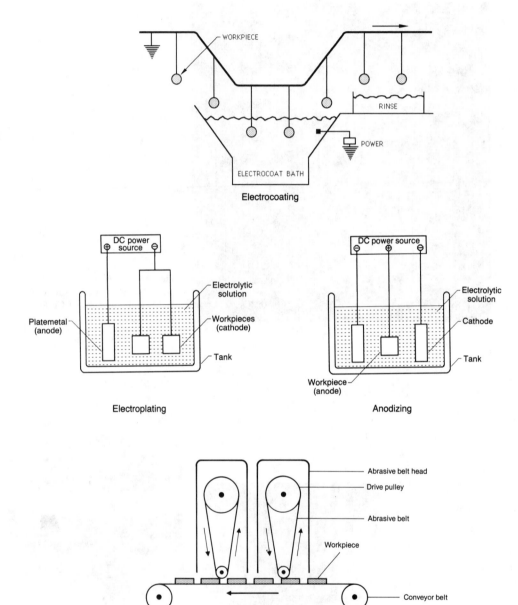

Figure 4.39   Representative material finishing methods.

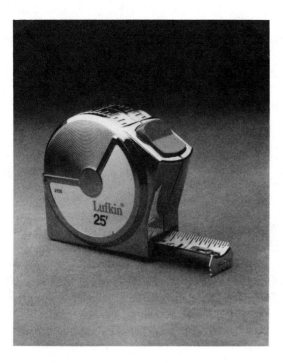

Figure 4.40 This attractive, durable, and user-friendly measuring tape features many finishing processes. Courtesy of Cooper Tools.

QUESTIONS AND ACTIVITIES

1. Prepare an analysis of the comparative processes used to manufacture the wood and aluminum baseball bats currently on the market. What are the performance differences between these two types of bats?

2. Select a common product, such as a fingernail clipper, a ballpoint pen, pliers, a lipstick case, a sandwich cookie, jeans, or tennis shoes. Prepare a product analysis report to include dimensioned sketches of the part pieces; identification of materials used and why; statements of the function of each part; descriptions of the processes used to produce each part; and a description of how the product is assembled. Also include a summary of how the product could be improved so that it is better and more easily manufactured. You may not know the precise answers to some of the aspects of this research activity, but you are encouraged to make reasonable guesses.

3. Many nontraditional cutting methods are *noncontact* by physical description. Why do you think this attribute is important in material processing?

4. Explain the difference between cutting by separation and cutting by removal.

5. Describe the reasons for, and the advantages or disadvantages of, the hot methods of material joining (welding, brazing, and soldering) with respect to the temperature required to effect the joining.

6. Describe the advantages and disadvantages of using stain as opposed to paint as an exterior finish for a house. Would the same hold true for exterior furniture finishes?

7. Describe the several eletrochemical cutting processes, and list some typical applications.

8. Describe the differences between the tooling and the machining speeds for wood- and metal-cutting operations.

9. Write your personal definition/description of the terms *mass reducing, mass conserving, mass increasing*, and *mass conditioning*.

10. List some ways of generating holes or openings in materials, and describe the attributes of each.

11. Investigate the frozen cakes and pastries found in the freezer section of your local supermarket. How are they mass produced, and what manufacturing processes can you identify?

12. Prepare a research report on the manufacture of jeans or sweatshirts.

13. Assemble into teams of 3 or 4 people and analyze the processes used to produce each of the badminton racquets shown in Figure A. The material used is 1/4″ aluminum tubing.

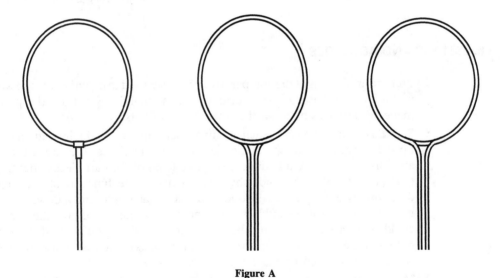

Figure A

CHAPTER 5 | Human Factors in Design

Human factors is the study of the interactions between people and the products they use and the environments in which they work and live. Its central theme is design for human use through the recognition of human capabilities, needs, and limitations. (See Figure 5.1.) Other terms used to describe this field of study are *engineering psychology, human engineering, ergonomics, bioengineering*, and *biomechanics*, among others. Historically, it is held that the term *ergonomics* was first coined in England in 1949 when the Greek words *ergon* (work) and *nomos* (study of) were combined to give license to this emerging discipline.

During the Industrial Revolution, there was an emphasis on finding efficient ways to manufacture goods to meet the demands of the consumers. As many more complex products were introduced, there was little concern for how these artifacts affected the user. In the 1940s, psychologists began to study the causes of the numerous aircraft accidents experienced by military pilots. It soon became apparent that the failure to match the sophisticated new equipment with the abilities of the human operator was a major contributor to these accidents. This was the beginning of what was to become known as *human factors engineering*. As interest in the interaction between operators and the products they used expanded, more individuals became involved with these studies. The discipline began to expand beyond the cognitive aspects of design and response to include the physical capabilities and limitations of the human operator. Outside the United States, many people called this new discipline *ergonomics*. Today, ergonomics and human factors engineering, or just human factors, are interchangeable.

A basic and current approach to this discipline involves the examination of

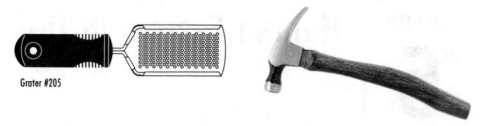

Grater #205

Figure 5.1 The food grater and this claw hammer are identified as ergonomically designed tools in the promotional literature. The handles are uniquely fashioned for human use. Grater Courtesy of OXO International.

human characteristics, behaviors, needs, and expectations, and the planning of strategies to incorporate this knowledge into design programs. Ergonomists are specialists who know, for example, how muscles and tendons are used by humans to grasp a tool or operate a cash register in a grocery check-out lane. Many industrial engineering programs require considerable experience in this field, and product designers should consult these experts as the need for their services arises. Four related factors or sciences that enable product planners to account for ergonomics in their work are anthropometry, physiology, anatomy, and psychology.

ANTHROPOMETRY FACTORS

The formal name for the technique used to express quantitatively the form of the human body is *anthropometry*. It is the science of human measurement and it deals with body sizes and proportions. The significance of this science lies with the fact that the comfort, performance, and physical welfare of people are enhanced to the extent that the things they use are designed to "fit" them. One of the major myths in product design is that there is an "average" person, when in reality no two people have the same physical structures. For many years, artifact design was centered on the concept that if an average or mean dimension was applied, the artifact would be satisfactory for most users. This was an erroneous and unfortunate assumption. Common male body dimensions are shown in Figure 5.2. If work spaces, equipment, or products were designed using the mean dimensions, more than 50 percent of the population would be too large or too small to be accommodated by them. One solution to this problem would be to plan things that could be adjusted to fit the user, such as moving the car seat forward so that the driver could reach the brake pedal. Barring this, an acceptable alternative is to strive for a 95 percent fit of the target population.

Male and female shapes, weights, and dimensions and other anthropometric data have been charted, and these provide an invaluable bank of statistics to aid planners in their goal of designing for human use. For example, the height of a chair seat and keyboard table for a word processor station can be determined from these

1985 Male*

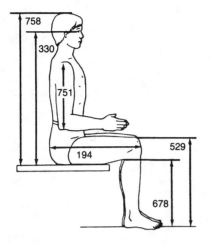
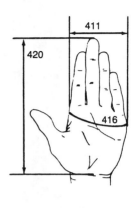

No.	Dimension	5%ile	Mean	95%ile
758	Sitting height	88..5 (34.8)	93.6 (36.9)	99.0 (39.0)
330	Eye height, sitting	76.4 (30.1)	81.3 (32.0)	86.5 (34.1)
529	Knee height, sitting	52.1 (20.5)	56.1 (22.1)	60.3 (23.7)
678	Popliteal height	40.4 (15.9)	44.0 (17.3)	47.8 (18.8)
751	Shoulder-elbow length	33.3 (13.1)	36.1 (14.2)	38.9 (15.3)
194	Buttock-knee length	56.4 (22.2)	60.8 (23.9)	65.4 (25.7)
420	Hand length	17.9 (7.0)	19.2 (7.6)	20.6 (8.1)
411	Hand breadth	8.3 (3.3)	8.9 (3.5)	9.6 (3.8)
416	Hand circumference	20.1 (7.9)	21.6 (8.5)	23.2 (9.1)

*Data given in centimeters with inches in parentheses.

Figure 5.2   Anthropometric data are shown on this chart.

measures, as can the heights of toilet seats, kitchen counters, sinks, and table saws, as well as the positioning of controls to accommodate an airplane pilot, truck driver, or lathe operator. The designers of the banking check desk shown in Figure 5.3, had a dual measurement challenge. They not only had to concern themselves with the size of a seated person but also the dimensions of a wheelchair. As shown, the result was a very usable piece of banking furniture.

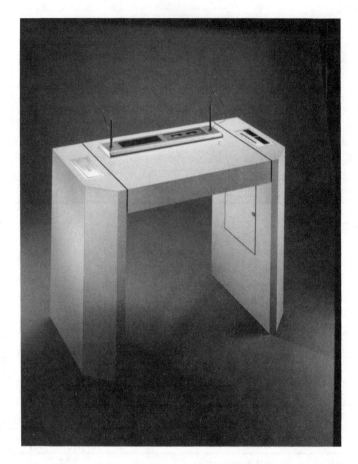

Figure 5.3 This check-writing desk for use in commercial banks was designed for a person in a wheelchair. Courtesy of Matel Inc.

Manufacturers of the very popular exercise machines must account for the differences in human anatomy in their products. Seats, backpads, footplates, and handles all must be positioned easily to accommodate the variations in body structure. Additionally, they have to be familiar with the human musculoskeletal system so that weights and force devices can be adjusted to provide the desired exercise without damaging soft tissues such as tendons and ligaments. (See Figure 5.4.) Convenience, safety, and efficiency of operation are important factors in this equipment. The in-house design staffs are typically comprised of creative and energetic engineers and mechanical technicians who work closely with outside consultants such as industrial designers, materials specialists, and exercise physiologists.

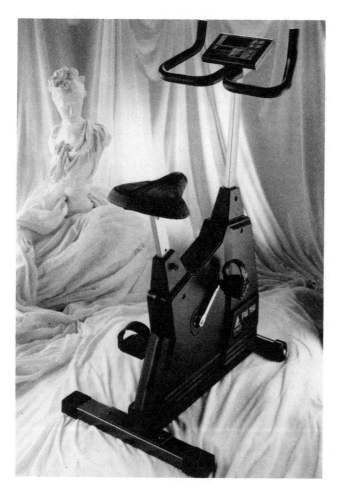

Figure 5.4 Quality fitness equipment must be adjustable to accommodate different user capabilities and body sizes. This is prize-winning exercise bike is a good example. Courtesy of Fitness Master.

PHYSIOLOGY FACTORS

Physiology is the science dealing with functions of living organisms and their parts. Especially important to the designer are the human visual, tactile, and auditory senses and how they relate to products and environments. These features relate to human work efficiency and safety.

Visual Senses

A number of factors affect the quality of human vision, or visibility. *Illumination* is the intensity or amount of light, and *brightness* is the difference in light intensity between an object and its background. *Glare* is light reflected directly from a surface. Additionally, *color* plays a role, as does the *motion* of the object being perceived. What all these factors mean is that visibility determines how well a product assembly task, for example, can be performed, and is therefore of especial importance in the design of work stations. Different tasks require different illumination levels and types, as shown in Figure 5.5. It should be noted that the task conditions also affect the illumination requirements. The rate at which light is emitted from a source is called *luminous flux*, and it is measured in *lumens*. *Illuminance* is measured in *footcandles (fc)*.

Aside from the matter of vision and work areas, product designers must assure that warning labels on vehicles and equipment are readable. The labels on luggage racks fastened to automobile trunk decks usually are pressure-sensitive plastic tabs with the printed warning: "Maximum Load 50 Pounds." Unfortunately, most of those tags are faded and unreadable within a year or so. This is also true of the information labels affixed in the engine area. These warnings are important because they relate to user safety and should be designed to last the lifetime of the automobile. There are literally hundreds of other instances of product visibility problems to which the designer must attend.

Task Condition	Type of Task or Area	Illuminance Level in fc	Type of Illumination
Small detail, low brightness, low contrast, prolonged periods, high speed, extreme accuracy	Sewing, inspecting dark materials	100	General plus supplementary, e.g. desk lamp
Small detail, fair contrast, speed not essential	Machining, detail drafting, instrument repairing, inspecting medium materials	50-100	General plus supplementary
Normal detail, prolonged periods	Reading, parts assembly, general office and laboratory work	20-50	General, e.g. overhead ceiling fixture
Normal detail, no prolonged periods	Washrooms, power plants, waiting rooms, kitchens	10-20	General, e.g. random natural or artificial light
Good contrast, fairly large objects	Recreational facilities	5-10	General
Large objects	Restaurants, stairways, warehouses	2-5	General

Figure 5.5 A chart of recommended illuminance levels for selected tasks. Note that footcandles (fc) is an obsolescent customary measure. The preferred metric unit is lux (lx), or lumen per square meter. One lx = 0.0929 fc.

Tactile Senses

Tactility is the sense of touch or feel. It relates to those sensitive nerve fibers near the surface of the skin that can detect contact pressure and temperature differences. This sense can be a feature in the design of control knobs according to size, shape, and texture to avoid recognition problems. For example, control lever knobs for aircraft have tactually discriminate shapes that bear a relationship to control function, such as a wing-shaped knob for flaps and a wheel-shaped knob for the landing gear. (See Figure 5.6.) Similarly, *kinesthetic* senses are based on nerves in the joints and muscles and relate to bodily movements. They can provide an awareness of arm or leg positions without any visual or tactile clues, such as a vehicle driver moving his or her foot to the brake pedal automatically, without any forethought. This information can help designers in positioning machine control levers, footplates, and knobs for safe and rapid operation.

Auditory Senses

Sound is an important source of information as well as a means of communication. Perceived sound can be an alarm signaling danger or the purr of a smoothly functioning machine. It is the sensation produced when the ear receives vibrations from the surrounding air. Noise is an unwanted sound and is important in the design of products and facilities because it can lead to temporary or permanent hearing damage. Noise also interferes with ordinary communication and can be sufficiently annoying to affect work performance as well as leisure.

Sound can be described as a combination of auditory frequency and intensity. The term *frequency* refers to the number of occurring vibrations (or "sound waves") per second, and is measured in hertz (Hz). One Hz equals one cycle per second. The human ear is sensitive to frequencies ranging from 20 to 20,000 Hz. As an illustration, the lowest tone (A) on a piano has a frequency of 27 Hz; the highest (C) is 4,000 Hz. People typically are more sensitive to higher frequencies than to low. *Intensity* relates to the pressure level of the sound waves, or amplitude, and is expressed in decibels (dB). The more powerful a sound is, the more intense it is.

The term *loudness* describes the degree of sound occurring as an interplay of

Figure 5.6 These aircraft controls were designed for tactile recognition to avoid confusion. The lever on the left is for flaps control; that on the right is for the landing gear.

frequency and intensity. It is the loudness of noise, sustained or repeated over a period of time, that can result in impaired hearing. Sound-level meters are designed to measure loudness in terms of dB weighted with a frequency response scale. There are three of these scales: A, B, and C. The A-scale is generally used because it most closely approximates the response characteristics of the human ear. The unit used to measure the loudness of sound is therefore the *dBA*, or decibel A-scale weighted. When frequency and intensity approach the range of 0 dBA, the sensation of hearing occurs, known as the *range of audibility*. A noise of 120 dBA is known as the *threshold of feeling*, and pain or hearing damage can result. Such noises must be controlled or prevented.

The detrimental effects of noise have been noted. The Occupational Safety and Health Administration (OSHA), a federal agency, has established standards for noise exposure in order to protect workers from such hazards. Such data are also valuable to designers of equipment and machinery for consumer and industrial use. (See Figure 5.7.) The matter of duration is especially important here. Intermittent noise causes less hearing damage than does continuous noise because the ears have an opportunity to recover before a new exposure.

Noise-reduction measures.   Protection in the workplace includes such measures as using quieter equipment, providing damping devices for noisy equipment

Duration Per Day Hours	Sound Level dBA
8	90
6	92
4	95
3	97
2	100
1.5	102
1	105
0.5	110
0.25 or Less	115
Noise above 115 dBA is not permitted. Noise below 85 dBA is considered to have zero effect.	

Figure 5.7   Permissible noise exposure levels are specified to protect workers from hearing damage.

and work areas, separating people from noise by barriers, and requiring workers to use personal ear protection. Such remedies, singly or in combination, are all aimed at keeping noise within the approved limits. Ear protection, such as plugs and muffs, can be reasonably enforced among industrial employees, but can only be recommendations for the private citizen who purchases a piece of powered machinery for use at home. Table saws, lawn mowers, and chainsaws are noisy, and prolonged use can be both annoying and hazardous. (See Figure 5.8.) It is important for designers of such equipment to reduce noise where possible, and to write safety precautions into owner manuals. The aims obviously are both humanitarian and directed at reducing product liability claims. In the final analysis, the best ear protector is one that is worn.

Vibration. Though it is a different form, vibration is often associated with noise. It occurs as an oscillating or other periodic motion of a rigid or plastic body, and can be either of a whole body or localized form. *Whole body* vibrations are associated with the operation of construction equipment or heavy machinery. *Localized* vibrations are produced by hand-held tools such as chainsaws and air tools in the range of 40 to 65 Hz. The results of excessive local vibrations are "dead" fingers due to lack of sensation in the hands, or "white" fingers caused by impaired circulation. Preventive measures include handles covered with damping materials, and larger handles for a looser, more comfortable grip without loss of necessary machine control.

As a final comment on the subject of noise, it must be said that too little noise also can affect work performance. When noise is totally absent, as in a very quiet work area, transient noises such as the ticking of a clock, the dripping of water, or

Noise	Sound Level, dBA
Jet Airplane	120
Jack Hammer	115
Chain Saw	110
Rock Concert	110
Table Saw	100
Power Lawn Mower	95
Outboard Motor	95
Motor Bike	95
Vacuum Cleaner	79

Figure 5.8 Representative noise levels for selected products.

nearby conversations can be annoying. These situations may require the introduction of sound-masking systems such as background music. Furthermore, a city park with a water fall that creates a sound of about 75 dB will mask unwanted traffic noise. This calculated introduction of sound creates what is commonly called "white noise," and has applications in many environments.

Other Senses

Human performance also is affected by *thermal* conditions. The thermal environment is determined by a combination of air temperature, air velocity, humidity, and radiant heat—elements that interact to form an environmental "envelope." Heat exchange between the human body and the surrounding air is a function of air velocity. Humidity, or the water vapor in the air, determines the rate of sweat evaporation from the skin tissue surfaces. One goal of proper work area design is to minimize the adverse consequences of hot environments. This is accomplished in three ways:

1. Modify the environment to minimize heat exposure.
2. Provide the worker with the opportunity to adapt to heat and permit the human body to function physiologically at an efficient level.
3. Provide personal protection and local cooling when appropriate environmental changes are not possible or when physiological adaptation does not take place.

In addition to thermal concerns, other factors include the elimination of noxious odors and dust from the workplace, primarily through the provision of adequate general exhaust systems and personal face masks. Health, safety, and productivity are once again the important aims.

ANATOMY FACTORS

Anatomy is the science of the structure of the human body, to include the skeleton, muscles, nerves, blood vessels, and various organs. The physician must have a knowledge of these in order provide proper medical treatment. The designer must understand select anatomical functions in order to plan satisfactory products and environments.

The human frame, or musculoskeletal system, is comprised of 208 bones and numerous muscles, tendons, and ligaments, all of which are used in the performance of physical work. There are two major *bone* groups that give the body its structural shape. The long bones of the upper and lower extremities and the spine provide the basis for human activity. The other group provides protective cover to vital organs, such as the skull and ribs. Bones are connected and given stability at their joints by *ligaments*. Bundles of muscle fibers make up the *muscles*, and the ends of each blend into strong fibrous tendons that attach the muscle to the bone. The unique property of muscle is that it can contract to apply mechanical leverage at the joint and thereby perform work. The larger the muscle, the greater the force that can be applied with it. Muscle fibers are made to contract by nervous impulses incoming from the brain.

As suggested earlier, bones and muscles working in unison act as levers. The forearm, for example, has its fulcrum at the elbow, and one of the upper arm muscles provides the force to bend the forearm. The mechanical advantage produced by this lever movement makes human work possible. The study of these motions is called *kinesiology*, and there are several classes of such motions:

1. *Flexion:* bending or decreasing the angle between body parts
2. *Extension:* straightening or increasing the angle between body parts
3. *Hyperextension:* moving a member beyond its normal limits
4. *Adduction:* moving toward the midline of the body
5. *Abduction:* moving away from the midline of the body
6. *Pronation:* rotating the forearm so that the palm faces downward
7. *Supination:* rotating the forearm so that the palm faces upward

These representative motions result in certain kinds of movements used in human work, and they are important to the ergonomist who is planning a special type of production tool. A *positioning* movement would be used to place the foot on the brake pedal of a vehicle. A *manipulative* movement would be used to handle parts in an assembly operation. A *continuous* movement would be used to guide a board through a table saw. A *sequential* movement would be used to place a washer and screw in a wood panel and then use a power screwdriver to tighten it. A *repetitive* movement would be used to hammer a nail into a board.

Work Devices

Collectively, this whole process of body movement in response to the requirements of an operational task must be considered in the design of products, equipment, tools, and work areas. The aim is to plan a tool, as an example, that will function well, be easy to use, and will account for the fact that a human being with special capabilities and limitations will be using it for extended periods of time during a workday. A carpenter who spends eight hours pounding nails to secure plywood sheets to the roof of a house will experience backaches, hand cramps, and sore arms. The ergonomist can alleviate such ailments by designing special air-actuated nailing machines that can be wheeled into position by a person standing in an upright position.

On industrial production lines, tool use is enhanced by devices such as balancers, positioners, and retractors. A *balancer* (see Figure 5.9) is a tension-adjustable unit that requires a very light force to extend the support cable and position the tool, no force to maintain the position of the tool, and a very light force to

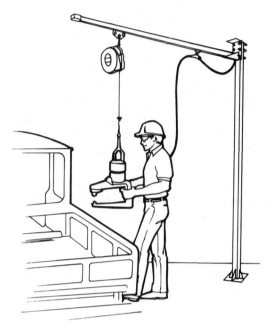

Figure 5.9 A balancer permits a worker to use a heavy tool with minimal effort. Courtesy of Aero-Motive Company.

retract the cable and position the tool. The unit shown features a tapered drum and a special spring configuration. This allows for any load for which the balancer is rated to be suspended within its travel range without drifting up or down when unattended.

A *positioner* is a tension-adjustable unit that requires a light force to extend the cable and position the tool, no force to maintain the position of the load, and a force equal to the weight of the suspended tool to retract the cable and reposition the tool. This unit features a straight drum and a unidirectional friction brake. (See Figure 5.10.) A *retractor* is a tension-adjustable unit that requires light force to extend the cable and position the tool, a light force to maintain the position of the tool, and no force to to retract the cable and reposition the tool. (See Figure 5.11.) This unit features a straight drum and spring configurations that optimize functional characteristics within specific working ranges.

Torque is the twisting action generated by a power tool to tighten a screw or nut by applying an equal and opposite force on the tool handle. All the torque applied to the fastener must be reacted by the operator holding the handle of the tool. With *torque arms*, shown in Figure 5.12, the torque transmitted by the tool is absorbed by the unit itself, and not by the operator. This neutralizes the undesirable tool forces and reduces the repetitive strain on hands, wrists, and arms, which can contribute to a number of painful afflictions. Workers can freely position and handle tools within a wide work envelope. Easily adjustable spring tension allows workers to adjust for tool weight and location preference for any type of operation.

These four units, and many others for similar applications, help to take the weight out of production tools and equipment, thereby reducing strain on workers'

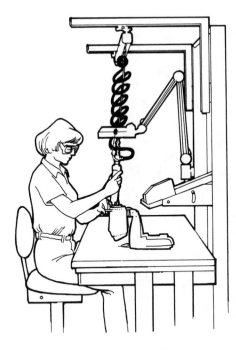

Figure 5.10 A positioner eliminates the necessity to exert force to maintain the working level of a production tool. Courtesy of Aero-Motive Company.

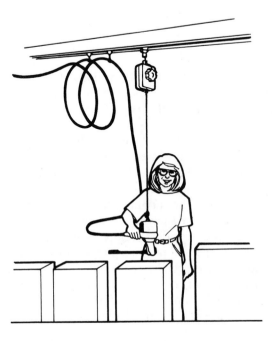

Figure 5.11 A retractor requires only light force to lower and raise a heavy tool. Courtesy of Aero-Motive Company.

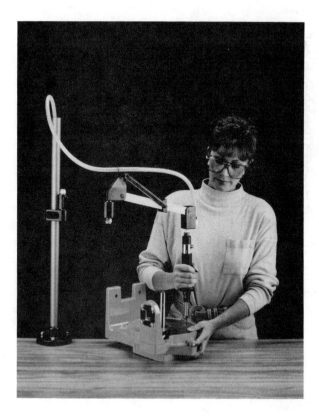

Figure 5.12 A torque arm reduces the tool twisting force experienced by a production worker. Courtesy of Aero-Motive Company.

upper bodies, hands, and wrists. Properly suspended power tools can be positioned and operated with minimal effort for longer periods of time. Constant tension controls allow operators to set and maintain tool positions at comfortable work heights, whether the tools be power nut-runners or meat processing saws. This equipment also reduces tool damage, keeps the workplace safe and properly organized, and increases production.

Ergonomically designed workstations are planned to allow for operator differences in height, weight, reach, and strength—essentials for their comfort, efficiency, and productivity. The station shown in Figure 5.13 is a good example. It provides people with adjustment options for work surfaces, chairs, component bins, tools, power, lighting, and related accessories. The upright system allows proper placement of all components via articulating arms and shelving designed for primary horizontal and vertical reach zones. This creates a work envelope that reduces unnecessary movement and places the worker in the best possible position for the tasks at hand.

Hand-held power tool design is especially challenging. A power screwdriver must be designed to fit the hand without compromising the function of the tool, as shown in Figure 5.14. The clever handle configuration of this tool allows it to be used for drilling, nut-running, or screw-driving, in a range of work positions. (See Figure

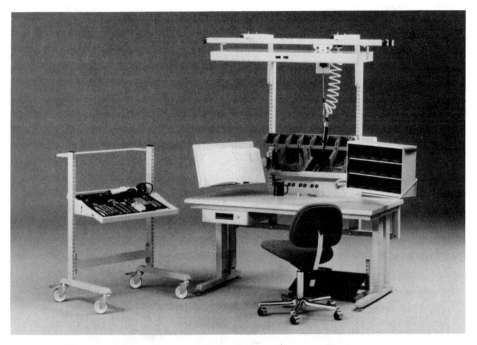

Figure 5.13 All components of this work station can be adjusted to meet the needs of a product assembler. Courtesy of Aero-Motive Company.

5.15.) Note also that it operates by a hand squeeze pressure, and not by a tiring and less efficient finger or thumb pressure.

A tool handle is important because it acts as the interface between a worker's hand and a tool, and all forces associated with a task are transmitted by the hand through the handle to the tool. The ideal handle must be one designed ergonomically to place a minimum of strain on the operator. The optimum position for transmitting forces between hand and arm is to ensure the wrist is kept straight. It is also important to reduce, as far as possible, the twisting and bending forces applied to the wrist. Tools also should be designed to have a high power-to-weight ratio. Heavy tools not only put more load on the wrist but they also make precision tasks more difficult. As there are so many different working positions and tool functions, tools feature a variety of handle shapes.

Bow handles (see Figure 5.16) are used to transmit high feed forces without causing a torque in the wrist. To achieve this, the handle should be designed to ensure that the feed force is transmitted through the center line of the arm and wrist. The natural angle of a hand grip is approximately 70 degrees and is independent of hand size. By designing the handle to this angle, the muscles stabilizing the wrist are in their optimal position and strain is reduced. The limitation of this design is that the weight of the tool is in front of the hand. If used as a single-handed tool, it would put

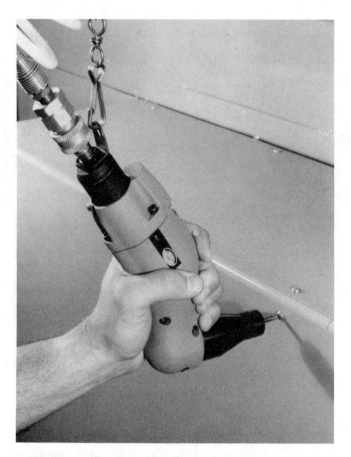

Figure 5.14 Power tools such as this screwdriver must be designed to fit the human hand. Courtesy of Sioux Tools Inc.

an undesirably high bending force on the wrist. In contrast, *straight* handles offer the user more directive control over tool operations. The diameter of a handle that when held, allows the strongest grip and causes minimum hand strain, has been found to be 38 mm for men and 34 mm for women. Obviously, design considerations must include the probable sex of the worker.

   Pistol handles are used to shorten the length of the tool and reduce the bending load on the wrist. (See Figure 5.17.) Keeping the distance between hand and workpiece as short as possible allows precision tasks to be performed efficiently. For tasks requiring high feed forces, the handle needs to be angled at 70 degrees. Where only twisting forces occur, the center of the tool weight should be on top of the handle to lessen the bending force on the wrist.

   Triggers are devices that actuate power tools, and the type is closely related to the handle design and tool function. (See Figure 5.18.) Triggers can be designed so

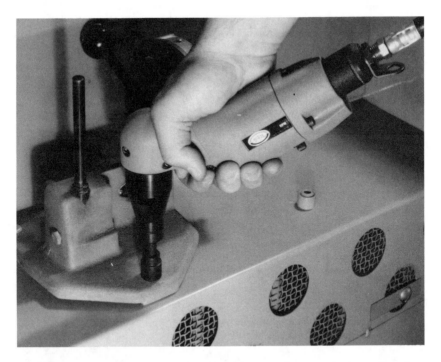

Figure 5.15   This nutrunner tool is designed for convenient operation in a variety of work positions. Courtesy of Sioux Tools Inc.

that constant hand pressure is required for operation, generally for safety reasons as with a chain saw, or they may be flipped on by a touch of the finger. Constant pressure can be tiring—a fact that can attested to by anyone who has operated a spray-paint can for prolonged periods. *Finger* triggers are used often on pistol handle tools where the trigger can be operated without affecting the grip on the handle. The trigger

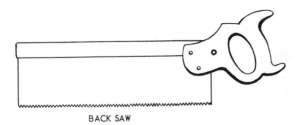

BACK SAW

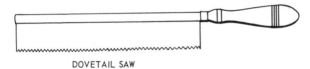

DOVETAIL SAW

Figure 5.16   A backsaw has a bow handle designed to deliver a controlled force. The straight handle of the dovetail saw permits a more measured and controlled cutting action.

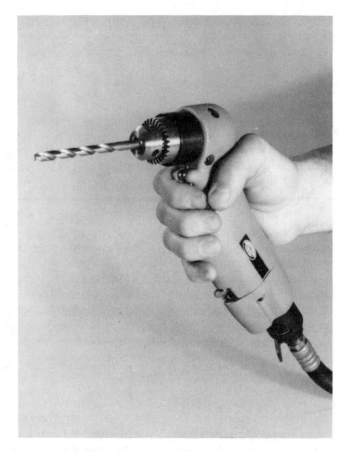

Figure 5.17 Pistol handles offer a comfortable tool grip to reduce wrist loads. Courtesy of Sioux Tools Inc.

should be designed to ensure that it is pressed with the middle part of the finger. If the tip of the finger is used, tendon damage can result. The finger trigger gives precise control over the speed of the tool. To give the worker the right "feel" when using the tool, it should be designed so that to increase tool speed, the pressure on the trigger must be increased. These triggers are used for frequent operations, for precision tasks, and when the tool needs to be positioned before it is started.

Strip triggers provide tool operation with low force because all the fingers are used, thus reducing hand strain. However, with this type of trigger, the grip on the tool is less stable. It is ideal for nonprecision repetitive functions such as blow-gun applications and simple assembly. *Thumb* triggers, conversely, can be operated while maintaining a firm grip on the tool. They are particularly suitable for tools such as chipping hammers where high feed forces are used. In this application, the trigger operation becomes part of the feed force. *Push* start triggers are used primarily on

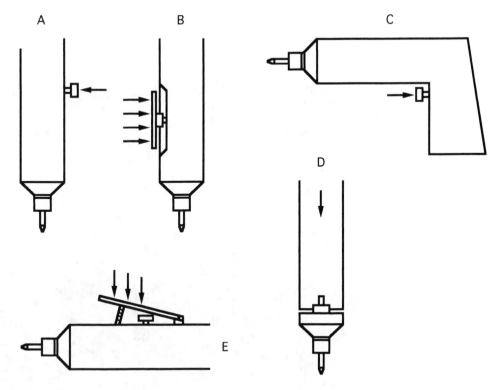

Figure 5.18 Note the differences in trigger designs to accommodate the fingers and the hands. A = Thumb; B = Strip; C = Finger; D = Push Start; E = Lever.

straight screwdrivers in repetitive high-volume operation. The trigger is activated by putting a feed force on the tool and a stable grip is maintained on the handle at all times.

Lever-start triggers are integral parts of the handle grips and offer great stability. They are especially suitable for long-cycle operations and for tasks that place high forces on the tool. They enable the tool to be positioned prior to start-up; in addition, a safety device can be built into the trigger without affecting its operation. The lever-start triggers are used on grinders, nutrunners, drills, and screwdrivers.

To summarize this discussion of handles and triggers, it should be stated that hand tools are used for cutting, finishing, and assembling, and must be held and operated by people. Consequently, they must be designed commensurate with human physical capabilities and limitations. People have natural holding positions for tools and implements that are both comfortable and efficient. It is one thing for a person to use a tool for brief periods in a home workshop setting, and quite another for the worker who must use it for eight hours, day after day. The aim of ergonomics is to account for this difference in duration in order to provide optimum work conditions for reasons of health, safety, and productivity.

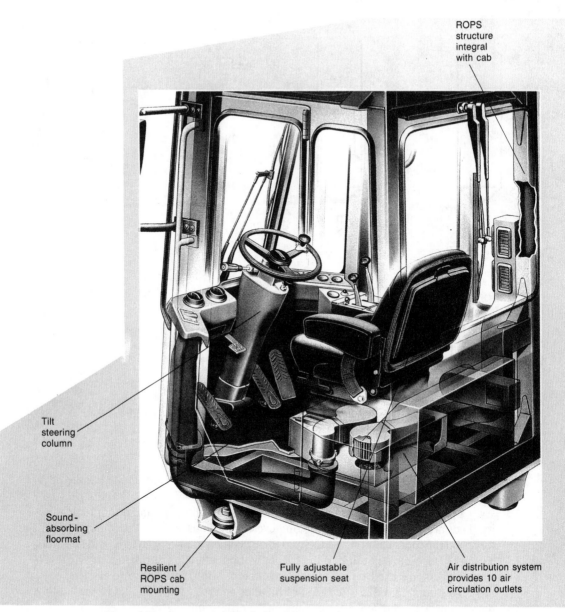

Figure 5.19   An ergonomically designed heavy-equipment operator compartment. Courtesy of Caterpillar Inc.

Much of the previous discussion was directed toward workplace design in a factory setting. Other kinds of workplaces are equally important. For example, the cabs of heavy vehicles (such as farming and earth-moving equipment) must be ergonomically designed for operator comfort, safety, and performance. The compartment shown in Figure 5.19 is commendable because of its enhanced visibility, tilt steering column, adjustable padded seat, air distribution system, shock-resistant structure, and convenient controls. Users of this equipment must work long hours under the adverse conditions of severe noise and vibration, dust, heat, and rough terrain, and their jobs are made easier with properly designed operating compartments.

A dental office is a totally different kind of work setting that is hygienic, quiet, bright, and comfortable. The equipment matches the physical environment to provide dentists and technicians with apparatus designed to rigorous ergonomic standards. The elegant reclining and completely adjustable chair by the Fimet Oy in Finland, shown in Figure 5.20, expresses patient comfort. It also offers comfortable and efficient working positions for dental personnel.

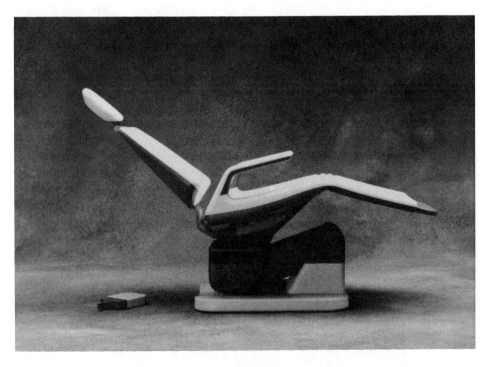

Figure 5.20   This very pleasing design for a dental chair provides a comfortable work station for both patient and doctor. Courtesy of Design Forum Finland.

PSYCHOLOGY FACTORS

Human job performance also is affected by the mental state of the worker. *Psychology* is the science of human nature and behavior. Industrial psychologists are concerned with those societal, cultural, mental, and emotional conditions such as boredom, isolation, fatigue, and confusion, and their effects on human performance. The physiological factors described earlier are contributors to the human mental state. Persistent noises, glaring lights, vibrations, foul odors, dust, and exertion most assuredly do affect the human outlook, and consequently the quality of work. Basic human needs include security, stimulation, and identity. Their absence results in anxiety, boredom, and anonymity. This matter of psychologic concern is difficult to identify and manage. Here, the role of the ergonomist is to provide a healthy, pleasant, caring work situation wherein people have a sense of active participation in their jobs and are motivated to be worthy contributors.

One example of the human influence on machine controls pertains to the design of the familiar membrane or touch switches. (See Figure 5.21.) Some of these control buttons require only a passing touch of the finger, whereas others require a slight pressure to actuate them. Plant mangers have found that some employees dislike the nonpressure switch because of the high possibility for error, as the operator touches the wrong switch while searching for the correct one. There also was a displayed feeling of need for personal control of the machine by the operator pressing the switch firmly in order to actuate it. The addition of an audible click when pressure contact was made was felt to be even more desirable. In this instance, the issue was resolved by the simple expedient of changing to audible pressure switches.

Figure 5.21 A common and familiar membrane switch. Courtesy of Memtron Technologies, Inc.

CUMULATIVE TRAUMA DISORDERS

According to current industrial safety data, musculoskeletal problems rank first among disease groups in frequency, and they affect nearly one-third of this nation's workforce. Collectively identified as cumulative trauma disorders (CTD), these injuries usually affect the hands, wrists, arms, elbows, shoulders, neck, and lower back, and result from the repeated use of these body parts. Aside from the primary issue of human suffering, the cost to industry in worker compensation, insurance claims, and lowered productivity amounts to billions of dollars annually. Methods of reducing these injuries have been described earlier, and include the proper design of tools, equipment, and work stations. Some CTD forms common to the arm are described here. (See Figure 5.22.)

1. Tendonitis is a very common ailment caused by the inflammation of a tendon due to excessive use.
2. Epicondylitis is elbow tendonitis and is the irritation of the tendinous attachment of the finger muscles. It is called *tennis elbow* if it involves the finger extensor muscles on the outside of the elbow, and *golfer elbow* when associated with the finger flexor muscles on the inside of the elbow.
3. A ganglion cyst is a tendon disorder resulting in an enlarged blister that generally appears as a lump on the wrist.
4. Carpal Tunnel Syndrome is a very common result of continual and forceful wrist actions, such as manipulating cash-register keys. Inflamed finger tendons running through the carpal tunnel of the wrist compress the median nerve to cause pain, numbness, and swelling, and the eventual erosion of the thumb muscle.
5. DeQuervain's tendonitis causes severe pain at the base of the thumb when moving the little finger toward the thumb or when closing the fist.

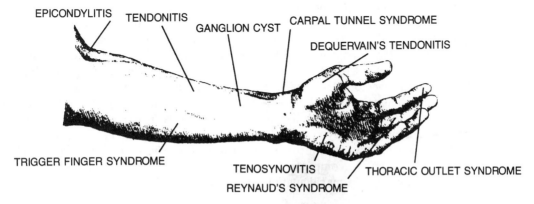

Figure 5.22 A diagram of common CTD problems associated with the human arm.

6. Trigger finger syndrome is caused by the excessive movement of a finger against sharp edges. The afflicted person can flex the affected finger, but the finger has to be passively straightened. A clicking noise generally accompanies this movement.

7. Reynaud's syndrome, or vibration-induced white fingers (VWF), is caused by the regular and prolonged use of vibrating hand tools. Workers suffering from this affliction experience pain and numbness when their hands are exposed to cold temperatures.

8. Tenosynovitis is an inflammation of the tendon and its sheath.

9. Thoracic outlet syndrome results from the compression of the nerves and blood vessels.

Employment injuries can be prevented by proper tool design based on the tool function and use, work station design, and job design. Constant reaching, pulling, squeezing, pushing, lifting, and turning in the performance of a job or task is exhausting, boring, and debilitating. Proper work tools and stations are important, but the best remedy for CTD afflictions is employee rotation to other jobs that require different work motions and offer a change of environment. Also, prolonged exposure can be limited by providing short but frequent recovery pauses, or the familiar work breaks. One enlightened approach to work management is to require that industrial engineers and other job planners perform certain production tasks for a normal work day. The real problems will soon become evident and corrective remedies will be taken.

HUMAN FACTORS IN OFFICE DESIGN

Up to this point in this chapter, the discussion of human factors has centered primarily on the demands of production work in factory settings. However, the work that goes on in modern automated offices also has ergonomic demands, the most important of which is perhaps the video display terminal (VDT). These terminals—which are comprised of a display screen, a keyboard, and a central processing unit—have rapidly replaced the use of typewriters and other office machines.

The *display screen* is the output device that shows what the computer is processing. Display screens can be monochrome (green, white, or orange on a black background) or color. An enhanced color screen can provide 16 colors out of a possible 64 at any time. The *keyboard* is the input device that allows the user to send information to the "brains" of the computer. Keyboards are commonly used for data entry and inquiry. The keyboard is similar to a standard typewriter keyboard but with additional special keys and functions. The *central processing unit (CPU)* is referred to as the "brains" of the computer. It is the center of operation for all the computer processing and it performs calculations and organizes the flow of information into and out of the system.

The VDT operates at high voltages, but the power supplies generating these voltages produce very little current. All data-processing equipment, including VDTs, must meet stringent international safety standards in this regard. However, in the wake of the expanding use of VDTs, concerns have been expressed about other potential health effects. Complaints include excessive fatigue, eye strain and irritation, blurred vision, headaches, stress, and neck, back, arm, and muscle pain. Other problems include general physical discomfort, CTD, and possible exposure to electromagnetic fields. Research has shown that these symptoms can result from problems with the equipment, work stations, office environment, or job design, or from a combination of these. Some of the most common stresses, their related health effects, and their means of prevention are discussed briefly in the following sections.

Eyestrain

Visual problems such as eyestrain and irritation are among the most frequently reported complaints by VDT operators. These visual symptoms can result from improper lighting, glare from the screen, poor positioning of the screen itself, or copy material that is difficult to read. These problems usually can be corrected by adjusting the physical and environmental setting where the VDT users work. For example, work stations and lighting can and should be arranged to avoid direct and reflected glare anywhere in the field of sight, from the display screen, or surrounding surfaces.

Changing focus is one simple way to give eye muscles a chance to relax. The employee need only glance across the room or out the window from time to time and look at an object at least 20 feet away. Other eye exercises may include rolling or blinking the eyes, or closing them tightly for a few seconds. Operators also can reduce eyestrain by taking "vision breaks," which may include periodic exercises to relax eye muscles. The National Institute for Occupational Safety and Health (NIOSH) recommends a 15-minute rest break after two hours of continuous VDT work for operators under moderate visual demands, and a 15-minute rest break after one hour of continuous VDT work where there is a high visual demand or repetitive work tasks.

Fatigue and Musculoskeletal Problems

Work performed at a VDT may require sitting still for considerable time and it usually involves small frequent movements of the eyes, head, arms, and fingers. Retaining a fixed posture over long periods of time requires a significant static holding force, which causes fatigue. Proper work station design is very important in eliminating these types of problems. Some variables of work station design include the VDT table, chair, and document holder. An individual work station should provide the operator with a comfortable sitting position sufficiently flexible to reach, use, and observe the display screen, keyboard, and document. Some general considerations to minimize fatigue include posture support (back, arms, legs, and feet) and adjustable display screens and keyboard. Tables or desks should be vertically adjustable to allow for operator adjustment of the screen and keyboard. Proper chair height and support to the lower region of the back are critical factors in reducing fatigue and related musculoskeletal complaints. Document holders also allow the operator to position and view material without straining the eyes, or neck, shoulder, and back muscles.

The type of task performed at the VDT may also influence the development of fatigue. In designing a work station, the types of tasks involved should be considered when determining the placement of the display screen and keyboard. (See Figure 5.23.) Operators are subject to the potential risk of developing musculoskeletal and repetitive motion disorders. Carpal tunnel syndrome (CTS) is common to VDT operators. CTS usually can be reduced by stopping or limiting the activity that aggravates the tendons and median nerve (e.g., data/keyboard entry), by maintaining proper posture, or, as a last resort, by having surgery. For correct posture, VDT

Figure 5.23 This VDT work station is well designed to account for many of its problems related to employee safety, comfort, and efficiency. Courtesy of Miles Inc.

operators should sit in an upright position at the keyboard, with arms parallel to the floor. The wrists and forearms also may require support, depending on the tasks involved.

Radiation

Another issue of VDT concern is whether the emission of radiation, such as x-ray or electromagnetic fields in the radio frequency and extreme low frequency ranges, poses a health risk. Some workers, including pregnant women, are concerned that their health could be affected by electromagnetic fields emitted from VDTs. The threat from x-ray exposures is largely discounted because of the very low emission levels. The radio frequency and extreme low-frequency electromagnetic fields are still at issue despite the low emission levels. To date, however, there is no conclusive evidence that these low levels of radiation pose a health risk to VDT operators. Some workplace designs, however, have incorporated changes (such as increasing the distance between operator and terminal and between work stations) to reduce potential exposures to electromagnetic fields. Since the possible effects of VDT radiation continue to concern operators, the issue is still being researched and studied.

Lighting

The matter of human vision, which was discussed earlier, is important in VDT work. Lighting should be directed so that it does not shine into the operator's eyes when looking at the screen. Further, lighting should be adequate for the operator to see the text and the screen, but not so bright as to cause glare or discomfort. The basic lighting factors that must be controlled to provide suitable office illumination and avoid eyestrain are quantity, contrast, and glare.

Quantity refers to the work surface illumination provided by light fixtures and natural daylight, usually around 50 to 100 fc. High illumination "washes out" images on the display screen, and therefore illumination levels in VDT settings should be somewhat lower, about 28 to 50 fc.

Contrast is the difference in luminance or brightness between two areas. To prevent the visual load caused by alternate light and dark areas, the difference in illuminance between the VDT display screen, horizontal work surface, and surrounding areas should be minimized. Most VDT tasks do not require precise visual acuity, and indirect or diffuse lighting is appropriate to provide a more uniform visual field. Diffuse lighting has two main advantages—there tend to be fewer hot spots, or glare sources, in the visual field, and the contrasts created by object shapes tend to be softer.

Glare is a harsh, uncomfortably bright light dependent on the intensity, size, angle of incidence, luminance, and proximity of the source to the line of sight. Glare may be the result of direct light sources in the visual field (e.g., windows), or reflected light from polished surfaces (e.g., keyboards), or from more diffuse reflections that may reduce contrast (e.g., improper task lighting). Glare may cause annoyance, discomfort, or loss in visual performance and visibility.

In many cases, the reorientation of work stations may be all that is necessary to move sources of glare out of the line of sight. The proper treatment for window glare includes baffles, blinds, draperies, shades, or filters, and the face of the display screen should be at right angles to windows. To limit reflection from walls and work surfaces visible around the screen, these areas should be painted a medium color and have a nonreflective finish. Anti-glare filters that attach directly to the surface of a VDT screen can help reduce glare, as can the newer model keyboards with anti-glare matte finishes.

DESIGN FOR HUMAN FACTORS GUIDELINES

Worker comfort, performance, and safety are mutually inclusive aspects of the people/workplace interface. The guidelines listed here are intended to aid in the process of designing equipment and facilities for people at work.

1. The person should be able to work in either a standing or seated position at will.
2. Work should be performed below the level of the heart. Where this is not possible, frequent rest periods to relieve upper-arm stress are necessary.
3. Work should be performed in the most natural and least awkward positions without undue body and limb stretching, bending, or twisting.
4. Workplace environmental conditions should include proper lighting, ventilation, noise reduction, and temperature.
5. Handtool design should be driven by the natural methods of grasping and controlling, and by the limb strength capabilities.

QUESTIONS AND ACTIVITIES

1. Write your definition of *human factors* and describe how it relates to tool and workplace design.
2. Explain the significance of anthropometry to tool design.
3. Define *tactility* and cite some examples of its importance in control design.
4. Explain the difference between sound and noise.
5. Identify some common measures used to reduce noise in power tools and in the workplace.
6. Explain the difference between noise and vibration.
7. Explain the relationships among bones, tendons, and ligaments.

8. Select one of the motions related to kinesiology and cite an example of it in terms of tool use.

9. Give specific production tool operation examples of the following movements: positioning, manipulative, continuous, sequential, and repetitive.

10. Explain the functions of balancers, positioners, retractors, and torque arms in assembly operations.

11. List the various styles of tool handles and describe their applications.

12. Describe the positive and negative features of the several kinds of triggers.

13. Identify some of the affects of mental state on worker productivity.

14. Refer to Figure 5.22. Select one CTD and explain its relationship to a specific tool operation. How could proper tool design help to alleviate the affliction?

15. Describe some of the human factors associated with office design.

Applied Ergonomics

The subject of human factors was examined in the previous chapter, and the theoretical framework for the relationships between product and user was established. There was an emphasis on the importance of, and the strategies for, becoming involved with these factors in the early stages of product design. This chapter will examine some specific areas of the product-user relationship and the considerations that can lead to subsequent quality design solutions. The topics include serviceability, security, vandalism prevention, safety, usability, the physically disadvantaged, and the environment.

DESIGN FOR SERVICEABILITY

Serviceability factors in product design influence the performance and availability of products in use. At best, the goal is to develop a product that requires little or no service or maintenance. Barring this optimum condition, needed maintenance should be accomplished easily and economically. Several important terms relate to this concept.

Maintenance is the act of keeping products in, or restoring them to, an operating condition. As such, maintenance includes both preventive and corrective actions. *Maintainability* is a design attribute that assures that maintenance can be performed with a minimum of cost, inconvenience, and effort. It is a before-the-fact action. Maintenance occurs after the fact.

Another related term is *reliability*, which simply refers to the probability that a product will continue to perform its intended function over a prescribed useful lifetime and under unintended, adverse operating conditions. A ballpoint pen is a reliable product. It usually works. The relationship to maintenance and maintainability is obvious. The most reliable product cannot continue to perform if it has not been adequately maintained or if it is subjected to operational abuse. *Availability* refers to continued product performance without excessive down-time, and is a factor of reliability and maintainability. It implies the readiness for use. People assume that a product such as a copying machine will be usable when it is needed. If it is not, people are inclined to blame the manufacturer even if the failure was due to faulty maintenance.

The problem with attempting to achieve a maximum measure of reliability is the issue of economics. For a given design, cost often increases as reliability is improved. More expensive materials may be used, or more costly methods of fabrication may be implemented. The design goal is to deliver a product that will achieve a reliability level satisfactory to the customer, while holding production costs to an acceptable figure.

Approaches to the broad areas of maintenance and reliability have been quantified by providing mathematical analyses of these factors. This permits some valuable measurement units. For example, the reliability inherent in a product can be commonly stated as its mean-time-between-failures, or MTBF. The unit for expressing the maintainability of a given design is its mean-time-to-repair, or MTTR. Several of the references listed at the end of this book (Chapter 11) provide detailed information regarding these calculations and their uses.

Product performance also is influenced by the care with which a thing is made. This is called *quality*, and it is an attribute built into the manufacture of a product. Quality can be defined as those features of a product that bear on the ability of that product to satisfy a customer. Although there are a number of approaches to this issue of product satisfaction, the Taguchi Method is, in the author's estimation, one of the more interesting, effective, and useful. It also is one of the most demanding and complex. According to this method, quality is the loss imparted by the product to the user from the time the product is delivered. Customers suffer loss whenever they buy a faulty product because they are deprived of its use. In quality control (QC), the focus is on those losses that result from the deviation of the product's functional characteristic (what is expected of the product) from its target value (the ideal state of product performance from the user viewpoint).

For example, assume that the functional characteristic of concern is the force required to turn the faucet handle on a kitchen sink. The target value of the force is five pounds. Any losses due to the deviation of the force from the five-pound target are losses due to functional variation. The undesirable features that cause such variations from target values are called *noise factors*, and there are three types:

1. *Outer* noise is a function of environmental conditions, such as changes in temperature and humidity, or the presence of dirt. A copying machine may jam because of moist paper or because of dirt in the system.

2. *Inner* noise is a dysfunction between parts caused by the deterioration of product components. The jamming in the copy machine may be caused by a worn paper-feed mechanism.

3. *Between product* noise is caused by variations between product parts supposedly made according to rigid manufacturing specifications. The jam may be a result of an improper paper-feed mechanism assembly or adjustment.

An automotive seatbelt that fails to retract fully and gets caught when the car door is being closed is an example of noise—so are ceramic tiles that crack during firing, leaky faucets, and excessively noisy home furnace fans and refrigerators. The aim of QC is to manufacture products that are robust with respect to the three noise factors. The term *robustness* implies that product functional characteristics are not affected by such noise or nuisance variations. Noise source can be ascribed to off-line (occurring outside the actual production line) manufacturing activities, such as product design and process selection, or to on-line activities (occurring during or after product making), such as production and customer service. Obviously, if quality can be assured at the off-line stage, it will minimize the need for on-line reworking. The Taguchi Method, complex as it is, combines quality engineering and statistical techniques to experimentally reduce the causes of functional variations. One of its greatest advantages is to permit the identification of that specific stage of manufacturing that may be the source of a product fault. Readers who wish to pursue the topic further are invited to examine the related books listed in Chapter 11.

For example, a large grocery supermarket was recently opened for business, and it soon became apparent to customers that the automatic swinging entry doors were showing obvious signs of wear and tear. The problem was that the employees charged with the responsibility of retrieving the shopping carts from the parking lot were stacking a dozen or so in a long line and crashing them into the automatic doors to move the carts into the store. (See Figure 6.1.) The immediate result was largely cosmetic: The clear polycarbonate plastic door pane was being scarred and scratched, as were the metal push bar and the door frame, all of which resulted in a very unattractive unit. The cosmetic problem soon would degenerate into a structural one, resulting in door failure and possible injuries. The solutions to the problem are several:

1. Reposition the door push bar to a level matching that of the shopping cart, or add more bars. This would prevent the gouging of the plastic window, but not necessarily the door frame and jamb.

2. Redesign the shopping cart so that the height matches that of the push bar. Also, consider a plastic cart that would not damage the door. (A plastic cart would also be quieter.)

3. Train the employees to refrain from stacking too many carts and to patiently await the full opening of the door before moving the carts through the door. Encourage them to be aware of potential damage and to be careful.

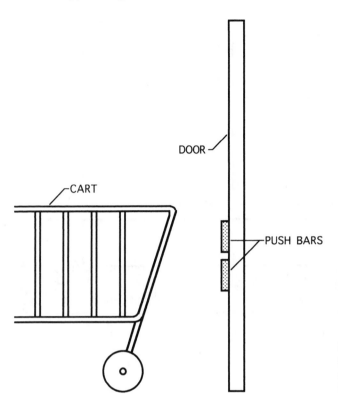

DOOR

CART

PUSH BARS

Figure 6.1 It is evident that this grocery cart will cause damage to entry doors because of improper placement of the door push bars.

The store management replaced the plastic sheet at a considerable cost and added another push bar, but the doors still show some scuffing. The point is that a design error led to a maintenance expense. With some study at the design stage, the problem could have been avoided.

A second example is of a slightly different nature. A certain seashore beach hotel has painted tubular steel railings lining the balcony walkways that run the length of the building. The railing support struts are set in the concrete floors of the walkways. (See Figure 6.2.) The atmospheric conditions are such that the salty ocean spray collects at the bases of the struts, gradually deteriorating them and the surrounding concrete, and ultimately causing a product failure. Each of these failed structural points now requires a costly maintenance program, as follows:

1. Break away the crumbling concrete at the strut base, using pneumatic tools and manual hammers and chisels.
2. Cut and remove six inches of the rusted steel strut base with an oxyfuel torch.
3. Arc weld a replacement steel piece to the bottom of the strut.
4. Grind and polish the weldment with electric abrasive equipment.

Figure 6.2 Hotel railing structure, showing area of support strut deterioration.

5. Construct a wooden containment form at the strut base area to receive the concrete to secure the newly welded strut.
6. Mix and pour the fresh concrete into the form.
7. Remove the wooden form after the concrete has set, and polish the new concrete to match the texture and appearance of the original floor.
8. Clean and then repaint the steel strut and the new concrete to match the original construction.

There are 150 of these strut points, and each involves the field repairing and restoring operations just described. This is a good example of a very time-consuming and expensive design error. One might be tempted to dismiss it as a construction error, except that one should be reminded that the perceptive designer would have anticipated this potential failure and addressed it when the railing structure was being planned in the first place. Also, the designer's grasp of materials science was wanting. There are any number of more robust design solutions possible, and the readers of this book will have an opportunity to participate in a project to provide one. (See "Questions and Activities" at the end of this chapter.)

Vacuum cleaners for both home and institutional use should have soft, resilient bumpers on the noses to protect door trim and baseboards from scars, dents, and scratches. Most do not, and so, in time, these wooden pieces become unsightly and require costly maintenance. Designers should use the cleaning equipment they design in order to become aware of such design faults, and correct them before the units goes into production.

Another common device that invites failure is a popular type of liquid soap dispenser found in public restrooms. This model requires the user to pull a lever forward in order to deliver a few drops of soap into the palm of the awaiting hand. (See Figure 6.3.) The problem is that this unit is designed to be fixed to the wall with thick double adhesive tape strips. With time and required repeated pulling, the tape fails, and the unit loosens and eventually falls away from the wall. The designer had a simple solution to this potential maintenance problem: design a dispenser lever that must be pushed rather than pulled to deliver the soap. Unfortunately, this obvious solution was ignored.

The several types of panels used to separate the individual cubicles in public restrooms and shower stall areas, or in industrial welding or spray booth facilities, are generally fastened to masonry walls with screw anchors, as shown in Figure 6.4. Note that the panels generally will fail in time as normal wear and tear loosens the anchors and pulls them away from the walls. This will happen even if the front end of the panel is supported by a vertical post. There are two reasons for this condition: (1) the design of the connector bracket is faulty because the torque drags the panel down and away from the wall, and (2) such masonry anchors are notoriously unreliable because they are so often installed improperly. These anchored panel arrangements have been used for years. Someone should design a new system.

One could argue that if the soap dispenser unit had been installed properly (that is, by assuming that the wall and unit were clean and dry), the structural adhesive tape would not fail. The same argument may be valid in the example of the

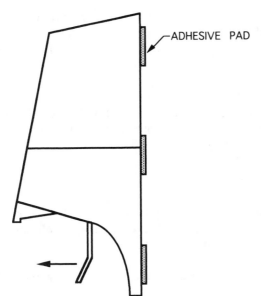

ADHESIVE PAD

Figure 6.3 This liquid soap dispenser would have been better designed with a push lever rather than a pull lever to avoid potential product failure.

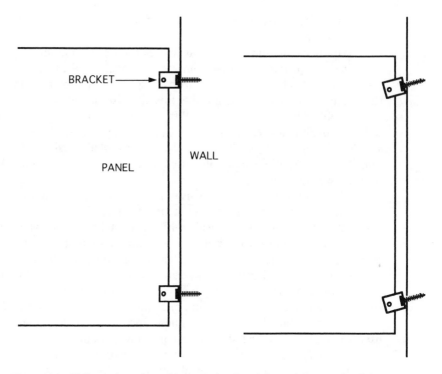

Figure 6.4 Wall panels such as this example often fail, as shown on the right, because of poor design and faulty installation.

wall panel anchors described earlier. The fact is, however, that many installers are careless or poorly trained and do not follow prescribed installation procedures. Designers should consider such human frailties at the earlier design stages, determine any possible failure points, and design accordingly.

The common push bars used to open large institutional doors have a mechanism that eventually fails with normal use. (See Figure 6.5.) This not only makes door opening difficult or impossible but it also can pinch and bruise the fingers. This recurrent condition can be corrected by redesigning the mechanism, thereby solving both a maintenance and a safety problem.

Design for Serviceability Guidelines

Although no exhaustive checklist can be compiled for serviceability design, or for any of the other design topics featured in this chapter, some general statements are in order. The following guidelines can be helpful:

1. Provide access openings of sufficient size and lighting to adequately perform maintenance tasks.

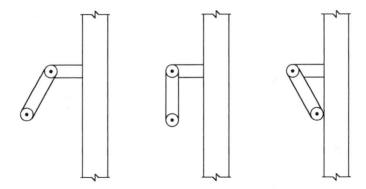

Figure 6.5 The problem of excessive wear in push bar door openers can lead to maintenance and safety problems. Fingers can easily be pinched because the mechanism has failed.

2. Place frequently pulled service units on rollers, slides, or hinges.

3. Provide ready access to all test, adjustment, and service points.

4. Assure that cases and covers open without interference.

5. Design cases that lift off of units, instead of units out of cases.

6. Design maintenance parts of a size to be handled by one person.

7. Provide adequate worker protection against electrical, pressure, exhaust, and contact hazards.

8. Provide safe and convenient hand-holds.

9. Design components that require a minimum number of fasteners and parts.

10. Provide easy access to fasteners, with adequate space for tool manipulation.

11. Use captive fasteners, when possible.

12. Design snap-on cases and covers rather than those requiring threaded fasteners.

13. Design units utilizing quick-change replacement modules.

14. Minimize the number of parts in any product.

15. Provide sensory clues to product malfunctions, such as warning lights, horns, and displays.

16. Prepare user-friendly installation, assembly, operation, and maintenance manuals.

17. Specify standard, readily available, interchangeable replacement parts.

18. Simplify product use or equipment operation through good control system design. This may prevent any improper product use that could contribute to the need for maintenance.

19. Involve maintenance specialists in all routine product planning operations.

20. Anticipate everything that could go wrong with a product, and design such potential faults out of the product. This can be accomplished by thorough testing. Designers should personally try out everything they design. (Generally they do not.)

21. Consider the fact that maintenance may be undertaken by semiskilled technicians or by completely unskilled consumers, and make design modifications as necessary.

22. Avoid complex disassembly procedures. Keep things simple for maintenance personnel who may be working under adverse conditions and deadlines.

23. Incorporate design features that will hold the possibility of maintenance error to a minimum.

DESIGN FOR SECURITY

Security can be defined as the condition of being protected from danger or not exposed to danger. Security design ranges from planning systems to assure freedom from vandalism, tampering, eavesdropping, danger, or industrial espionage, to protection against unauthorized entry, access, or removal. Examples of such systems include police patrols, guard dogs, fences, doors, lights, locks, alarms, safes, and tamper-proof packaging. The mailbox in Figure 6.6 looks secure and is secure

Figure 6.6 This mailbox is constructed of durable aluminum with a strong locking mechanism. The simple forms and bold graphics contribute to its pleasing appearance. Courtesy of Danish Design Centre.

because the designer set out to make it so. A variety of plastic heat shrink sleeves are available to protect the handles of sporting equipment and the closures of drink bottles and medicine containers. (See Figure 6.7.) The key to the effective use of such sleeves is that they protect, are tamper evident, and are convenient to remove.

Security becomes a design issue whenever current product design encounters problems relating to freedom and protection. A firm engaged in making doors and windows is immediately faced with the problem of guarding its products against unauthorized entry. In short, the designs have obvious security implications, and the solutions must lead to the provision of adequate mechanical or electronic locking systems. The approach should be one of determining how an intruder might possibly gain entry through a door, and designing devices to thwart such illegal efforts. Security design, much like medical device design, is a highly specialized technology. Companies engaged in security design must avail themselves of the latest social and technical research, and common sense, in order to design effective, fool-proof security devices.

One appropriate example is the design of a hardened steel plate to totally cover the area of the door lock mechanism. It is installed flush with the door and jamb surfaces, and extends over the jamb to deter access to the deadbolt, as shown in

Figure 6.7   Shrink sleeves can protect products and are tamper-evident. Courtesy of Halpak Plastics, Inc.

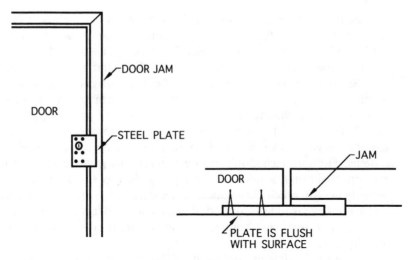

Figure 6.8 A door jamb plate will help to thwart the efforts of a burglar. The simplest solution is often the best solution.

Figure 6.8. If jimmying the door lock is a problem (which it is), the designer must invent a device to foil the intruder (which this will).

Design for Security Guidelines

1. Consult marketing and security personnel to ascertain the prevalence and nature of security problems with specific products.
2. Review literature related to reported security concerns applicable to products manufactured by your company.
3. Anticipate the possibilities of security problems with a product under consideration, and design to overcome these potential problems.
4. Study available catalogues and literature, and consult with sales representatives to learn of the new security technology pertaining to your company's line of products.
5. Field test all security designs to ascertain their worthiness.

DESIGN FOR VANDALISM PREVENTION

Vandalism is the deliberate mischievous or malicious damage of property. It is not a recent phenomenon resulting from current technical and social conditions; in fact, its name derives from those Germanic peoples of the fifth century (Vandals) who sacked Gaul, Rome, and Spain. Vandalism is a willful, not an accidental, act, usually directed at public property because such holdings are not considered to belong to anyone in particular. Vandals might be reluctant to harm the belongings of someone known personally but would have no compunction at trashing the property of a total stranger. It is a serious condition that leads to the deterioration of facilities and neighborhoods, causes personal and public inconvenience, has human safety implications, and is costly. There are several characteristics of this negative behavior that can provide insights for product designers.

1. It is damage to property belonging to another person, generally unknown to the vandal.
2. It is damage that must be repaired by someone.
3. It is damage that generally occurs in areas of restricted surveillance.
4. It is made easier because of the increased mobility of vandals (e.g., automobiles, motorcycles, all-terrain vehicles, and snowmobiles.)
5. It is fostered by the increased presence of coin-operated vending, copying, banking, and telephone equipment.
6. It is frequently paired with theft, such as smashing a vending machine to get at the coins.
7. It often is perpetrated by perfectly respectable and moral individuals who are occasionally bent on property destruction. Not all hoodlums are vandals, nor are all vandals hoodlums.

The designers of vandalism-prone products are challenged to develop approaches to combating this serious issue, beginning with a study of research reports dealing with the sociology of vandalism. Of equal importance is the collection of data pertaining to the specific product design task. For example, if the task is vending machines, the design group should meet with the people who locate, install, and service the machines. Why are they placed where they are? Could a safer, better policed area be found? What are the most frequent maintenance problems? How many are caused by vandals? Would the use of better materials or a different locking system lessen the prevalence of destructive acts? These are sensible, relevant questions whose answers may lead to effective design solutions.

A researcher once noted that the kinds of materials used in a bus-stop shelter influenced the way vandals responded, which may suggest a kind of "psychology of materials." The formidable appearance of a sturdy, secure, and attractive shelter of concrete, or a heavy, flat, stainless steel housing for a vending machine coin box may deter vandalism as well as theft, thereby lending credence to the notion of materials psychology. It is a thought worth considering.

A possible solution to the endemic problem of vandalizing lights in pedestrian and vehicle passageways is shown in Figure 6.9. The special lights were planned to conform to the steeply sloping sides of the passageway. Incidentally, the sides were designed to prevent vandals from crawling to them and to prevent trash from collecting on them. The tough plastic light lenses are unbreakable and hard to reach without special equipment.

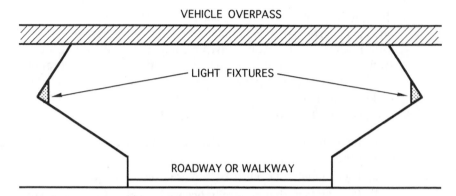

VEHICLE OVERPASS

LIGHT FIXTURES

ROADWAY OR WALKWAY

Figure 6.9 A good solution to the problem of vandalizing safety lights. Provision has been made for both adequate illumination and protection.

The flush-mounted telephone (see Figure 6.10), based on a single sheet of stainless steel, is a hands-free, reliable, and durable communication device, and it does possess something of the materials psychology described above. It is resistant to severe abuse, and, since there are no external parts to be damaged, the primary

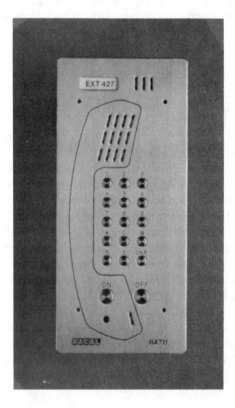

Figure 6.10 A single sheet of flush-mounted stainless steel and clever design resulted in this vandal-resistant RA711 telephone. Courtesy of Racal Acoustics Limited.

function of communication is not easily lost. This vandal-resistant telephone is a good example of a creative analysis of all the shortcomings of traditional phones, with special attention to the matter of vandalism. Although the instrument was designed to meet the needs of clients currently experiencing financial loss through vandalism, this phone is so easy to install and so versatile that there are many other uses. For example, it can be adapted for use by the hard of hearing and for others who find it difficult to hold a receiver.

Another example of designing for vandalism prevention derives from the common problem of street sign damage. This issue has economic, safety, convenience, and esthetic implications. By personal experience, the author lives on a street where the street sign generally has been bent askew or broken from its post by vandals. (See Figure 6.11.) This causes inconvenience to the motorist (or can be a matter of life or death for an ambulance driver) looking for the street; results in considerable expense (about $100) for the city to replace or repair; and detracts from

Figure 6.11 A common street sign vandalism problem.

neighborhood esthetics. The author discussed the matter with his design for manufacturing class and assigned the problem to them. Several practicable solutions
resulted (along with some questionable ones, such as coating the sign edges with
broken glass or razor blades, or running a powerful electric charge to the structure),
but one stands out. One design team proposed that a tough coil spring be attached as
an interface between the sign and the post, as shown in Figure 6.12. This is a simple,
practical solution that would deter vandals, for the sign would merely spring back to
normal position when bent. The plan included a universal fitting that would accom-

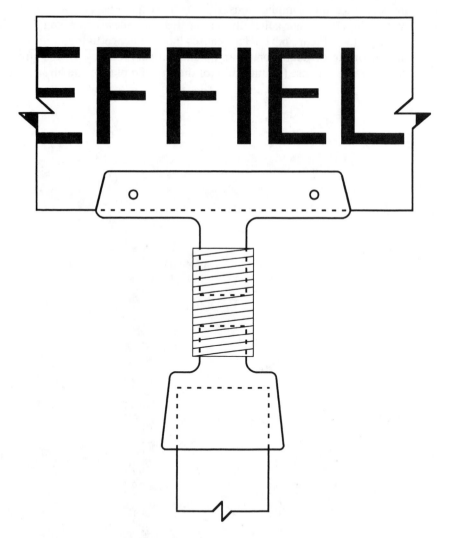

Figure 6.12   The addition of a steel spring as an interface between a street sign and
post will prevent sign damage.

modate the typical round, square, or U-shaped signposts. The same kind of inventiveness can lead to equally sound solutions to many vandal-related problems.

Design for Vandalism Prevention Guidelines

Typical of the facilities and structures subject to acts of vandalism are signs, lights, vending machines, telephones, public toilets, public transport, room furnishings, and recreational facilities. The damage can include writing on walls, breaking lights, slashing bus cushions, wrecking furniture, plugging toilets, or bending street signs. A list of suggestions leading to possible solutions follows:

1. Provide for built-in units in public restrooms. Soap dispensers and towel containers are more vandal-proof if they are built into the wall, not projecting from it.
2. Use glazed tile in washrooms to discourage graffiti. It is more difficult to write on and easier to clean. (The problem is that felt-tip marker designers keep inventing pens that will write on any surface.)
3. Place building signs at an elevation that would preclude vandals removing them.
4. Design mail slots instead of fragile mailboxes.
5. Design lighting systems that would discourage vandals. (Refer to Figure 6.9.) Note that the proposed structure emits sufficient light while presumably discouraging vandals.
6. Select materials appropriate to the product, and consider a psychology of materials.
7. Investigate new technologies in the search for vandal-resistant products.
8. Review the literature dealing with vandalism and vandal-resistant products to learn of the newer approaches to problem resolution.
9. Involve experts in vandalism prevention early in the design process.

DESIGN FOR SAFETY

The single-most important concern to manufacturers is that their products do not endanger the people who use them. This condition can be more reasonably assured if safety is built into a product at the conceptual stage and based on a before-the-fact philosophy. Fortunately, this ideal is becoming a matter of course by many design teams through appropriate safety analyses encouraged by both altruism and possible liability suits.

Safety can be defined as the freedom from those conditions that can cause injury to people or damage to equipment. As such, safety relates both to work environments and to products. This present discussion will be concerned with those factors dealing primarily with the design of safe products.

Aside from the important issue of protecting users from product hazards, manufacturers also must concern themselves with the equally important issue of product liability. (See Chapter 10.) Stringent product safety laws have established legal obligations for the makers of products—a fact that provides further inducement to the production of things that are safe.

Types of Product Hazards

The purpose of a product safety analysis is to identify potential product dangers. Once identified, these dangers may be eliminated or mitigated through normal product design procedures. There are six basic types of hazards:

1. *Entrapment* injuries are caused by the nip points created by machine parts moving toward each other. Examples include meshing gears, belts and pulleys, chains and sprockets, cams and levers, and closing doors. Parts of the body can be drawn into these nip points and be pinched or crushed.

2. *Contact* or tactile injuries occur by the accidental movement of the body against sharp points, hot surfaces, or electrically charged units. Typical sources of these hazards are moving saw blades and cutters, gas engine mufflers, soldering equipment, and electric switches.

3. *Impact* injuries result when a part of the body is struck by reciprocating or rotating machine parts. Spinning lathe chucks, collars with projecting bolt heads, and impact tools are common sources of such blows.

4. *Ejection* hazards are associated with particles of debris thrown by a moving machine part. Bits of wood are routinely ejected from rapidly rotating saw blades and router cutters. Similarly, ceramic chunks can fly like sharp missiles from a cracked grinding wheel. All such dangers can result in serious (even fatal) eye, head, and hand injuries.

5. *Entanglement* of hair or loose clothing in rotating shafts or workpieces is a common hazard. Parts need not be moving rapidly to cause serious bodily damage. One example is a piece of clothing getting caught in a rotating drill chuck or a fan blade.

6. *Noise and vibration* hazards can result in temporary or permanent hearing loss or in the loss of sensation in the limbs. Operating a gasoline-powered lawn mower without ear protection can eventually affect one's hearing.

A further point should be made regarding these hazards. Some are *inherent*, inasmuch as they occur routinely in the course of normal machine operations, and are simply unexpected accidents. As described below, a chainsaw is an inherently dangerous machine. Other hazards are of a *contingent* nature because they result from absent or improperly installed machine guards, the failure of safety devices, malfunctions due to the faulty assembly of product components, or poor maintenance procedures. While unexpected, they all have common bases of operator, equipment, or design error. Additional information on hazards can be found in Chapter 5.

All hazards can be at least partially overcome through reasonable product analysis procedures. Some call for the installation of proper machine guards or other safety devices, others call for effective warning labels, and still others call for adequate personnel safety and maintenance training, including excellent operation and service manuals.

Safety Devices

Situations occur when it is impossible to totally eliminate a product safety hazard. The solution is to provide safety devices to minimize the occurrence or severity of a use-related accident. *Safety devices* are elements added to a design to control hazards. The gasoline-operated chainsaw is a case in point. (See Figure 6.13.) This machine is inherently dangerous because of the rapidly moving, sharp-toothed chain that provides the cutting action. This factor, coupled with noise and vibration, can easily result in a serious accident. A safety device common to quality saws is a chain brake, designed to stop the chain immediately in case of a kick-back or bind. A brake-switch interlock stops the motor when the chain brake is pushed forward with the back of the left hand. The brake can also serve as a hand guard. Other devices or features include a right hand chain guard, a trigger lock-off to prevent accidental acceleration and starts, a shield to protect hands from contact with a hot muffler, and a sawdust chute to direct flying chips away from the operator. Manufacturers and designers must attend to situations where hazards cannot be eliminated but can be controlled.

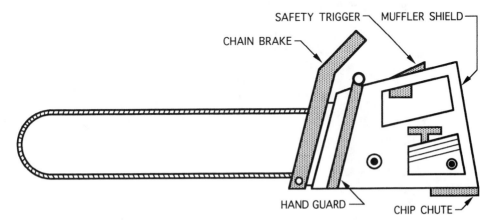

Figure 6.13   The manufacturers of chainsaws have made these inherently dangerous tools safer by designing in a number of safety devices and features.

In a totally unrelated example, the simple act of sharpening household tools such as knives or scissors poses a possible consumer safety problem. Many people are careless or unskilled and often end up with a cut finger or hand. The unique sharpener in Figure 6.14 is safe, efficient, and easy to use. The convenient hand-hold keeps the fingers away from the sharpening action, and the method of using the tool is clearly visible. This is a good solution to a potential problem.

The popular and very usable craft knife shown in Figure 6.15 has been for many years a tool of choice for artists and hobbyists. It is easy to use, reliable, and convenient. However, a redesigned model has some improvements that make it an even more desirable cutter. (See Figure 6.16.) The soft plastic barrel, available in several striking colors, provides a comfortable and sure grip. The addition of a simple anti-roll nut at the end of the chuck tightener offers the added convenience of preventing the tool from rolling off a work surface. This is a decided improvement over the round, rolling models (see Figure 6.15), which are a source of considerable aggravation. A potential safety problem has been resolved by moving the chuck

Figure 6.14 A safe and convenient scissors sharpening device. Note the positive hand-hold to protect the user. Courtesy of Design Forum Finland.

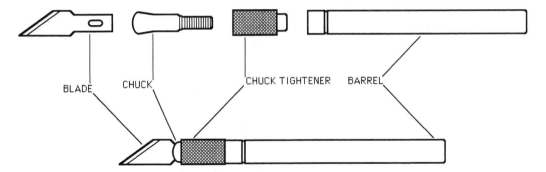

Figure 6.15 A typical craft knife has many household and hobby uses.

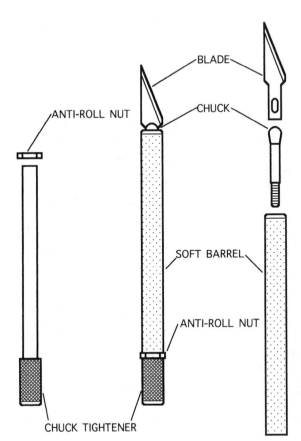

Figure 6.16 Note the several improvements in the redesign of the common craft knife, making it a safer and more convenient tool.

tightener to the end of the barrel, thereby keeping the fingers away from the sharp blade. A plastic protective cap is also provided as an added safety feature. This knife is a well-conceived, usable, and attractive product.

The special hook latch is an excellent example of a simple yet effective safety device. (See Figure 6.17.) Its purpose is to conveniently secure open hooks so that they will not separate from chains, rings, or trolleys while in use. Note that the side tabs on the latch prevent the fingers from entering the danger zone, thus providing an additional user safety feature. This product is a logical result of the designer's careful examination of the common safety hazards in such hooking devices, and eliminating them in the early design stages.

The familiar key used to lock a drill in a drillpress chuck also can be dangerous. The key is necessary in the operation of the machine, but must be removed after use. If left in the chuck inadvertently, and the machine is turned on, the high speed of the revolving chuck could throw the key into the face of the operator and cause a serious accident. One clever designer studied the problem and devised a key that has a pressure spring that permits chuck tightening with a normal finger push, but that

Figure 6.17 This latch is easily and safely managed to secure the hook to an industrial lifting device. Courtesy of SIPCO Products, Inc.

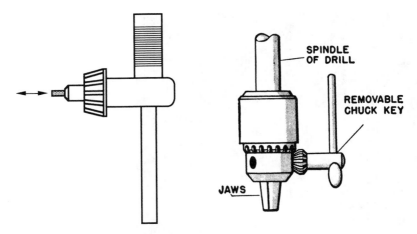

Figure 6.18 This drill press chuck key will self-eject after use. This action prevents it from being accidentally left in the chuck during operation, as shown in the inset.

ejects the key when the pressure has been released. (See Figure 6.18.) This simple modification has eliminated a potential hazard.

The striking welding hood shown in Figure 6.19 provides a convenient, effective, and reliable eye and face protector for the welder operator. Many of these models also have auto-darkening filters in place of the traditional fixed shade filters. This permits the human welder to maintain an uninterrupted view of the workpiece, without having to raise and lower the welding hood.

Quality control factors also affect product safety by way of possible manufacturing errors. Faulty weldments, incorrectly installed O-rings, poor worker training, and employee fatigue are among the conditions that may contribute to making below-quality products. These, in turn, may lead to product failures and accidents. This is a management problem, not a designer problem, and must be dealt with by vigilant production control staffs. However, the diligent design team will study the contemplated product plan, identify possible manufacturing errors, and account for them.

Design for Safety Guidelines

1. Anticipate all potential safety hazards in a product, and design accordingly for elimination or mitigation.
2. Design a product to eliminate the possibilities for assembly errors that may result in a hazardous product.
3. Recognize that products may be used by careless people who will not read the owner's manual and may be used incorrectly; create design features to overcome this fact.

Figure 6.19 This welding hood offers excellent eye and face protection to the user. Also, it is nice to look at. Courtesy of Atlas Welding Accessories.

4. Practice safeguarding by location, where machine hazards are positioned so they will be inaccessible to the operator during normal machine operation.

5. Review literature dealing with safety hazards inherent in the products of your company.

6. Employ benchmarking techniques to identify and emulate the outstanding product safety features and procedures in competing companies.

7. Involve safety specialists early in the design process.

DESIGN FOR USABILITY

A product is usable if it plays on innate human tendencies to perform according to some natural pattern. The term implies convenience, ease of use, functionality, logic, and directness, as contrasted to confusion, difficulty, misdirection, misperception, and aggravation. Usable artifacts give people great pleasure; unusable products are annoying. Public address audio systems are notoriously unreliable and difficult to use. Ballpoint pens are reliable and easy to use. The term *user-friendly* has become a vernacular expression relating to the ease of computer operation, and in a way sums

Figure 6.20   This computer is a user-friendly machine.

up the entire realm of usability. Such computers are in fact easier to operate, more forgiving, and provide visual clues to extricating an errant user from some confusing deadend. (See Figure 6.20.)

Donald A. Norman, in his book *The Psychology of Everyday Things* (see Chapter 11), refers to this phenomenon of utility as "natural mapping," and suggests that it infers the proclivity of people to act or react in some natural, logical manner. For example, the typical action of a right-handed person in reaching for the hardware to open the heavy outside door to a building is to grasp it with a clenched fist, and not with the extended fingertips. (See Figure 6.21.) This is natural, perhaps because through experience people have learned that this is the most comfortable, efficient, easiest, and often the only possible way of opening a heavy door.

The portable radio in Figure 6.22 is very usable, as personally attested to by the author, who owns one. Every control feature provides an immediate visual or aural response, without any need to try to deduce what (if anything) is happening. You turn the tuning knob and the indicator moves to the desired station. The volume control is positive and simple. The product has few controls, and each is direct, logical, and immediately responsive. For each action, there is a readily apparent reaction. There is a feeling of confidence that one is in command of the machine.

Another popular and very "high-tech" radio was considered by the author, but

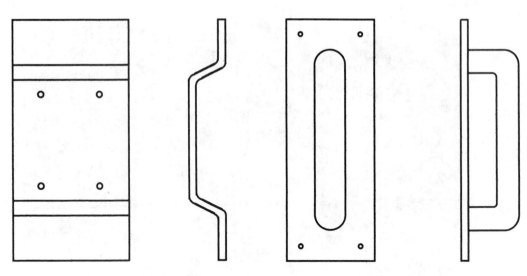

Figure 6.21 The door pull on the right is much easier to use than the model on the left. It provides for a natural gripping action whereas the other does not.

the salesperson had to spend five confusing minutes trying to explain how it worked, and finally gave up in exasperation. The designer of that machine was determined to include every bit of available electronic technology, to the point that the radio was a confusing array of balancers, scanners, seekers, programmers, replayers, and wake-up sequences—a good example of technology unbound. The designer of that radio approached the task diligently, but forgot to try it out personally or to watch someone else use it, to see if it worked. It was not very commendable, and it went unpurchased.

Not all usable or unusable products are electronics related, however. A cross-country skier has a feeling of profound satisfaction toward a sure and convenient method for attaching a boot to a ski. Optimally, one should only have to press the boot toe into the binding without any further physical exertion. To be required to bend down and wrestle the two members into a lock position, generally under adverse weather conditions, is not a satisfying experience. While on the subject of skis, one manufacturer has added to the enjoyment of winter skiing by designing a toe-warming, portable, battery-operated heater that fits any ski boot. (See Figure 6.23.) This unit weighs only 75 grams (2.5 ounces), and the rugged battery case is injection molded of a polycarbonate ABS plastic resin.

Household clothes washers and dryers generally have four adjustable footpads to level the machines and to prevent undue vibration during operation. (See Figure 6.24.) Through normal use, one or more of the pads loosen and throw the machine out of balance. Invariably, the one needing adjustment is inaccessibly located at the rear bottom of the appliance, making readjustment inconvenient, if not impossible.

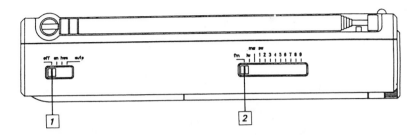

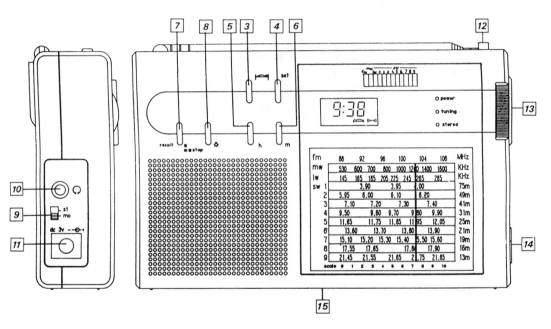

Figure 6.22 A very user-friendly portable radio, whose controls provide visibility to
the user. Courtesy of Grundig AG.

This is not a situation requiring a professional service call. It is assumed that the
owner will attend to this minor malfunction. Clearly, the designer should have
considered the potential problem and devised a method for easy operator adjustment. Such annoying service matters lead to customer dissatisfaction.

These few examples offer some valuable clues to designing usable products.
Thinking designers must consider the anthropometric, physiological, and psychological capabilities and characteristics of humans (introduced in the previous chapter)
as they plan solutions to design tasks. Some points worth considering are discussed
next.

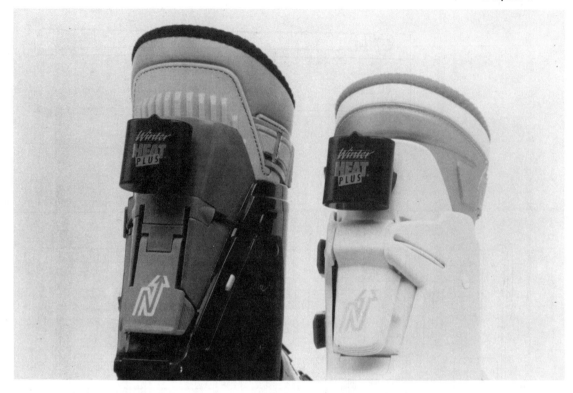

Figure 6.23 A rugged, lightweight, and effective ski boot heater adds to the pleasures of winter skiing. Courtesy of Alpine International Corporation.

Visibility factors involve the assurance that users clearly perceive how they must manipulate product controls in order to cause something to happen. They should not have to guess, engage in trial and error, or search their memories or the owner manual in order to determine which button to press next. Controls should be logical, positioned properly, and clearly marked. The radio controls in a current automobile (Figure 6.25) are confusing. The arrows on the tape deck do not provide the proper clue as to which button to push for rewind or for fast-forwarding. Visibility is absent. To compound the felony, this radio has 30 different controls the driver has to contend with, many of which are equally unfriendly. Contrast this with the very visible and easily managed automatic bank teller machine shown in Figure 6.26.

Another aspect of this factor is the human response to visual cues. The several warning lights and gauges on an automobile dashboard signal that the engine is overheating, or the oil level is low, or the fuel tank needs refilling. Many pieces of industrial equipment having similar visual attention-getting devices, all designed to protect the operator or the machine, or to indicate clearly some remedial action to be taken. They are especially effective and necessary in noisy, intrusive environments.

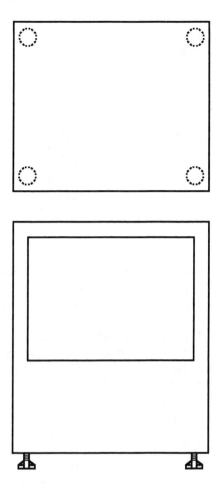

Figure 6.24 The adjustable feet on this clothes dryer can be a constant source of user irritation. They can go out of adjustment, causing the dryer to go out of balance and resulting in a violent vibration.

The inclusion of these systems makes all equipment operation more satisfying, more efficient, and safer.

Auditory factors take advantage of the fact that humans hear changes in tones, respond to warning buzzers, or awaken to the sounding of an alarm. Furthermore, they are more confident that they have caused something to happen in response to pressing a key if they can hear it, as with operating a computer keyboard. Sound can provide information that is difficult to convey by any other means or that an operator does not know. Some examples include the rattle or the rushing air noise that indicates that a car door is not closed; the whistle of a teakettle to remind one that the water is boiling; that scratching sound emerging from the front automobile wheels to alert the driver that the brake discs need replacing; the change in pitch when the vacuum cleaner hose is clogged; the highway rumble strips to remind drivers of an approaching potential danger area; and the satisfying purr of a properly tuned

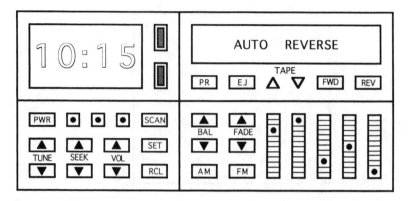

Figure 6.25   An automotive radio and tapedeck should not be as confusing as this one.

automobile engine. Surely, one of the most annoying and potentially dangerous situations is following an automobile whose turning signals continue to blink eternally. The driver of that errant vehicle has failed to respond to the visual dashboard light, indicating that the turn is complete and that the signal can be turned off. The solution is to provide a loud aural signal with the visual, so that the driver is sufficiently annoyed and will turn off the signal. One could argue that such a signal should turn off automatically. Generally, they do, but not always. Hence, the need exists for sound signal backup. Furthermore, aural signals are imperative in work situations where an equipment operator is frequently away from the immediate work station or where visibility is poor. The operator can hear warning signals that cannot be seen. Humans perceive sounds and respond accordingly—a fact that should be considered by product designers.

Tactile factors relate to feel or touch. The membrane switch problem described in the preceding chapter is a good example. Recall that machine operators in those situations were more reassured that electronic instructions had been properly relayed if they felt and heard the depression of a control key or switch. Also, the controls on an electric blanket must be effectively operated by touch, or perhaps sound, because it is generally dark when they are being adjusted. Because of the tactile feature of embossed toilet paper, it is easier to lift off the roll and is therefore more "usable." The inventive designer must attend to this product requirement. Consider also the people who have sight-impairments and their special needs for Braille dot codes on elevator control buttons and other equipment.

A skilled woodworker can "feel" the sharpness of a chisel shearing through a piece of walnut, or "feel" the blade of a tablesaw while ripping a piece of oak. In both cases, there may be a feel of unusual resistance, and, for the tablesaw a change in sound. These are clear indications of tools that need sharpening. When using a pencil sharpener, one can feel when the wooden pencil is sharp. The matter of tactile capabilities possessed and practiced by humans can be an important factor in designing usable products.

Figure 6.26 Automatic tellers facilitate consumer banking needs and generally are as simple and convenient to use as this example. The controls and visual clues clearly indicate how the user is to interact with the device. Courtesy of Diebold, Incorporated.

A final element relates to human *physical* endowments and directly relates to the ergonomic factors (again described in the preceding chapter.) Such human measures obviously control the sizing of objects to fit the human frame in such a fashion as to make them more usable, as shown in Figure 6.27. Note how the air-driven screwdriver's angle head and molded shape provide a comfortable, reassuring, safe grip, and convenient access to fastener installation in tight places. Further examples include the proper positioning of control levers and switches to match human reach; the provision for the easy adjustment of the height of a bicycle seat or

Figure 6.27 An ergonomically designed industrial screwdriver that is comfortable and convenient to use in product assembly operations. Courtesy of Sioux Tools Inc.

the seat, back, and footrests on physical exercise equipment; the assurance that the push-button on a can of spray paint is placed for proper index-finger operation, pointed in the right direction, and needing only a modicum of pressure; designing hair-dryers and hand drills with handles set at an angle for convenient operation; and making chairs of a comfortable work or leisure height.

The squeegee in Figure 6.28 is a clever, efficient, and reliable solution to the problem of removing soap and water residue from residential shower stalls. Through ergonomic studies, the designers determined that the traditional T-shaped squeegee was not the most efficient for a wiping motion in a confined area, and that a cylindrical handle would keep the wrist from tiring. The solution was a handle that feels soft to the touch and two wavy flexible blades, like double windshield wipers, all made of inexpensive extruded plastic. When not in use, it stands against the tile, a piece of functional art. Other uses include industrial cleaning, commercial window washing, drying boats, and even wiping down horses after a thorough hosing.

The finned projections at the base of the electrical plug in Figure 6.29 serve as a better grip when removing the plug from an electric outlet, as well as preventing the cord from being torn out of the plug. This makes it a more usable product, and one featuring designed-in maintainability.

Figure 6.28 This elegant Cleret squeegee is a very usable and efficient tool. Courtesy of Hanco.

There is a basic user-unfriendly problem with many consumer-packaged goods. The issue is the proliferation of confusing sizes, or "funny numbers." A good example is the toothpaste tube. A perusal of local market shelves will reveal the following ounce-sizes: 0.75, 0.85, 1, 2.7, 3, 4, 4.3, 4.5, 4.6, 5, 6, 7, and 8.2, and perhaps others. There may be some good reason for these sizes, but it escapes the consumers. In their estimation, this is perceived as a marketing gimmick directed at frustrating cost-comparison shopping.

One practical solution to this problem is to adopt the metric system, whereby toothpaste will be packaged in only four sensible sizes: 25 milliliters (ml) to replace the one-ounce tube; 50 ml to replace the two-ounce tube; 100 ml to replace the four-ounce tube; and 150 ml to replace the six ounce tube. It is a user-friendly solution—

Figure 6.29 The fins on this electric plug serve the dual purpose of aiding in the plug withdrawal and of preventing the cord from tearing from the plug.

one that will thwart deceptive marketing practices and offer the consumer easily comparable package sizes. It also should be an economic benefit to manufacturers by reducing the numbers of package sizes.

Design for Usability Guidelines

1. Apply the principle of simplicity to all products to make them more usable.
2. Make use of natural mapping in all product designs.
3. Avoid all deceptive marketing practices. They are unjustifiable by any measure.
4. Test all products personally to debug them for unusability features.
5. Test all products for anthropometric fit.
6. Make use of the results of market analyses to determine the receptivity of new products, and modify them as required.

DESIGN FOR THE PHYSICALLY DISADVANTAGED

Increasingly, designers are becoming aware of both the problems and the potential of designing products expressly for the disabled and the elderly. This is one of the fastest-growing population groups, and the opportunities for addressing their special product needs are great. It is estimated that by the year 2000, about one-third of the U.S. population will be over 65 years of age or be disabled in some way. Further estimates indicate that 80 percent of persons over age 65 continue to live in private residences rather than in patient care facilities. With aging comes a range of health problems such as arthritis, heart disease, and hearing, sight, and orthopedic impairments, among others. These statistics should be of particular interest to the design community because this population is the group most affected by their environments. Interestingly enough, this is one of America's wealthiest demographic groups. It is not a matter of such persons being unable to afford products appropriate to them, but instead the difficulties they may encounter in finding such products.

There are several important pieces of federal legislation that bear on this issue, such as the President's Commission on Employment of the Handicapped, Sub-Committee on Barrier Free Design; The Older Americans Act; The Rehabilitation Act of 1973; and the Americans with Disabilities Act; among others. The Consumer Products Safety Commission also has regulations that refer to this group. The individual states generally have enacted special regulations, laws, and codes based on the federal acts, and in many cases extending them. Some of this legislation deals with barrier-free design, which means those architectural designs that eliminate the hindrances that deter physically limited persons from having access to and mobility in and around a building, structure, or improved area. Designers should become familiar with such legislation as a guide to their work.

Some readers may take issue with the author's grouping together persons who are old with persons who have disabilities, feeling quite properly that people may be old but not necessarily disabled. Coupled with this is the controversy associated with the word *disabled*, where some prefer terms such as *functionally limited, physically disadvantaged, impaired, exceptional, infirm aged, elderly frail,* and the like. The implication of this part of this chapter is simply that a sizable portion of our population possesses some degree of some deficiency and has special product and living space needs. Furthermore, designers of such products and spaces must be aware of these deficiencies and account for them in their creative efforts. There are four general groups of such disabilities:

1. *Physical or motor disabilities*, implying the loss of use of one or more limbs, or having some skeletal or motor coordination impairment that may require the use of wheelchairs, crutches, braces, or other special devices.
2. *Emotional diabilities*, as seen in those who are psychologically disturbed and who require special care and attention.
3. *Sensory disabilities*, involving visual, aural, or tactile impairments that generally require special aids and devices.
4. *Intellectual disabilities*, as evidenced by some degree of mental retardation or deficiency, and that may require care in an institution or regular counseling.

The manifestations of these diabilities, whether brought about by aging, accident, or birth, must be scrutinized in order to determine what special needs must be addressed when designing products for those who have disabilities. Obviously, a best way to determine these needs is to meet with these people to observe how they function in their environments, to learn what difficulties they may have with the seemingly simplest of tasks, and to ascertain how appropriate product designs could contribute to their welfare. Product planners also should bear in mind the emotional effects of aging and debilitation. The realization that one can no longer drive a car or dress or do household chores can be depressing and distressing. Too many special buttons and dials on a washing machine or a dryer can confuse and discourage an elderly user. The ordinary act of vacuuming the carpets can be made possible for many by designing a machine with a power assist and with few and simple controls. There should be an appropriate level of technology for all products, not just those designed for this special group. Products that are generally more functional and usable will be good for the entire market population. Some selected examples of such products are discussed next.

Electric cords and plugs pose a special problem for people who have fingers that are weakened from arthritis or muscular degeneration. Ordinary plugs are too small for efficient finger grip, even for persons with no dexterity ailments. The plug in Figure 6.30 features a large rear projection that can be easily and effectively grasped and pulled from an electric outlet. Contrast this with the plug in Figure 6.29.

A line of hand tools for the kitchen was conceived with the notion that a utensil appropriate for the elderly or the manually disabled would also better serve the

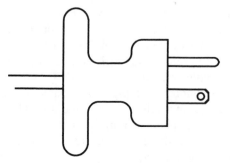

Figure 6.30 An electric plug designed for easy removal by persons with finger strength and dexterity problems.

general user. Based on gerontological and ergonomic research, a wide, thermoplastic rubber-like handle with soft grip fins on either side was developed as the universal contact point for the human hand. (See Figure 6.31.) The handle distributes the stress more evenly, alleviates the need for a tight-gripping action, and is comfortable and easy to hold regardless of the function of the tool. The tools are convenient, dishwasher-safe, durable, and visually attractive. There also are special scissors with oversized cushioned handles that are spring-loaded to gently reopen the scissors after each cut. (See Figure 6.32.) This easy action makes them ideal for persons experiencing weakness in their hands. A sliding lock switch, located conveniently on the upper handle close to the thumb, keeps the tool in a safe locked position when not in use.

Figure 6.31 This line of kitchen tools has handles of a convenient size and configuration for safe and easy use. Courtesy of Oxo International.

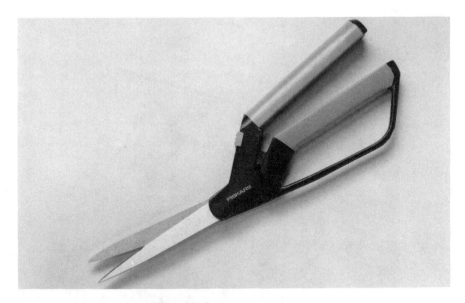

Figure 6.32 These attractive scissors were designed especially to make cutting easier for persons with hand muscle problems. Courtesy of Fiskars USA.

The barrier-free shower in Figure 6.33 is typical of the several styles available. This durable unit is constructed of rugged acrylic, with safe, smoothly rounded contours. It features molded corner soap trays, a wrap-around grab bar, wall brackets for a hand-held shower nozzle and hose, and a stainless steel seamless curtain rod. Some units also have a molded-in seat or a fold-up seat as an added safety feature. Conveniently placed water controls are nonprojecting for safety and are easy to manipulate. Toilet facilities are further enhanced by supplying a special highline toilet 18" high, with an elongated bowl, for more convenient seating and mobility. (See Figure 6.34.)

The special attention given to the needs of this population group is evident in the several products described above. Another example is the design of large tables for nursing homes. They should be round with a large central pedestal bases so that wheelchairs can be moved up to them. People should have eye-to-eye contact when socializing. Round tables facilitate this; side-by-side gang seating does not. Folding wheelchairs should be easy to erect for use; most of them are not—even for an ambulatory person. The special-powered doors for wheelchair users should be efficient and reliable, with controls conveniently placed.

Design for the Physically Disadvantaged Guidelines

1. Visit nursing homes, health care facilities, and other agencies dealing with physically disadvantaged persons to become familiar with the special product needs of this group.

Figure 6.33 A shower stall designed with safety and convenience features necessary
for older persons. It is equally effective for any user. Courtesy of Eljer.

2. Study the literature devoted to problems and solutions of the aged to gain data
   to guide the design activities of companies wishing to enter this product field.
   Joseph A. Koncelik's book (see Chapter 11) is very helpful.
3. Designers should personally use the products they create to ascertain their
   effectiveness. Also, designers should observe these products in use by the
   physically impaired, discuss their usability, and modify as required.

Figure 6.34 A highline toilet features a stool 18″ tall for safety and comfort. Courtesy of Eljer.

4. Design products taking into account the special safety needs of the disabled. Avoid furniture designs having sharp edges or projections that may contribute to a falling injury.
5. Remember the importance of independence, personalization, and control in the lives of the infirm aged. It is a psychological blow to become increasingly dependent on other persons or to be unable to perform simple tasks.

Design artifacts that are accessible, easy to control, and offer personal reinforcement.

6. Remember that products designed for this special group will be more usable
and more acceptable to the general population.

DESIGN AND THE ENVIRONMENT

Whereas ergonomics is the study of the relationships between humans and their
products, the issue of environment and design concerns itself with the interactions of
humans with the world in which they live. Very simply, the problem is that Americans
produce over 160 million tons of solid waste each year, or about 3 pounds per person
per day, which is the highest per capita rate among industrialized nations. The
responsibility of designers and manufacturers regarding this intolerable situation is to
plan and make environmentally benign products. Consumers also have a role in this
national effort and are in some measure responding to the challenge. But in the final
analysis, it is industry that must provide the leadership by taking the situation
seriously and reflecting this concern in their products. (See Figure 6.35.) Such waste

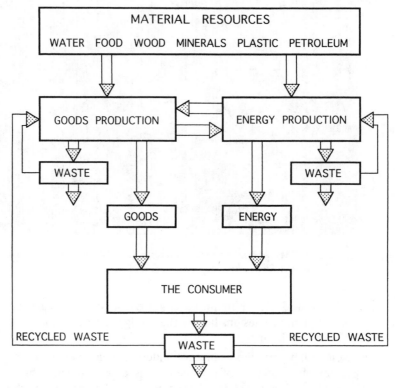

Figure 6.35   This illustration charts the flow of materials in the product manufacturing system.

consumes material resources, expends energy, and creates garbage that requires disposal. At the end of its useful life, a product should not contribute to the degradation of the environment.

Many approaches toward resolving this matter have been offered. *Source reduction* is practical because it decreases the amount of waste at the point of origin, generally at the manufacturing level. *Reuse* is desirable because it returns the product to the manufacturing stream with little if any loss, such as returning glass beverage bottles for refilling. *Repairing, rebuilding, restoring*, and *remanufacturing* are sensible approaches, inasmuch as they retain a basic structural unit while only requiring replacement parts to restore a product to a usable condition. Excellent examples are the many older buildings that have been restored to active use, rather than been razed. (It costs about one kilowatt of energy to tear down each kilogram of building material.) *Recycling* is of prime importance because it prevents the introduction of material into the waste stream, and it makes people feel better because they are actively engaging in a solution to the problem of trash. Each of these approaches must be met cooperatively by designers, manufacturers, and users in order for the issue to be resolved.

There are endless opportunities for designers to effect source reduction. The innovative cartonless case system Figure 6.36, attests to this. The two-row vertical tray cleverly doubles as a bulk shipping container and as an inviting countertop

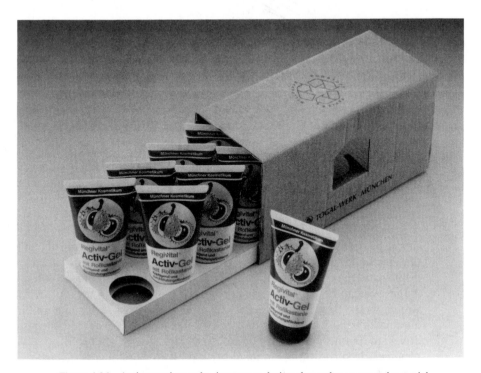

Figure 6.36 An innovative packaging system designed to reduce costs and material waste. Courtesy of IWK Packaging Machinery Inc.

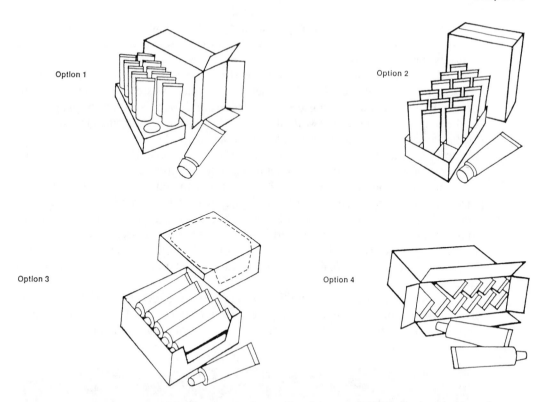

Figure 6.37   Additional tray packaging options. Courtesy of IWK Packaging Machinery Inc.

dispensing display. This is an ideal unit for tube-filled cosmetic, adhesive, paint, and food products. Individual product tubes are loaded cap end-first into a precut paperboard tray and fitted with a protective paperboard shipping sleeve. By eliminating unnecessary secondary packaging, the creators of this system offer an environmentally sound method to reduce packaging waste, while simultaneously lessening the expensive space requirements for shipping, storage, and shelving. Some other options for this system are shown in Figure 6.37.

Some manufacturers of quality flashlight batteries offer 4, 8 or 12 cells in one simple paper package, thereby using 91 percent less plastic and 44 percent less paper than in the familiar carded package. To be sure, the "blister" packs are easier to display and may be more attractive, but one must question the expense and waste. Adhesives can be used to stack cartons for pallet loading, to replace the common stretch wrap. (See Figure 6.38.) This unitizing system saves time and money, and eliminates solid waste. This endemic problem must be studied carefully and creatively to diminish gratuitous or excessive packaging. Other source reduction examples include substituting laminate brick packs for metal coffee cans, flexible drink pouches for glass bottles, bag-in-box wine for glass bottles, and flexible peelable

COLUMNAR STACKED CARTONS	CROSS STACKED CARTONS	SKID-LOCK®APPLICATION

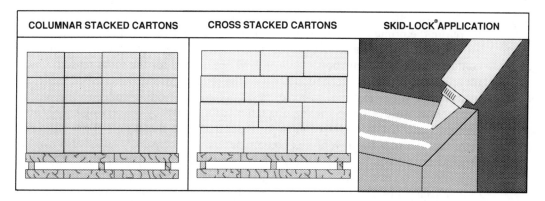

Figure 6.38 Material waste is eliminated by using adhesives instead of plastic wrapping in carton stacking and shipping. Courtesy of Glue-Fast Equipment Co., Inc.

membranes for metal can ends. All of these examples result in weight, volume, and material reductions.

Materials recycling is both popular and sensible, for it offers an opportunity for people to become committed and actively engaged in doing something positive about saving the environment. Also, recycling is conservation. Aluminum recycling can save 95 percent of the energy required to produce the metal from bauxite ore. There is a 50 percent savings in recycling steel and glass over making them from iron ore and silica. Making a part from plastic scrap saves 85 percent of the energy used in fabricating it from virgin resin. There are some other important and interesting features associated with the recycling of these materials.

Aluminum cans are lightweight and valuable; are easy to collect, crush, transport, and reprocess; and are therefore the most popular recycling material. Recycling is encouraged by deposit laws, reverse vending machines, and buy-back centers. These have resulted in a recovery rate of about 65 percent in nondeposit states and over 90 percent in deposit states. Does one need to say more about the need for bottle and can laws in all states? Altruism is fine, but experience shows that people need to be prodded into serious recycling by legal means. Organic coatings are removed in special furnaces, and about 90 percent of the recovered cans are used to produce sheet for new can bodies. Brass and copper are other nonferrous metals that are valuable scrap and therefore commonly collected.

Iron and steel scrap accounts for about 35 percent of the ferrous metal used in this country, most of which comes from manufacturing scrap. Recycled cans from deposit states and voluntary collection programs also provide scrap for steel making, but at very low recovery rates—around 2 percent. The tin on tinplate cans must be removed, at some considerable expense, because minute amounts of it can contaminate a melt and cause hard spots to form, which makes the sheet rolling process difficult. Reprocessing steel can scrap is complex and costly, and once recovered has a low value.

Glass bottles and jars constitute more than half of all glass products, and are therefore the prime target in glass recycling and reuse. Returnable beverage bottles are commendable because they prevent the wasting of large numbers of such valuable containers. Short of this, bottle deposit laws are helpful for the same reason. Once recycled, the glass container industry uses the crushed waste glass, called *cullet*, to mix in with virgin material to make new containers. Adding cullet to the mix is desirable because it lowers the melting temperature, thereby conserving energy and furnace wear. Colored glasses must be separated, and metal and plastic closures removed, in order to prevent mix contamination. Postconsumer cullet also can be used as an aggregate for roadways, brick and tile, sewer pipes, and glass-wool insulation. The effective recycling of glass suffers because of poor economics.

Plastics recycling offers simultaneously the most promise and the most problems, inasmuch as so much plastic is used in product manufacture and it is so susceptible to contamination. The Society of Plastics Industry (SPI) coding system mentioned earlier is helpful in separating the many kinds of plastic to facilitate their reuse as raw materials. Products made up of different plastics pose similar problems. For example, polyethylene terephthalate (PET) soft drink bottles often have high-density polyethylene (HDPE) bases for strength and metal caps, requiring costly and inconvenient disassembly before reprocessing. Recovery rates for plastics are up, but figures are misleading because of the many different plastics in use. It is estimated that by the year 2000, more plastics will be recycled annually than any other recyclable material.

Paper products have been recycled for many years, with the result that over 25 paper of paper and board currently is manufactured from secondary fibers. The difficulty in using paper scrap materials is classifying them according to fiber quality because they so frequently arrive as mixed scrap. Further, the economics is controlled by the cost of competetive virgin pulp, and the fact that reprocessing shortens and weakens the fibers and reduces their value. Corrugated box board is most commonly recovered because it is easy to collect and costly to landfill. Reprocessed board is used for cereal and shoe boxes, toweling, construction board, and insulation. Carpet underlayment is made from out-of-date telephone books. Various inks, coatings, glues, and laminations impede waste paper recycling because of high costs and environmental disposal issues—problems that continue to plague the industry.

In spite of these problems, the industry can promote recycling by making the task easier. For example, the Society of the Plastics Industry has established a voluntary coding system for plastic containers to identify them by material type. The codes consist of a triangle formed by three arrows with a number in the center and distinguishing letter codes under the triangle (see Figure 6.39), which are applied to the container bottom by imprinting or molding. As new recycling technologies emerge and new regulations regarding waste management occur, recyclers will be able to more conveniently sort the various plastics to earn the highest price for the reclaimed materials. At least 31 states have laws requiring such coding on plastic containers over 8 ounces.

Similarly, automobile manufacturers recently have joined with materials and

HDPE

Plastics Coding System

1: PETE (polyethylene terephthalate)
2. HDPE (high density polyethylene)
3. V (Vinyl)
4. LDPE (low density polyethylene)
5. PP (polypropylene)
6. PS (polystyrene)
7. Other

Figure 6.39 Plastic coding systems ease the efforts of recyclers.

recycling associations to form the Vehicle Recycling Partnership to find ways to make cars more recyclable.

Polyethylene stretch-sleeve labels for packages (Figure 6.40) allow the recycler to remove the labels easily, and without having to deal with the problem of residual glue. The combination of high-speed application, enhanced graphics, and environmentally friendly labels make this an excellent solution to the problem of packaging. Adhesive flaps on large mailing envelopes have largely replaced the metal bending tabs, again making recycling easier.

Designers are responding to the challenge of environmentally correct manufacturing by specifying the use of recycled materials. Industrial products include items such as the pallet used in material handling systems. (See Figure 6.41.) This structure is injection molded from 100 percent postconsumer waste polycarbonate, coming mostly from the automotive industry. Wildlife signs, benches, tables, marine walkways, lighting standards, parking lot car stops, planters, bird feeders, receptacle skirting, and barricades are among the many products made from plastic lumber. (See Figure 6.42.) This versatile material—made from recycled polystyrene foam coffee cups, food containers, milk jugs, and many other plastics—can be worked like wood but requires no finishing and is impervious to weather and decay. The design challenge here is to create attractive products of used material, such as the many trash containers, signs, and playground equipment one can find today. Many components in house construction are made from recyclables, such as reinforcement struts, insulation, building board, rubber paving pads, carpeting, and roofing.

At least one company is manufacturing a line of berry baskets and produce

Figure 6.40 This environmentally friendly sleeve label is applied to a nicely-formed plastic jug to create an attractive fabric softener package. The label is easily removed, and the jug is easier to recycle. Courtesy of Venture Packaging, Inc.

trays from 100 percent recycled PET plastic. The company has arranged with a municipal curbside recycling program to collect such specially marked packages, which are then repurchased and reprocessed into the packaging stream. The packages are clearly marked with a bright colored dot to help consumers and recyclers to identify and sort them easily. This is important. Recycling must be made as convenient and cost effective as possible for it to be successful. PET plastic also is being used as fibers for woven textiles, carpet backing, fiberfill for pillows, and molded parts for automobiles.

On the downside, consumers purchase significantly more disposable razors than durable styles due to their low price and convenience. The unit has a polystyrene handle and frame, which is a low code 6 on the plastics recycling scale, and contains an integral metal blade. The shavers are easy to manufacture but almost impossible to disassemble and recycle. Consequently, this product is a prime candidate for design reconsideration, leading perhaps to a more economical reusable model. Some 1.3 billion of these shavers enter the waste stream each year.

Figure 6.41 An industrial pallet is used to hold products to be moved with forklifts as a part of a materials handling system. This one is made of recycled plastic. Courtesy of Co-Ex-Co, Inc./Polyzin.

The use of more benign materials also is an attractive option. For example, a water-decomposable, 95 percent cornstarch packing filler is an environmentally sound alternative to the familiar puffy styrene plastic "popcorn." (See Figure 6.43.) The Eco-Foam™ is available in configurations for both light- and heavy-weight product support applications. This starch foam is formulated to decompose on contact when saturated with water. In a similar vein, there are biodegradable golf tees made from plant fibers, food by-products, and other water-soluble natural materials. The tees biodegrade in one day under normal watering and mowing conditions. One company has replaced the wood in the shell of an ordinary writing pencil with a rigid shell made from recycled paper. Other examples include the use of natural water-base solvents, plasma etching, and laser ablation to replace the traditional toxic materials for industrial parts cleaning, and the 1978 ban on the use of chloroflorocarbon (CFC) in aerosol cans.

One of the better methods of keeping products out of the waste stream is to repair and resell small electric appliances such as toasters, irons, mixers, lamps, and

Figure 6.42 Plastic planks are construction pieces made from recycled plastics. They have many uses, such as this sturdy, weather-proof, picnic table. Courtesy of American Container.

razors. All too often, the breakdown of these gadgets is due to something minor (such as a loose wire or fastener or a faulty cord) that can be repaired easily. Some social agencies such as Good Will or rehabilitation groups collect these, repair and rebuild them, and sell them to the public at low cost. Much of such effort is labor intensive and is an excellent way to help people engage in productive, environmentally correct work. Industry could encourage these programs by offering an incentive to the public to return such products to a retailer for a discount on a new article. These collected products could then be donated to some appropriate agency.

Design for the Environment Guidelines

1. Design products to be made from a single material. When components made from other materials must be added, they should be designed for easy disassembly.
2. Design plastic containers with wide mouths to simplify cleaning prior to recycling.
3. Practice materials conservation. The amount of product packaging can be easily and justifiably reduced through design.
4. Paper, metal, and glass contaminants lower the value of recyclable plastics. Therefore, labels, tags, and other attachments should be designed for easy removal, and caps should be made of the same container material.

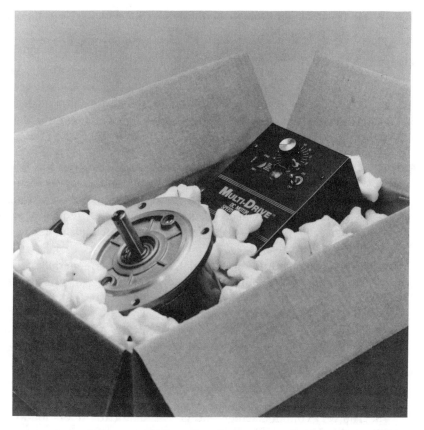

Figure 6.43   The packaging filler material used to protect this industrial part is made from a benign starch-base substance. Courtesy of American Excelsior Company.

5. Identify all plastic parts with the correct SPI coding symbol.

6. Specify ecologically friendly materials in all product designs. Examples include nontoxic inks and solvents, mercury-free household batteries, and unbleached papers.

7. Avoid "degradable" plastic materials in product designs. They are expensive, they add more bulk to landfills, they interfere with plastic recycling, they do not decompose faster than do comparable nondegradable plastics, and their use is not supported by any major environmental group and only a few plastics manufacturing groups.

8. Include information about proper disposal methods on all packages of industrial and household hazardous wastes, such as pesticides, cleaning products, and solvents.

9. Study current information about recycling and materials technology, and apply this knowledge to product designs.

10. Specify the use of recycled and recyclable materials in designs.

11. Design products for easy disassembly to facilitate the recycling and remanufacturing processes.

12. Design products for a second life, and design for durability.

13. Involve environmental specialists at the early stages of the design process.

QUESTIONS AND ACTIVITIES

1. Study the problem of the clothes dryer in Figure 6.24. Examine several such machines at an appliance shop to determine the frequency of the problem. Design a foot pad that will be less apt to go out of adjustment, and if it does, will be easier to readjust.

2. The junction boxes for buried telephone lines generally project from the ground and are regularly damaged by automobiles, snowplows, or vandals. Design a unit that will avoid the need for repair or that will make such repairs simpler.

3. Describe the components of a user-friendly motel room. Consider such factors as light and comfort control switches, bathroom facilities, and windows, among others.

4. Analyze the serviceability of a motel room, and propose solutions to perceived problems.

5. The very common screen and glass sliding patio doors found in homes and motels frequently do not function properly. Analyze these to determine the basic causes of the problems, and correct them.

6. A popular door stop device is made to flip up and lock to permit free movement of the door, and to flip down to hold the door open. The action of the device is controlled by the toe. A frequent product failure is the wearing of the snap lock so that the unit drags on the floor. Redesign this device to improve it.

7. Examine the toilet or restroom facilities in a college building, and list any examples of vandalism found there. How might these problem designs be made more vandal-proof?

8. Add to the lists of design guidelines for serviceability, security, vandalism prevention, safety, usability, the physically disadvantaged, and the environment. Base your work on personal experience, interviews, or library research.

9. According to your personal experience, what makes one computer more user-friendly than another?

10. In your own words, explain what is meant by the term *natural mapping*.

11. List examples of safety problems discovered in products you have used. How would you redesign the product to make it more safe?

12. What is the difference between maintenance and maintainability?

13. Examine the design for the physically disadvantaged guidelines and apply one or more to a design problem of your choice.

14. It has been noted that the kinds of materials used in products or facilities can often deter vandals from damaging them. Explain this and give some examples.

15. Refer to the faulty tubular steel railing design described in the serviceability section of this chapter. Propose some solutions to the problem.

16. Identify some examples of Taguchi "noise" in products you have used.

CHAPTER 7 | Computers in Design and Manufacturing

Included in the early chapters of this book were descriptions of the evolution of product making, from handcrafting to machine-aided to mass-production systems. The rationale offered for this progression included factors such as manual skills improvement, the discovery of new materials and the means to work them, the development of power systems, and the invention of methods for quantity production. As a further example of progress, modern manufacturing relies heavily on the computer to address the needs for high-quality, reliable, usable, and cost-effective artifacts. (See Figure 7.1.) The readers of this book are generally computer-literate, and so no detailed description of what computers are, how they work, and what they can do will be given here. The purpose of this chapter is to explain and illustrate the uses of computers in designing and manufacturing products.

COMPUTER-AIDED DESIGN

Computers have been used in design work for some time, and to a great extent have replaced the traditional board-drafting techniques in areas such as architecture, cartography, product planning, and graphics design. The use of computers to assist in the creation, modification, and analysis of a product plan is called *computer-aided design* (*CAD*). It is important to understand CAD as a technology, because it interrelates with so many of the other production systems. For example, CAD data are used to program assembly robots and automated machining centers, and to perform engineering analyses.

Figure 7.1   The familiar computer work station is an important manufacturing tool.
Courtesy of Applicon Inc.

A CAD system is composed of a graphics terminal on which a picture of the part being designed can be displayed. Designers enter the part data through a drawing program, and a keyboard is used to enter dimensions and other data. The part description is then stored as one of many such part descriptions in a CAD data base. The computerized part description is not a picture, but rather a graphic representation of coordinate points and geometric shapes from which a picture can be constructed. CAD offers many immediate benefits to the user. Parts can be drawn, rotated, scaled, and presented in three dimensions to enable designers to better visualize them; repetitive sections can be redrawn automatically; overlays can be easily shown on screen; and engineering drawings can be easily updated and printed. Garment parts can be designed and nested, and tool-cutting paths for metal structures can be programmed and verified.

The microelectronics industry probably has the most integrated uses of CAD. A new microchip can be designed on a CAD terminal. Once the design is in the CAD data base, the chip's performance can be simulated, the design can be modified if necessary, and the masks for the chip can be made, all automatically from the data entered at the CAD terminal. Although many other industries have not yet achieved this level of integration, it has become an embodiment of the computerized design-test-modify-test-fabricate model that can change the way manufacturing programs are organized.

Major CAD Functions

There are four basic categories of CAD functions or applications: geometric modeling, engineering analysis, kinematics, and automatic drafting. The results of these functions form a data base that can be integrated with computer-aided manufacturing schemes. (See Figure 7.2.) A typical case study can serve to illustrate the use of

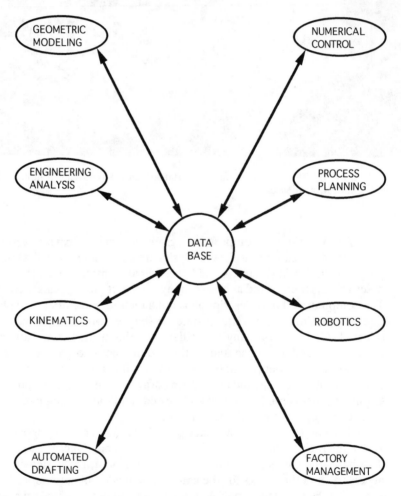

Figure 7.2   The primary functions of a CAD/CAM system, showing their relationships.

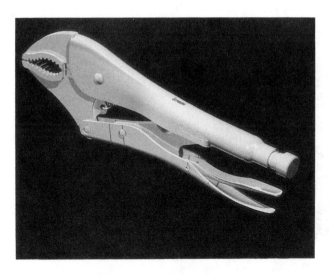

Figure 7.3 The standard model of this pliers has blunt, tough jaws good for general use. Courtesy of Applicon Inc.

these CAD functions in engineering design. The American Tool Company of Lincoln, Nebraska, undertook the redesign of their famous VISE-GRIP® pliers. (See Figure 7.3.) The jaws of the standard pliers are too large to fit into some narrow fittings and cannot reach the nut or bolt. (See Figure 7.4.) A new Dynamic Modeling™ system developed by Applicon Inc. was used to design a pair of long-nosed pliers to solve the clearance problem.

In *geometric modeling*, the designer constructs the shape of the object on the computer screen. This pictorial representation is converted into a mathematical model by the computer, which stores the information in the computer data base for later use. This model may be retrieved by the design engineer at any point in the

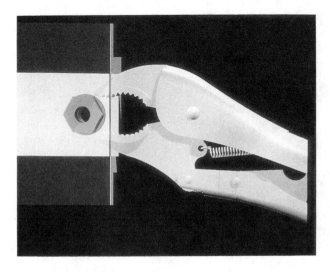

Figure 7.4 Note that the jaws of the standard pliers do not permit entry into tight places to reach a nut. Courtesy of Applicon Inc.

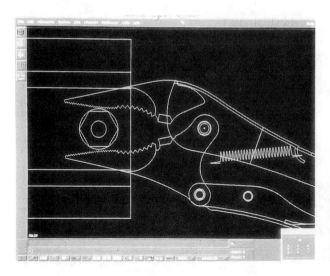

Figure 7.5 Wire frame modeling was used to create the long-nose plier jaws. The redesign utilized data from the original pliers, saving considerable time. Courtesy of Applicon Inc.

design process for refinement or analysis. Most basic models are wire frames that represent a part shape as a series of interconnected line elements, as used to define the plier jaws seen in Figure 7.5. The frame may depict a flat 2-D object or a full 3-D model of the pliers with sculpted features as shown in Figure 7.6.

Most CAD systems permit the user to move directly from a geometric model to *engineering analysis*. Simple keyboard commands can direct the computer to calculate area, volume, mass, or moment of inertia, for example. Perhaps the most powerful computer analytical technique is the finite element method. Here, the part geometry is reduced to a network of simple elements that the computer can use to calculate stresses, deflections, and other structural characteristics. For example, the

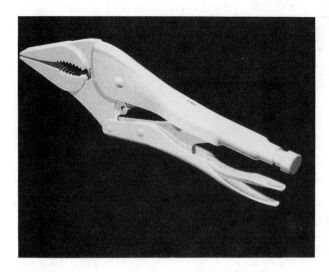

Figure 7.6 A 3-D model of the redesigned pliers. Courtesy of Applicon Inc.

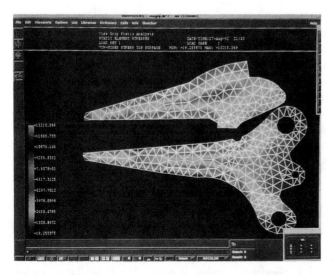

Figure 7.7 Finite element analysis was used to display levels of high stress in the plier jaw structure. Any areas of potential failure can be modified at this time. Courtesy of Applicon Inc.

strength of the plier jaws is determined by finite element analysis. (See Figure 7.7.) A judicious examination of this diagram can reveal the areas of high stress, which may indicate points of possible structural failure. Engineering design changes can be made to eliminate such faults before the part enters production. Before the advent of the computer, these element meshes had to be manually constructed—a tedious, time-consuming, and error-prone task. The computer makes analysis faster and better.

The *kinematics* feature of CAD systems permits the animated motion of hinged parts such as plier jaws, door locksets, front-end loader buckets, aircraft landing gear, and automobile hood linkages. (See Figure 7.8.) This allows the designer to assure that moving parts neither interfere with one another nor impact on adjoining

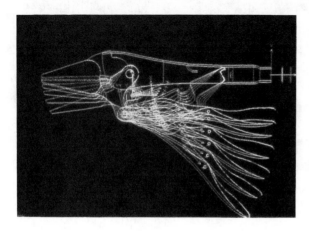

Figure 7.8 Kinematic modeling was used to visually display the full range of motion of the pliers assembly. This is helpful in design verification. Courtesy of Applicon Inc.

parts of a structure. There are several hinged parts on the pliers, and proper clearances and positions can be checked in the early design stages. Again, any obvious or potential fit problems can be eliminated. An earlier method of dealing with kinematic analysis was to construct pin-and-string-and-cardboard models to develop workable mechanisms by trial and error.

With *automated drafting*, detailed engineering drawings can be produced automatically from a data base. Most such drafting systems can generate up to six views on a screen, and have automatic scaling, cross-hatching, notation, and dimensioning features. CAD drawing can be five times faster than the traditional board method, and design changes can be made quickly and accurately. A detailed CAD drawing of the stub arm that connects the plier handles is shown in Figure 7.9.

Rapid Prototyping

Portions of Chapter 2 dealt with the subject of preparing experimental product models and mockups to test preliminary design results. These artifacts, called *prototypes*, are generally prepared by model makers employing such methods as numerical control (NC) machining or hand modeling from clay, wood, or plastic

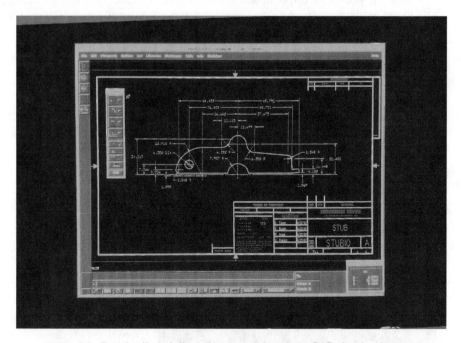

Figure 7.9 Part details and dimensions are an important CAD drawing feature. Shown here is the stub arm that connects the pliers handles. Courtesy of Applicon Inc.

foam. To reduce costs and speed up model-building operations, a number of more expedient prototyping techniques have emerged. These new technologies, whereby CAD plans are quickly transformed into solid objects and not merely 3-D drawings, are variously referred to as *rapid prototyping, free-form manufacturing, conceptual modeling*, and *desktop fabrication*, among others. The term generally used to define this concept is *rapid prototyping*, and it is an important CAD application with a number of distinct attributes. The most important of these is perhaps the enhanced visualization capabilities it gives to the designer who now can have a 3-D solid object to hold and measure, evaluate and verify. Other advantages include speed, accuracy, and the elimination of design flaws. There are several of these prototyping systems, the more common of which are described next.

With the *selective laser sintering* (*SLS*) process, developed by the DTM Corporation, powdered materials such as nylon, polycarbonate, and investment casting wax are transformed into solid objects, one thin cross-section at a time, using a modulated laser beam. The components of this system are shown in Figure 7.10. This method is not limited to generating models and prototypes; it also creates patterns, molds, and tools required for short-run production. A step-by-step description of the laser sintering system (Figure 7.11) follows.

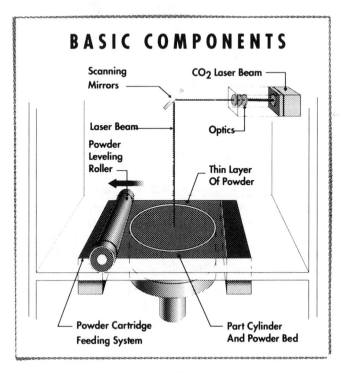

BASIC COMPONENTS

Scanning Mirrors

CO_2 Laser Beam

Laser Beam

Optics

Powder Leveling Roller

Thin Layer Of Powder

Powder Cartridge Feeding System

Part Cylinder And Powder Bed

Figure 7.10   The elements of the selective laser-sintering process. Courtesy of DTM Corporation.

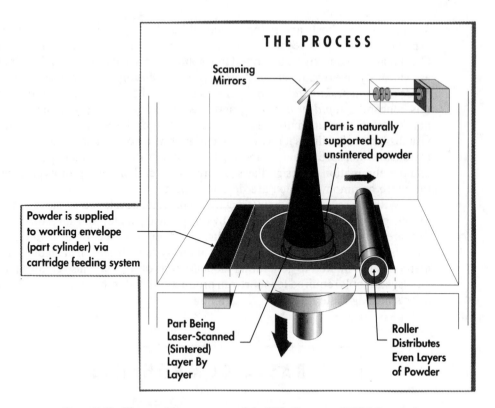

Figure 7.11   The operations sequence of the SLS. Courtesy of DTM Corporation.

1. As the process begins, a very thin layer of heat-fusible powder is deposited into a work-space container.
2. This layer is heated to just below its melting point.
3. The first cross-section of the object under fabrication is traced on the layer of powder by a heat-generating CO_2 laser. The temperature of the powder impacted by the laser beam is raised to the point of sintering, fusing the powder particles and forming a solid mass. The laser beam intensity is modulated to sinter the powder only in areas defined by the object's design geometry.
4. Another layer of powder is deposited into the work space, on top of the previous layer.
5. The process is repeated, with each sintered layer fusing to the layer below it until the part has been formed.
6. The finished part is retrieved and the excess powder removed. (See Figure 7.12.) Some parts may require postprocessing, such as sanding and epoxy coating.

THE PART

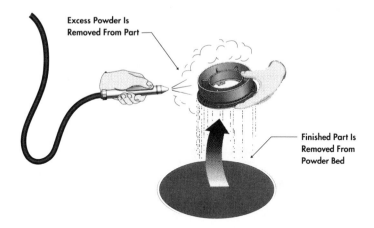

Excess Powder Is
Removed From Part

Finished Part Is
Removed From
Powder Bed

Figure 7.12 This completed SLS workpiece is being cleaned prior to use. Courtesy of DTM Corporation

This laser sintering system is versatile, accurate, and fast, and is widely used by a range of industries for fluid and motion control parts for airframes, housings, and handles for power tool products, testable circuit breaker components, and emergency vehicle spotlight assemblies.

Stereolithography was developed by 3D Systems, Inc. This process combines the technologies of computers, lasers, optical scanning, and photochemistry by taking advantage of photopolymers that change from liquid to solid when exposed to ultraviolet light. The patented StereoLithography Apparatus (SLA) converts 3-D computer image data into a series of very thin cross-sections, much as if the object were sliced into thousands of layers. (See Figure 7.13.) A laser beam then traces a single layer onto the surface of a vat of liquid polymer. The intense spot of ultraviolet light causes the polymer to harden precisely at the point where the light hits the surface. The hardened material is then lowered a small distance below the remaining liquid, recoating it with more resin. The laser then traces the next layer of the computer image model onto the now liquid surface. The process continues until the complete 3-D model is built, one layer at a time, from the bottom up. Typical layer thicknesses are from 0.076 to 0.381 mm (0.003 to 0.015 inch). The completed part is removed from the vat after the last layer is formed, and it undergoes a final curing with an intense ultraviolet light source. SLA product applications include medical bone implants, electrical discharge machining electrodes, nozzles for metal spraying, cellular phone cases, and automotive wheel rims.

The *laminated object manufacturing* (*LOM*) process is a development of Helisys, Inc., used to produce solid objects by successive deposition, bonding, and laser cutting of layers of sheet materials such as papers, plastics, or composites. The system

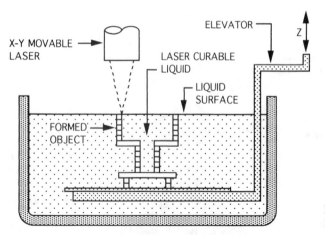

Figure 7.13 A diagram of the stereo-lithography apparatus.

employs CAD data inputted into the LOM machine computer. (See Figure 7.14.) A cross-sectional slice is generated and the laser cuts the outline of the slice. A new layer of material is then bonded to the top of the previously cut layer. A next cross-section is prepared and cut, and the process continues until all the required layers are laminated, whereupon the finished object is separated from the excess material. The resultant model is a durable multilayered structure of varying complexity. (See Figure 7.15.) These objects can be machined and finished for use as molding and casting patterns, and limited test or visual models for the automotive, aerospace, medical, computer, and electronics industries.

Since the materials used in the LOM process are in sheet form, and the system does not subject them to either physical or chemical phase changes, the finished parts do not experience shrinkage, warpage, or other deformations. LOM does not convert liquid polymers to solid plastic, nor does it convert plastic powders into sintered objects. Instead, it employs existing solid sheets that are glued with a hot roller and cut with a laser beam.

Other systems include the *solid ground curing* (*SGC*) process by Cubital Ltd., which forms layered objects with deep undercuts and thin walls in a solid environment; *fused deposition modeling* (*FDM*) by Stratasys, Inc., is a one-step thermoplastic extrusion and deposition technique to create a layered form on a 3-D wireframe CAD model; and *ballistic particle manufacturing* (*BPM*) involves depositing material in a controlled pattern to build a wax part.

To summarize this information on rapid prototyping, the several systems routinely couple CAD data directly with some type of CAM operation to produce accurate models and molds of varying complexities, quickly and economically. Each method has special advantages and applications, and designers should become familiar with these to simplify the preparation of their valued prototypes.

Figure 7.14 A typical laminated object manufacturing machine. Courtesy of Helisys, Inc.

Other CAD Applications

CAD systems can be employed to create accurate flat patterns for those product parts that are fabricated by folding sheet materials. For example, the layout for a sheet metal box shell is designed in a flat configuration, cut out, and then folded to form the box. Some of these software programs are capable of developing a pattern from a 3-D CAD model, of specifying the type of seam or joint, of determining the bend allowances according to the thickness of the sheet material, and then "nesting" the patterns on a sheet for economic cutting. An example of such a layout appears in Figure 7.16. This same design procedure is used to lay out cardboard toothpaste cartons, clothing patterns, television cases, and sheet metal ductwork.

An innovative application of the nesting theory is a method referred to as *tiling construction*. The table and chair group shown in Figure 7.17 is premised on the tiling concept, and was developed by Dennis B. Smith, a student at Western Michigan University. A *tiling* is a set of polygons that cover the euclidian plane without gaps or

Figure 7.15   Note the complex forms that can be produced with the LOM process.
Courtesy of Helisys, Inc.

overlaps, just as the single set square can tile a floor. The tiles—in this case, the legs, table top, and other furniture components—are nested within a single 4′ by 8′ plywood sheet such that, with saw kerf allowances, 100 percent of the material is utilized. (See Figure 7.18.)

All components on the board are adjacent to at least one other component, so that any change in the shape of one component will affect the shapes of those surrounding it. An indispensable tool that allows one to overcome this design challenge is the computer. This permits the designer to alter one component while the computer simultaneously alters all affected components. The computer permits the quick implementation of the many design layout changes that must be made before the furniture is perfected, without having to redraw any areas not affected by the changes. As the tiles are being developed, they can be rapidly assembled on the computer into a 3-D model of the furniture. This ensures that all components fit properly before the prototype is constructed.

Several restrictions are placed on the possible forms of tiling furniture. The components must serve both as structurally efficient parts of the assembly and as tiles in the plane. But because these two purposes work toward mutually exclusive ends, the furniture component must embody a synthesis of these two ends. Therefore, tiling constructions have an invisible logic driving their form rather than the usual function/form paradigm of furniture design. The final assembled pieces appear in Figure 7.19.

The geometric model of a product part developed on a CAD system can serve as the basis for generating a cutting tool path for a milling machine that will produce

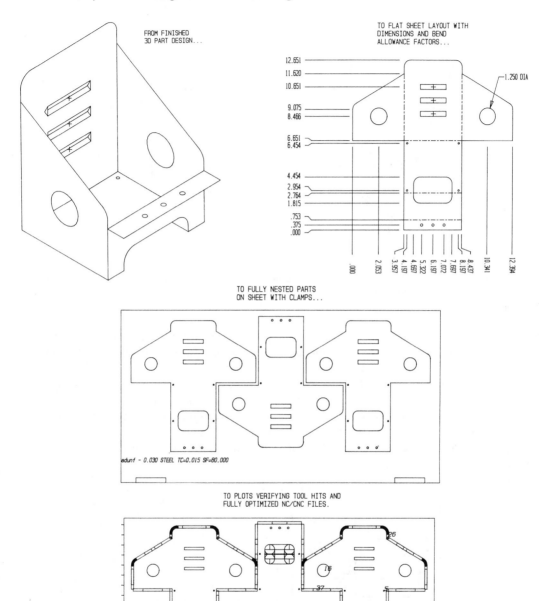

Figure 7.16 Sheet metal part nesting programs can be developed from 3-D wire-frame objects. Note the sequence of events. Courtesy of Anderson-O'Brien, Inc.

Figure 7.17 An example of a furniture group developed according to tiling construction theory. Courtesy of Dennis B. Smith.

the part. This is especially necessary for those tough machining tasks, such as sculpted parts, or those with multiple surface levels. (See Figure 7.20.) After the part has been modeled, the data can be used to create NC instructions to fabricate the part and to provide the necessary link to computer-aided manufacturing systems.

As a parting thought regarding CAD in design and manufacturing, it should be noted that much creating continues to be done by humans working at drafting tables using pencils and triangles. (See Figure 7.21.) This is the situation now and it will continue for some time. Much architectural and graphic design layout and detailing is

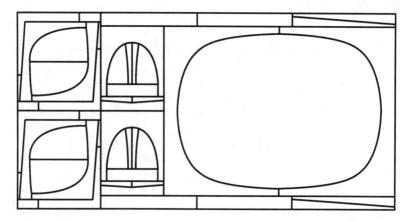

Figure 7.18 The nested parts of the tiling constructions. Courtesy of Dennis B. Smith.

Figure 7.19   The completed tiling construction furniture pieces. Courtesy of Dennis B. Smith.

performed on CAD with its staggering benefits and options. However, creative, speculative design thinking generally takes place at a drawing bench or a sketchpad.

COMPUTER-AIDED MANUFACTURING

The use of computers to monitor and control the processing of a designed part and to support its related production operations is called *computer-aided manufacturing*

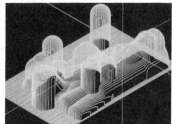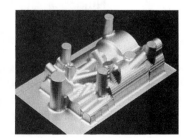

Figure 7.20   This example of tool-cutting path generation was done with CAD. Courtesy of Applicon Inc.

Figure 7.21 Human drafters at work in project and detail engineering design. Despite the latest technical resources that are available, the human being is always the creative force. Courtesy of GROB Systems, Inc., Bluffton, Ohio.

(*CAM*). It has as its major functions the numerical control of machine tool operations, process planning, robotics, and factory management. The aim is to provide an efficient manufacturing environment, and the computer is an integral part of this.

Numerical Control

Numerical control (*NC*) refers to the method of directing a machine tool to produce a part according to prerecorded, coded numbers, letters, and symbols. In the past, these instructions usually were stored on either punched paper tape or on magnetic tape and fed to the machine. When the job changed, the program of instructions changed. This change capability gave NC a desired flexibility, for it is much easier to write changes into the control program than to modify the production equipment. More advanced and current systems employ computer numerical control (CNC), where the machine tool contains a microcomputer that stores the information and controls the machine. The most sophisticated system is the distributed numerical control (DNC) scheme where several CNC machines are linked to a central computer.

A typical NC turret lathe machining example is shown in Figure 7.22. The workpiece is mounted manually in a chuck where it is drilled, bored, reamed, recessed, counterbored, and tapped in a sequence of NC programmed operations. The turret swings automatically from one numbered operation to the next, all according to the NC instructions. Some chucked workpieces may require only the

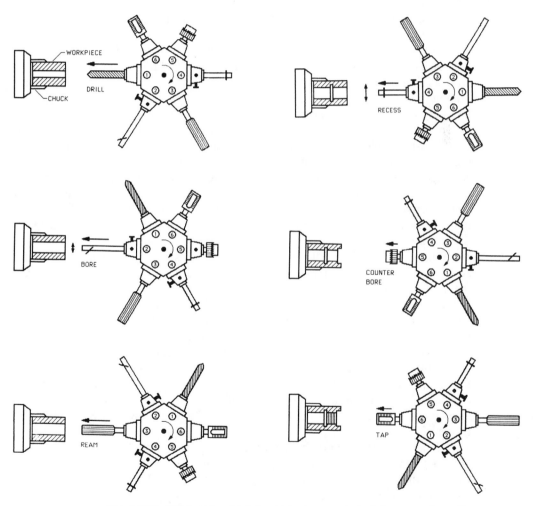

Figure 7.22   The turret on this lathe rotates to present a tool, then moves in to effect the necessary cutting operation on the chuck-mounted workpiece. An NC program regulates this series of six cutting operations.

drilling, reaming, and tapping operations, and the program could be written to do this. The NC direction of operations is logical, efficient, and automatic.

Process Planning

Process planning is the determination of how a specific part will be made in a factory. Whereas NC controls the operation of a single machine, process planning is the detailed breakdown of part production steps, and then selecting the optimum method of achieving each step. This includes determining the best manufacturing process to use, as well as that sequence of operations that will assure the most efficient measure of time and cost.

An important aspect of process planning systems is a concept called *group technology (GT)*, which is a way of organizing parts having similar shaped features into families to permit standardized fabrication steps. GT basically involves the examination of a group of related product parts to identify common attributes, grouping them according to similarity, and then using these common features to enhance the efficiency of part manufacturing. For example, if six part pieces all require drilled holes, the six could be directed as a group to a drilling machine for processing. GT is in common use as a retrieval technique based on part families and existing data bases for standard tooling and automated fabrication processes.

A fundamental unit in any computerized material processing system is the *machining center*. Such a center is in fact a highly sophisticated version of a common machine tool such as a mill, lathe, or drill press. This special equipment differs from the ordinary in that machine operations can be performed automatically, with little or no human participation. All process functions are computer controlled, from the actual cutting, to tool changing, to the movement of workpieces to and from the machine. These automatic machines are called horizontal or vertical machining centers, turning centers, or drilling centers, among others.

A typical *horizontal machining center* is shown in Figure 7.23. A key to the operation of such a center is the automatic tool changer, or ATC. One common model has three main parts, as shown in Figure 7.24. The chain-type tool magazine is a movable linkage unit with pockets to hold the machining tools. The tools can be drills, taps, reamers, or a variety of milling cutters. A tool-changer arm is mounted on a tool-changer column, and the arm pivots to select a needed tool from one of the chain pockets. The column then swings around to the machine head spindle to install the tool.

In a typical tool-change sequence, the computer reads ahead in the workpiece NC machining program as the cutting is taking place. It commands the tool magazine to move the next-needed tool to the pick-up position where it is grasped by the changer arm. Upon completion of the present cutting operation, the computer issues a command to retract the workpiece from the machine spindle area. Now, the tool change can take place. The changer column swings to the machine head position. The

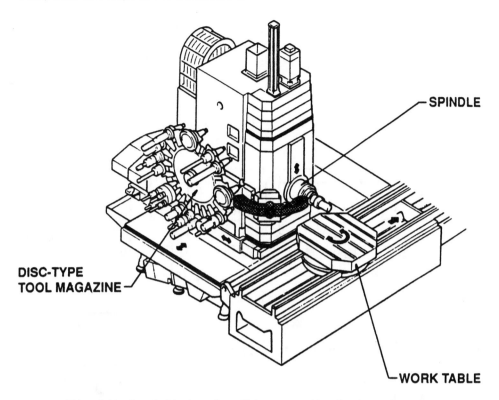

SPINDLE

DISC-TYPE
TOOL MAGAZINE

WORK TABLE

Figure 7.23 A typical horizontal machining center with a disc-type tool changer. Vertical machining centers also are in common use. Courtesy of GROB Systems, Inc., Bluffton, Ohio.

arm pivots to retrieve the just-used tool from the spindle, and again pivots to install the next-needed tool. The column now swings to the magazine position, and the arm pivots to return the just-used tool to the proper pocket. The workpiece moves back into position and the cutting continues. This process is repeated as the computer issues a command for another needed tool and the necessary chain movement, until the part machining has been completed.

Many industrial parts can be machined on these centers. One example is a cast aluminum automotive transmission case. (See Figure 7.25.) The operations include milling all the way around the large open end, changing tools, spot drilling six holes, changing tools, drilling those six holes and two others, changing tools, and reaming the two additional holes.

A *turning center* differs from an ordinary lathe in that it has one or two turrets that can automatically present a needed tool to do one of a series of turning or related operations. (See Figure 7.26.) The turret can be mounted at the end, in front, or

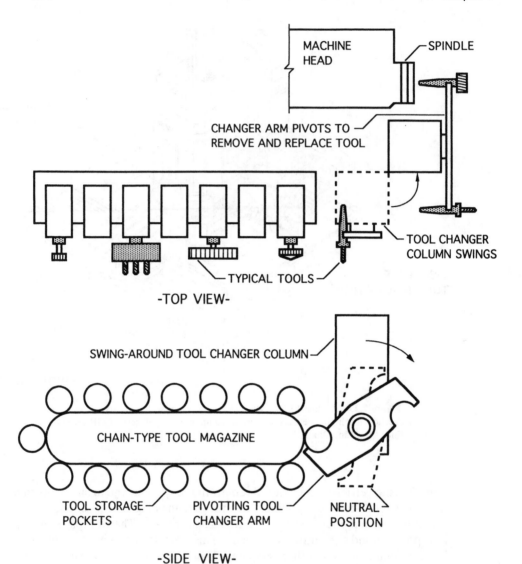

Figure 7.24   The operational sequence of an automatic tool changer for a horizontal milling center.

behind the workpiece. The tools are rotated into cutting position by the computer-controlled turret. The workpiece generally is placed in the headstock chuck by a robot, or it can be installed manually. A typical turned part is shown in Figure 7.27. Note the many automatic facing, sizing, bevelling, grooving, and shouldering operations necessary to turn the part. The part is shown being grooved in Figure 7.28.

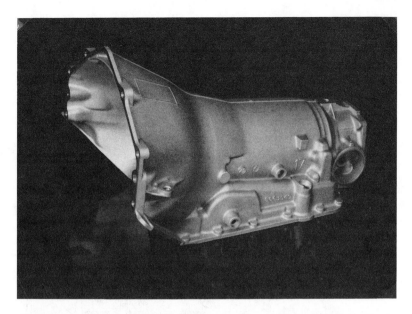

Figure 7.25 A number of machining operations were used to produce this aluminum transmission case. The work was done on a milling center. Courtesy of Cincinnati Milacron

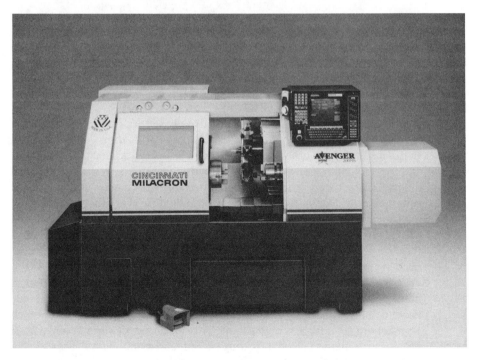

Figure 7.26 A turning center is a special type of lathe that can perform numerous lathe-related machining operations. Courtesy of Cincinnati Milacron.

Figure 7.27   The cast workpiece on the left was turned to the final shape on the right on a turning center. Courtesy of Cincinnati Milacron.

A number of machining centers, usually serviced by a robot and dedicated to the production of a family of parts, is called a *manufacturing cell*. (See Figure 7.29.) In this example, an automated guided vehicle (AGV) is used to deliver workpieces to and from a material station. A robot loads a workpiece into a turning center where the part is machined. The robot then moves the workpiece to the milling and drilling centers for the necessary machining. Completed parts are then moved to the measuring center where they are checked for accuracy. In operation, the various centers of the cell are all filled for continuous machining, and the robot can move workpieces from center to center and back again, as required. A layout for a more complex, and perhaps more typical cell is shown in Figure 7.30.

The term *flexible manufacturing system* (*FMS*) is used to describe an arrangement of machines, material transport devices, and a common computer hardware to produce parts randomly from a select family. It is a special type of machining cell whose main advantage is high productivity and product variety, and a good example obtains with methods for fabricating sheet metal components. These also are known as *flexible fabrication systems*. The theory of sheet metal FMS is that several types of

Figure 7.28   In this illustration, the cast workpiece is mounted in a turning center chuck. A grooving operation is shown. Courtesy of Cincinnati Milacron.

case products (e.g., television sets, video cassette recorders, and tuners) can be nested on a metal sheet. The cases are the family, for they all are sheet metal boxes that must be cut and bent to shape. (See Figure 7.31.) The system is designed to unload a sheet, feed it to the punch press, and separate the cut parts, all under computer control.

A typical FMS line is shown in Figure 7.32. Note the transfer mechanism where stacks of metal sheets are stored. In operation, a stack is moved forward to the

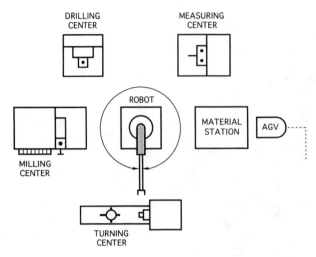

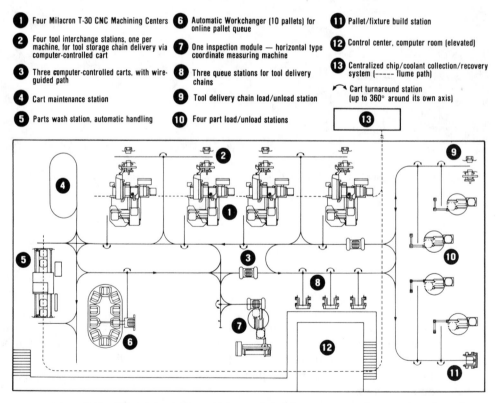

Figure 7.29 A diagram of a manufacturing cell. The robot moves the workpieces from center to center, and the automatic guided vehicle (AGV) carries workpieces to and from the cell.

① Four Milacron T-30 CNC Machining Centers

② Four tool interchange stations, one per machine, for tool storage chain delivery via computer-controlled cart

③ Three computer-controlled carts, with wire-guided path

④ Cart maintenance station

⑤ Parts wash station, automatic handling

⑥ Automatic Workchanger (10 pallets) for online pallet queue

⑦ One inspection module — horizontal type coordinate measuring machine

⑧ Three queue stations for tool delivery chains

⑨ Tool delivery chain load/unload station

⑩ Four part load/unload stations

⑪ Pallet/fixture build station

⑫ Control center, computer room (elevated)

⑬ Centralized chip/coolant collection/recovery system (----- flume path)

⌒ Cart turnaround station (up to 360° around its own axis)

Figure 7.30 This schematic diagram shows a layout for a more complex manufacturing cell. The equipment is tightly grouped for efficient, centralized part processing. Note the line paths for the wire-guided AGVs. Courtesy of Cincinnati Milacron.

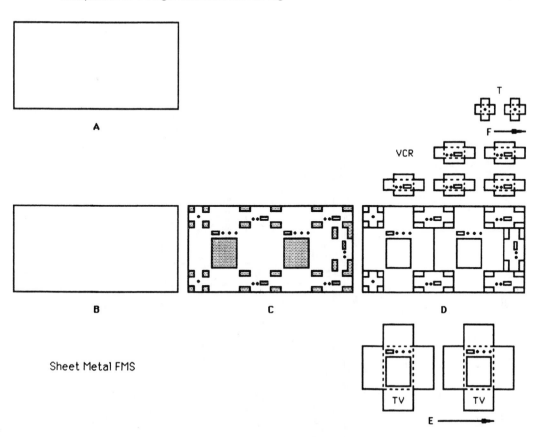

Figure 7.31 A case goods example of a sheet metal flexible fabrication system. The metal sheet is in storage at "A." It moves to the pick-up station at "B." It moves to the punching center at "C," where the shaded areas are cut away. At position "D." the linear cuts are made to separate different parts. The television, video cassette recorder, and tuner cases then move to the transfer areas, "E" and "F," where a conveyor takes them to a bending center.

pick-up area where suction pad grippers select one sheet and carry it to the punching center. (See Figure 7.33.) The sheet is positioned and held securely with finger clamps. Rotary punch and die sets (Figure 7.34) are mounted in the punching center, the punch above the table and the die below. The computer directs the punching of the sheet according to the nesting program. The punch moves down through the metal sheet and into the die to effect the cutting. The sheet moves in any horizontal direction under the punch. When the parts have been punched to shape and are separated by shearing, they drop off the table and onto a material conveyor, which moves them to the bending center. (See Figure 7.35.) The flexibility of the machine lies in its ability to cut any shape, randomly, within a family.

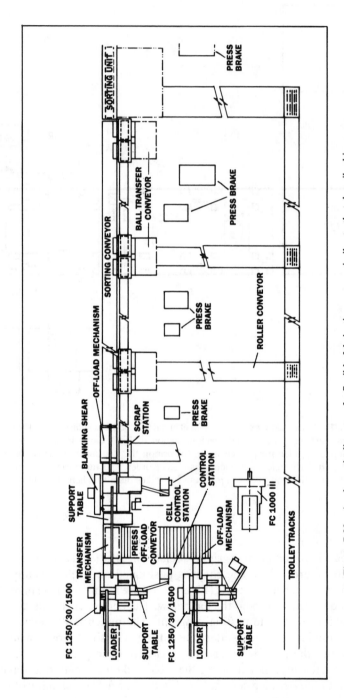

Figure 7.32 A schematic diagram of a flexible fabrication system similar to that described in Figure 7.31. Courtesy of Strippit, Inc.

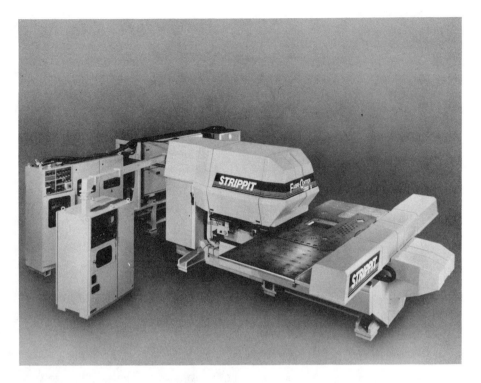

Figure 7.33 This is an example of a sheet metal punching center. Metal sheets also could be cut by plasma arc or laser processes. Courtesy of Strippit, Inc.

Robotics

A *robot* is a mechanical device that can spray paint, assemble parts, move workpieces in hostile environments, and perform a range of similar manual tasks. It is an integral element in many automated fabrication schemes and manufacturing cells because it can do this work efficiently and without tiring. A robot can arc weld metal chair frames with the same consistency of quality at five o'clock in the afternoon as it can at eight o'clock in the morning. Few human operators can maintain this level of quality—nor should they be expected to.

The term *robot* has its roots in the word *robota* from the 1921 play "R.U.R." (Rossum's Universal Robots) by Czechoslovakian Karel Capek. He coined the word to mean a manufactured mechanical worker, efficient but without feeling. The machines in his play very accurately resembled humans but were worked harder and longer, and they finally rebelled. Since then, there have been many interpretations of what robots are and how they are classified. One generally recognized definition is that of the Robotics Institute of America, which suggests an industrial robot to be "a programmable, multifunctional manipulator designed to move material, parts, tools,

Figure 7.34 Typical rotary punch and die heads used on punching machines. Note the many different tool shapes in the heads, which rotate to punch the desired cut according to the computer program. Courtesy of Strippit, Inc.

Figure 7.35 This Trumpf TC 120 Sheet Metal Fabricating Machine separates individual parts from the sheet and automatically deposits them in a bin under the machine. Courtesy of TRUMPF Inc.

or specialized devices through various programmed motions for the performance of a variety of tasks." Distilled to its basics, a robot is simply a machine that can be directed to do work accurately and tirelessly. Some robots are smart and can be computer-programmed to follow a very detailed sequence of operations and to make decisions. Others are less smart, pick-and-place manipulators, controlled by human operators to perform simple, routine tasks. Whatever their precise function and level of sophistication may be, they figure importantly in current manufacturing schemes.

Modern industrial robots have four basic parts. (See Figure 7.36.) The *base* supports the robot proper, and it can be fixed to a floor, attached to a movable platform, or hung from an overhead gantry. The *manipulator* is the action end of the robot—in reality a working arm capable of anthropomorphic waist, shoulder, arm, wrist, and hand movements. The hand unit is called the *end effector*, and it can be configured to grasp, paint, weld, or assemble as required. They may be mechanical grippers, electromagnet attractors, or vacuum-driven suction cups programmed to simulate human movements. (See Figure 7.37.) The *power unit* energizes the manipulator and is operated variously by electric motors, hydraulic actuators, or pneumatic units. Electric power units are fast, quiet, clean, accurate, and expensive. Hydraulic units are heavy, slow, dirty, precise, and powerful. Pneumatic or compressed air

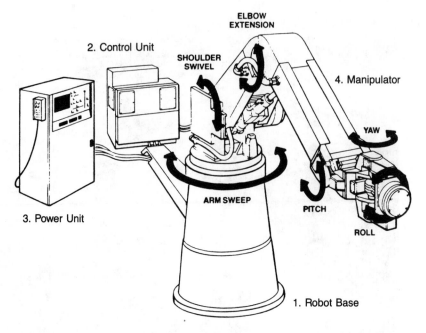

Figure 7.36 The parts of a typical jointed arm robot. Courtesy of Cincinnati Milacron.

systems are lightweight, inaccurate, fast, weak, and inexpensive. It is obvious that an air-driven unit would be better for small, pick-and-place part placement operations than would be a hydraulic unit.

Finally, robots have a *control unit* that directs the movement of the manipulator, and sometimes of the base. There are two types of robot control systems. Smart robots have *closed-loop* or feedback, systems, where signals sent by the controller to the manipulator are fed back to the controller to indicate whether or not the proper action has taken place. If not, the situation can be corrected by further controller signals. The term *servo robot* often is used in reference to such feedback methods, and they can be sufficiently sophisticated as to sense the joint region of two metal parts to be welded, begin the weld, complete it, and check for faulty welds. (See Figure 7.38.) In contrast, in an *open-loop* system, there is no sensor that measures manipulator performance in response to a controller signal. Although there is no feedback in such nonservo robots, they are simple, fast, reliable, and inexpensive, and they are valuable tools for light, repetitive tasks. For example, the robot in Figure 7.39 is transferring an automotive seat bottom to a fixture where nuts are welded to it.

Types of Robots.   All robots do not look alike, nor do many of them look like C3PO or R2D2 from the popular film *Star Wars*. They have different processing assignments and are designed to perform in closely defined areas. The outer limits of

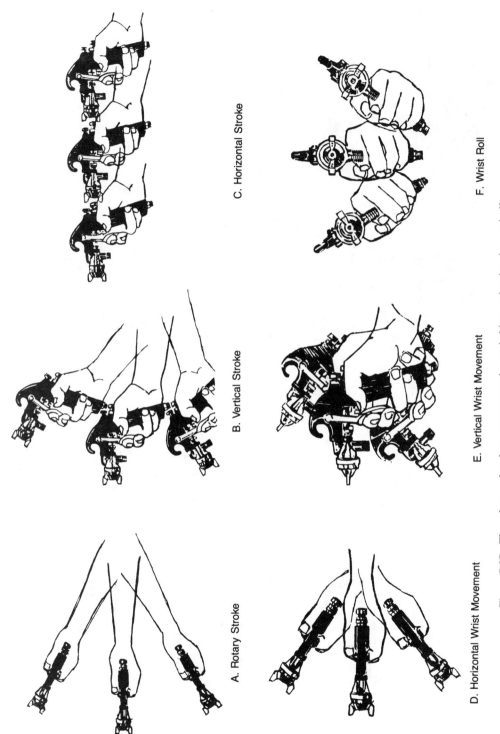

C. Horizontal Stroke

F. Wrist Roll

B. Vertical Stroke

E. Vertical Wrist Movement

A. Rotary Stroke

D. Horizontal Wrist Movement

Figure 7.37 These human hand movements can be copied by a robot hand or end effector. Courtesy of Binks Manufacturing Company.

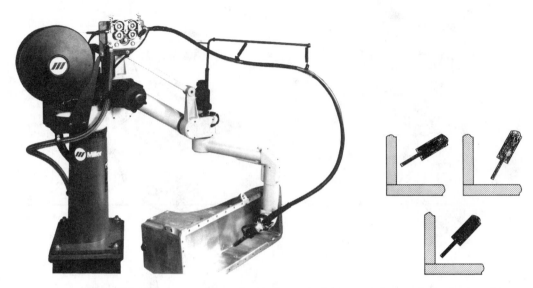

Figure 7.38 This smart welding robot can locate the joint area and proceed to make the weld under computer control. Courtesy of Miller Electric Mfg. Co., Appleton, Wisconsin.

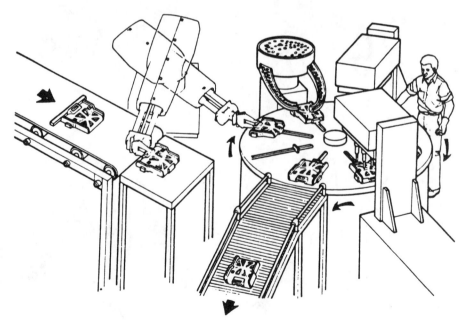

Figure 7.39 This robot is moving seat frame parts to a welding assembly area. Courtesy of General Motors Corporation.

the places a robot can reach with its end effector is called its *work envelope*. This area is determined by a robot's axes of operation. Each of the X, Y, and Z axes have both a minimum and a maximum dimension that clearly specify the work envelope in relation to the center line of the robot base. By standardizing these dimensions, it is possible for engineers to have a common reference when specifying a particular robot for a given application.

The axes of movement for a *rectangular* robot are shown in Figure 7.40. This robot moves in the vertical, horizontal, and transverse axes in simple straight-line motions, forming a rectangular work envelope. The range of these robots can be extended by mounting them on a movable track or a gantry. The vertical and horizontal motions for a *cylindrical* robot are similar to those of the rectangular robot, except for the base movement. Instead of a straight-line motion, the cylindrical robot rotates about the base, forming an arc with a corresponding work envelope.

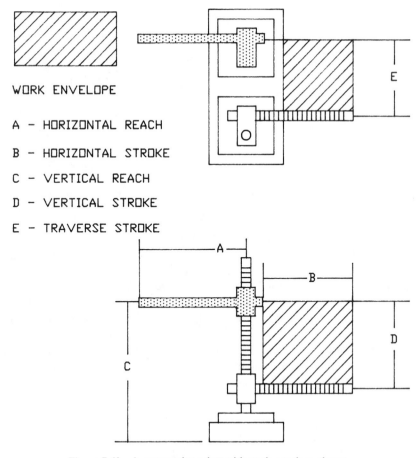

WORK ENVELOPE

A – HORIZONTAL REACH

B – HORIZONTAL STROKE

C – VERTICAL REACH

D – VERTICAL STROKE

E – TRAVERSE STROKE

Figure 7.40 A rectangular robot with work envelope shown.

WORK ENVELOPE

A – HORIZONTAL REACH

B – VERTICAL REACH

C – VERTICAL STROKE

D – HORIZONTAL STROKE

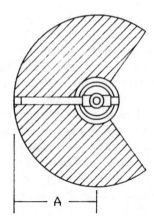

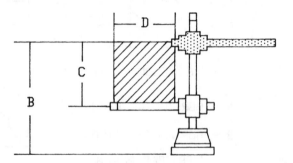

Figure 7.41 A cylindrical robot with work envelope shown.

(See Figure 7.41.) A *spherical* robot's reach and sweep (rotation about the base) motions are the same as with the cylindrical robot. The vertical motion results from a pivoting action of the robot arm, and forms an arc in the vertical plane. The work pattern of this robot is indicated in Figure 7.42.

A fourth type of robot is the *jointed arm*, or articulated, robot, shown in Figure 7.43, which forms a horizontal swing pattern similar to the cylindrical and spherical models. A unique motion in the vertical and horizontal reach planes is formed by the combination of angles that are possible between the various joints of the arm. This is the most sophisticated and expensive type of robot, and is typified by those used to spray-paint automobiles.

Factory Management

Factory management extends the basic CAM functions (numerical control, process planning, and robotics) to coordinate the operations of the entire production facility.

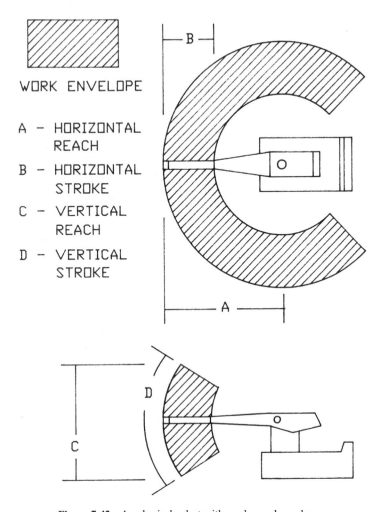

WORK ENVELOPE

A – HORIZONTAL
REACH

B – HORIZONTAL
STROKE

C – VERTICAL
REACH

D – VERTICAL
STROKE

Figure 7.42 A spherical robot with work envelope shown.

Examples of these coordinate operations are inventory control, materials planning, personnel assignments, and work scheduling. The ideal situation is considered to be hierarchical in nature. At the highest level, there is an administrative structure to oversee the complete production cycle from raw materials to finished goods by controlling the part- production job shops. The lower-level control centers manage the interface of operators and machines that make up the individual work centers. The goal is a sensible mix of humans and equipment to assure smooth and efficient product manufacture. The key to such success is continual computer and verbal communication between the various centers and the prompt feedback to aid in event planning and resource allocation.

One of the troublesome problems of this production network is that the various computer-based devices cannot always communicate with each other. They do not

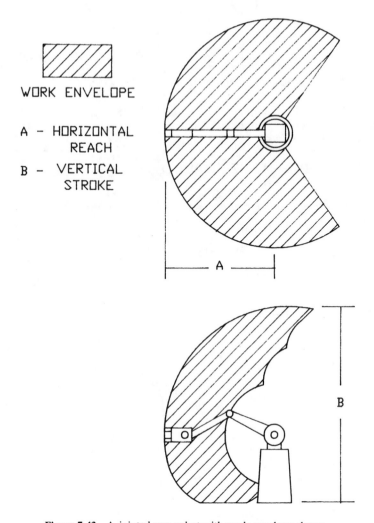

WORK ENVELOPE

A – HORIZONTAL REACH

B – VERTICAL STROKE

Figure 7.43   A jointed arm robot with work envelope shown.

have a common manufacturer and are therefore incompatible. To rectify this situation, a common set of procedures is required so that the equipment in the computer network can read the data exchanged between them. One such intercommunication program is the *manufacturing automation protocol (MAP)* promoted by the General Motors Corporation and other organizations to assure the compatibility of computer-based devices in their plants. This communications protocol will enhance factory management operations.

The important issue of materials movement and storage is addressed by *automated guided vehicles (AGV)* and *automated storage and retrieval systems (AS/RS)*. According to the Materials Handling Institute, *AS/RS* is defined as a combination of

Figure 7.44 This is an example of a small AS/RS, while others are several stories high. Such systems employ guided and manual vehicles to assure the prompt and accurate movement of product materials. Courtesy of Jervis B. Webb Company, Farmington Hills, Michigan.

Figure 7.45   Automated guided vehicles transport materials, parts, and finished products in manufacturing plants. Courtesy of Jervis B. Webb Company, Farmington Hills, Michigan.

equipment and controls that handles, stores, and retrieves materials with precision, accuracy, and speed under a defined degree of automation. These systems consist of a series of storage aisles and racks, where both automated and manually operated vehicles move, store, and retrieve raw materials, parts, and completed products. (See Figure 7.44.) The AGV used to move materials, parts, and products generally are battery powered and are guided by wires embedded in the floor or by reflective paint. Sensors on the vehicles can follow the guide wires or the paint, and a computer directs the route of the vehicle and the material it carries. (See Figure 7.45.) The manufacturing cell shown previously in Figure 7.30 utilizes AGV and a wire guide system. Both AS/RS and AGV are sophisticated material-handling methods to accompany computer-controlled manufacturing center and cell operations. The aim is to assure that the right materials get to the right station at the right time.

Figure 7.46 Conveyors such as these move product components to assembly areas. They are common features in all production facilities. Courtesy of Jervis B. Webb Company, Farmington Hills, Michigan.

There are, to be sure, many other more common methods of material transport, such as forklifts, handtrucks, and conveyors. Many are under the control of human operators, others are semiautomatic. These are valued devices in both conventional and computer-based production schemes. (See Figure 7.46.)

COMPUTER-INTEGRATED MANUFACTURING SYSTEMS

The factory of the future will utilize computer technology in all operational and information processing functions of a manufacturing organization, but in a manner

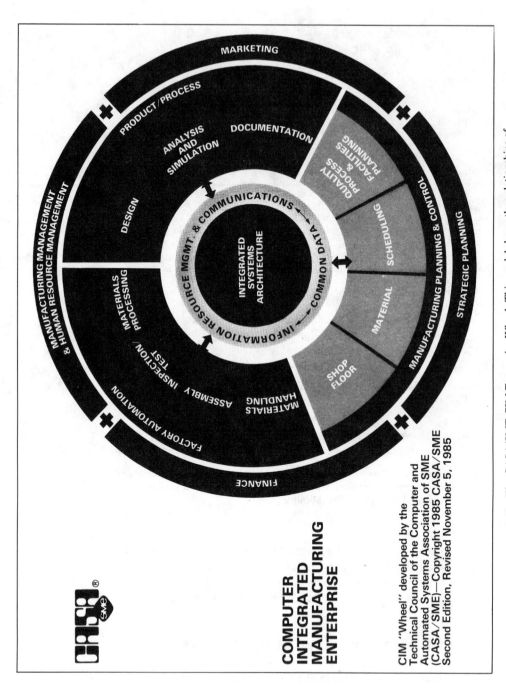

Figure 7.47 The CASA/SME CIM Enterprise Wheel. This model shows the relationships of the various production elements. Courtesy of the Society of Manufacturing Engineers. The Enterprise Wheel is copyrighted by CASA/SME of Dearborn, Michigan.

different from the CAD/CAM systems described earlier. There, it was shown that factory automation involved mechanical, electronic, and computer devices to operate and control production. The newly structured factory will feature a complete integration of every function, not just those immediately associated with product design and manufacture. The term used to describe this elegant concept is *computer-integrated manufacturing (CIM)*, and it can be defined simply as the total coordination of all production elements through the use of computers.

The CIM Enterprise Wheel (Figure 7.47) was designed by the Computer and Automated Systems Association to provide industry with a common vision of computer-integrated manufacturing. The wheel is important in understanding CIM and should be carefully studied. Note that the outer ring identifies the business functions of manufacturing and human resource management, finance, strategic planning, and marketing. The next ring includes the product/process, factory automation, and manufacturing planning and control elements. The central core of the CIM system includes the information resource management component. The efficient computer linking of these primary units is what gives CIM its unique character. There are few instances of such total integration in U.S. industry today, but it is a goal that is being sought and that is attainable. The current situation features individual CAD systems, manufacturing cells and centers, materials handling units, and the like, as islands of automation, devoted to converting raw materials into products, linked but not completely integrated. The factory of the future will be a new concept, a new approach to manufacturing, a new way of doing business.

QUESTIONS AND ACTIVITIES

1. How have computers influenced the fields of product design and manufacturing?
2. Describe CAD and list its major functions.
3. Write your personal definitions of *geometric modeling, engineering analysis, kinematics,* and *automated drafting*, and give examples of each.
4. Describe how you have used the computer in any personal designing activity.
5. Select a product of your choice and explain how computers may have been used in its design and manufacture.
6. Define the term *rapid prototyping* and indicate how it has revolutionized the preparation of engineering models and mockups.
7. List the differences between the SLS, SLA, and the LOM processes.
8. What are the attributes of nesting in the production of 3-D folded and flat parts?
9. Investigate the tiling construction process and use it in the design of a wooden bird house or bird feeder.
10. Describe CAM and list its major applications.
11. Prepare a research report on the origins of NC, and the course of its transition from paper tape to CNC to DNC.

12. Study the NC turret lathe example in Figure 7.21 and search the appropriate periodicals for other examples of such automatic machines.
13. Describe group technology and indicate its role in planning for the automatic processing of materials.
14. Describe the differences between a machining center and a machining cell.
15. Prepare a research report on the different configurations of automatic tool changers.
16. How does an FMS relate to a machining cell?
17. Write your personal definition of a robot, and describe its four basic parts.
18. List the four types of robots and give an example of how each can be used in manufacturing operations.
19. Describe the roles that MAP, AS/RS, and AGV play in modern production.
20. Explain what CIM is and give your personal assessment of its significance in future manufacturing.

Design for Producibility

As stated in the Preface to this book, the philosophy of design for manufacturing implies that all aspects of product planning be attended to as routine matters, to include the ease with which the product can be made. Designers (both engineering and industrial) may be good artifact innovators but all too often are not conversant with current material processing methods. This matter is significant because it contributes to market success. No matter how elegant a product form may be, and regardless of its performance and quality attributes, if the fabrication costs are excessive, the product will be economically unfeasible and probably will never reach the marketplace. Therefore, the design team must handle product-making issues up front, while the artifact is in its early planning stages and while important decisions regarding its manufacture can be considered, debated, and changed.

Producibility is defined in many ways by many field experts, but this author suggests that basically it is a measure of the efficiency with which materials can be processed in order create a product. It is a response to the question, How can a specified material be best worked to achieve an artifact having a specified geometry? The graphics planned for a perfume carton may feature such a complex array of shaded dots and colors as to render the idea impractical. The vast, sloping glass windscreen on a classy automobile may be too difficult to make and install to meet current industry safety and reliability requirements. The rubber shockmounts needed to attach a fiberglass chair shell to its steel frame may pose impossible structural engineering problems. All of these are elegant design concepts, but challenging to translate into feasible products. None of them measures to the fundamental concept

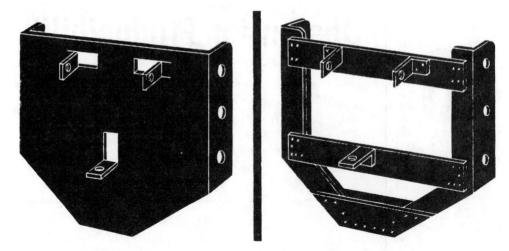

Figure 8.1   The steel support bracket on the left is more producible because it is a one-part part product. The example on the right is more complex because it has several parts that must be joined by spot welding.

of producibility, which encompasses the art of making parts easily and well, and then assembling the parts with equal facility into a usable product.

Study the small steel bracket examples in Figure 8.1. At issue is the need for a sturdy support for a mechanical device, and either of the two would suffice. However, one is comprised of eight separate pieces requiring multiple spot-welds to close assembly tolerances. The other is a one-piece redesign employing a series of progressive blanking, lancing, piercing, and bending operations, and is far more producible than the other. It is easier and less expensive to make, and exemplifies the producibility concept. Those factors that define this concept are described next, and will aid those involved in planning commercial products to design them to be easily made.

GUIDES TO DESIGN FOR PRODUCIBILITY

Years of experience in devising methods for assuring the producibility of artifacts has led to a number of proven guides or rules. Many relate specifically to product assembly, and others to general producibility theory. Although not totally infallible and not applicable in all situations, they are of some value to designers. Some general points include the following:

1. Reduce the number of product parts. Fewer parts result in fewer things to make, fewer process errors, and fewer pieces to build into the assembly program.
2. Use modular components wherever possible. Subassemblies are easier to control.

3. Strive for Z-axis assembly. Gravity is a natural assembly tool and designers should take advantage of it.
4. Use standard parts if at all feasible. Off-the-shelf items are cheaper and more available.
5. Eliminate or reduce the number of threaded fasteners in a design. They are troublesome for both human and robotic assemblers. Adhesives or snap-fits are preferable. If screws are necessary, specify those that provide for easier tool engagement, such as phillips-head.
6. Use self-aligning and self-nesting parts where practicable in a design. They are especially necessary in automated assembly systems.
7. Use symmetrical parts to simplify assembly. They also are easier to make.
8. Design specifically for automation in both material processing and assembly at the outset, and not as an afterthought. Failure to do so generally results in costly redesign.
9. Use human assemblers wisely. The human hand is an ideal assembly tool.
10. Match the design and the material to the forming process. All three warrant serious study to assure quality part pieces.
11. Design parts that are impossible to assemble incorrectly. Acknowledge the fact that humans and machines make mistakes.
12. Design for ease of fabrication. Simplify part designs so that they can be made quickly and economically.

PROCESS DESIGN FACTORS

Chapter 4 of this book dealt with the many ways of transforming raw materials into usable products. These are called processes, and generally are grouped as cutting, forming, assembling, and finishing techniques. Designers must be concerned with material processing because there are right ways and wrong ways of designing for them. Technical books on the subject include information on the special design considerations for metal extrusions, castings, and forgings, for example. A typical listing for extruded part design might run as follows.

1. Avoid deep narrow grooves, long thin legs, and undercuts.
2. Avoid sharp corners, bevels, and chamfers.
3. Avoid abrupt changes in section.
4. Avoid unnecessary hollow shapes.
5. Avoid knife edges, as they tend to present a wavy appearance.
6. Provide generous fillets and rounds for steel extrusions.
7. Design sections so that they have a uniform wall thickness.
8. Use webs for better dimensional control.

9. Use symmetrical designs for semi-hollow areas.

10. Use rib stiffeners to reduce twisting.

The drawings in Figure 8.2 illustrate some applications of these considerations. Note that the purpose of these design factors is to improve the quality, strength, and economics of extruded parts. Similarly, the cast part manufacture is enhanced by providing fillets and rounds, and avoiding undercuts, as shown in Figure 8.3. Design features such as these are applicable to any molded parts, such as sand castings, die castings, and plastic injection methods. Holes sheared in metal or plastic can be a

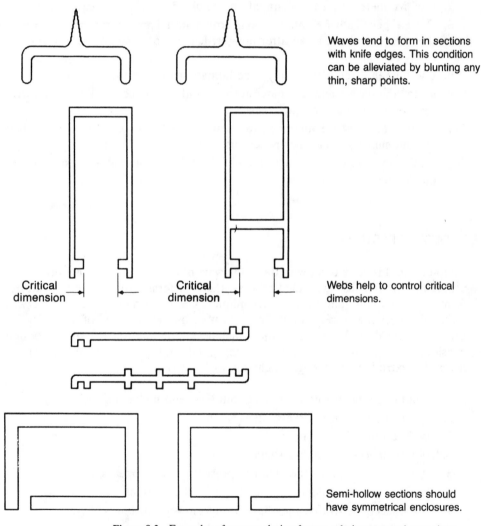

Waves tend to form in sections with knife edges. This condition can be alleviated by blunting any thin, sharp points.

Critical dimension

Critical dimension

Webs help to control critical dimensions.

Semi-hollow sections should have symmetrical enclosures.

Figure 8.2 Examples of process design factors relating to metal extrusions.

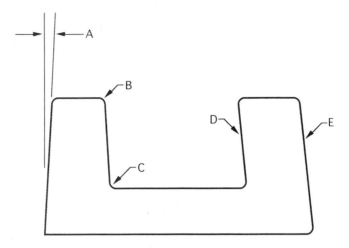

Figure 8.3 Cast and molded part quality is enhanced by providing in their design the features shown here. Draft (A) and (E) provide a taper to facilitate pattern or part withdrawal; rounds (B) provide for cleaner corners; and fillets (C) prevent the tearing of inside corners. Undercuts (D) are to be avoided because they make pattern withdrawal difficult or impossible.

problem. Punched holes leave little or no exit burrs. If this is a product requirement, this is the hole-generating process to use. (See Figure 8.4.) Extruded and pierced holes do leave such a burr or bushing, which may or may not be desirable. For example, the bushing left by the extrusion holing process may provide extra material for a tapped hole. There are many ways of generating holes, and designers must know the peculiarities of each and specify that which best meets the product needs.

There is a whole technology devoted to design for quality finishing. Electroplating, powder coating, and spray painting, among others, all have special requirements in order that consistency and reliability of finish can be maintained. For example, applying a spray coat to the part in Figure 8.5 is frustrated by the inability of the finish to cover the sharply defined interior surface and sharp corners. The provision for tapered interiors, fillets, and rounded corners improves finish quality. The desk

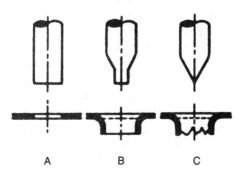

A   B   C

Figure 8.4 Different hole-generating processes result in different hole qualities. These three examples clearly illustrate those differences. (A) Punch; (B) Extrude; (C) Pierce.

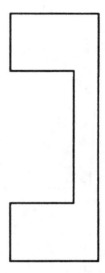

Figure 8.5 The design of parts to be spray-painted can contribute to the quality of the finish. Deep interior surfaces, undercuts, and sharp corners do not permit full coverage.

accessories in Figure 8.6 are made of brushed aluminum with a protective coating. The gentle surface makes these finishing operations easier.

Wooden rods can be bent to form contoured arm and leg structures, but there are limits to the degree of bend that the wood can tolerate, as exhibited in the chair in Figure 8.7. Experience and experimentation can lead the designer to specify feasible bends or to call for a structure made of several parts, each having bends of lesser degree. There is a logic to process design factors that demands such experience and experimentation in order that producibility may be achieved.

Many general-purpose knives are made of a one-piece blade and shaft, two halves of the handle to fit against the sides of the shaft, and two rivets to secure the handle assembly. (See Figure 8.8.) The production of this knife requires the shearing of the blade using punches and dies; the shaping of the two handle halves; the preparation of the rivets; and the sharpening, assembly, and finishing of the completed product. Many operations and considerable material handling and fixturing are required. Conversely, the elegantly simple knife in Figure 8.9 is cleverly made of one piece of stainless steel. It still functions as a knife, but it is easier to make by simple punch and die shearing, with no assembly required. It is therefore conceptually more producible, but this fact by no means detracts from the quality and usability of the hundreds of knives made from a number of separate parts. Indeed, specially shaped handles can be designed for surer and more comfortable holding, thereby justifying the design of a more complex form. The important point here is that the designer of the one-piece knife achieved a more producible product.

The knife is a simple example. The production of a ballpoint pen—a common, useful, reliable, and inexpensive writing tool—is more complicated. (See Figure 8.10.) The accuracy in making the pen tip is the key to its reliability and writing quality. As a rule, the seat that holds the roller ball must have an accuracy of plus or minus 0.002 mm (0.000787 inch). The drilled holes range in size from 0.5 mm

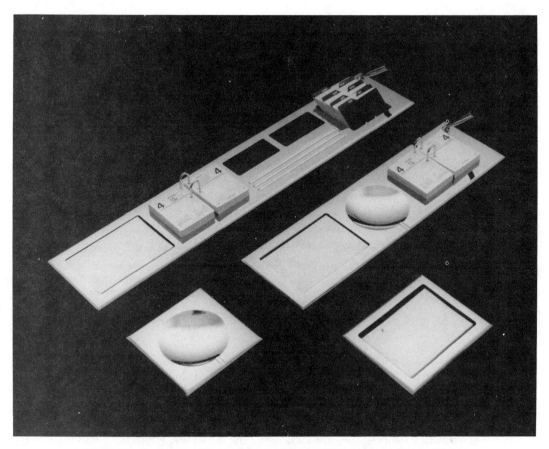

Figure 8.6 Proper attention to design for finish quality usually results in quality finishes. This attractive set of desk accessories supports this process design factor. Courtesy of Matel Inc. Designed and engineered by David Bar-or.

(0.0196 inch) to 1.2 mm (0.0471 inch). The tip must also be made inexpensively because the pen generally is discarded when it runs out of ink. A thousand of these tips cost about $6.00.

The brass tip is made on an automatic rotary transfer machine, which is a combination drilling machine and lathe shown in Figure 8.11. As illustrated in Figure 8.12, the machine is capable of many other operations, such as spinning, broaching, and planishing (end-turning or end-facing), among others. A tip is made in a working cycle of 12 steps or operations, at the rate of 100 pieces per minute. (See Figure 8.13.)

The action begins in Step 1, where the pre-cut brass workpiece measuring 11.5 mm (0.45 inch) long by 2.25 mm (0.10 inch) in diameter is loaded in a special collet chuck. The chuck holds the workpiece so that it can be worked from both the upper- and lower-tool positions. This feature makes it unnecessary to turn the workpiece over repeatedly, resulting in faster, more efficient and more accurate

Figure 8.7 Wood bending is a widely used furniture construction method. It requires a knowledge of the possibilities and limitations of the process to produce the satisfactory result shown here. Courtesy of Kinnarps AB, Stockholm

machining. Each part repositioning invites error. The other steps of the working cycle are listed in Figure 8.14, which should be examined along with Figure 8.13. Note that some operations are done on the upper half the the brass workpiece; others are done on the lower half. The efficiency of these machine operations is obvious. A similar process is used to make stainless steel tips for more expensive, reusable pens. After completing the tips, they are attached to the ink tubes and inserted into the barrels. In this pen tip example, the designers were aware of the capabilities of the equipment and designed the part to match these capabilities to achieve maximum producibility.

Many knowledgeable product designers carefully examine other types of machine process characteristics and design for them. Study the forms of the plastic kitchen appliances in Figure 8.15. They are paradigmatic of the many elegant German designs for such products and they have a pleasant technical look about them. It is almost as though the designer began with the question, What product forms can best be created by the plastic forming equipment in our factory? One obvious answer is, Clean cylindrical shapes. This answer provides a point of departure from which the designing proceeds. The basic shape unit for the coffee maker, coffee grinder, tea pot, and coffee thermos is a cylinder, with appropriate handles, spouts, and controls sensitively added. These forms in effect create a German style, and this philosophy is one to be considered by all designers.

Similarly, a knowledge of stainless steel sheet operations led to the design of the simple, ergonomically correct, and easily produced surgical staple extractor shown in Figure 8.16. (The companion surgical stapler was described in Figure 2.6.) This is a good example of efficient production with only two sheared and formed body parts

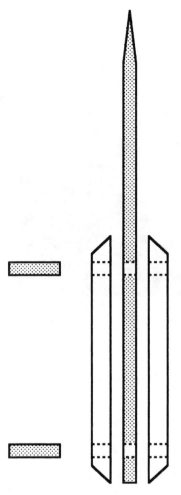

Figure 8.8 An example of the parts needed to produce the typical household knife.

and an assembly rivet. The ring handle and textured top lever provide a sure and comfortable grip, and an integral spring offers superior feel and control. The unique tool design permits staple withdrawal painlessly, perpendicular to the skin. All these attributes are accounted for by a designer skilled in process theory. The remover also lends itself to CAD design and nesting for added expedience. The body parts stretchouts are shown in Figure 8.17.

Tube bending technology has progressed from manual bend-and-fit methods to those under total computer control. For example, the complex configurations for aircraft hydraulic tube layouts have such tight space tolerances as to preclude any human efforts at making the bends. These same automatic systems can be used to good advantage in designing tubular furniture. One example in Figure 8.18 is made from cut and welded linear tube pieces, while the other features bent side structures,

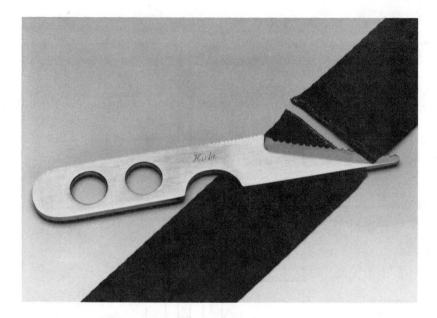

Figure 8.9 A very producible utility knife—one piece, very functional, and attractive. Courtesy of Urodesigns.

which are more economical—fewer parts to assemble, fewer weldments to possibly fail.

DESIGN FOR ASSEMBLY

The basic theory of design for assembly, or DFA, is that all unnecessary parts be eliminated from a product design or be re-created as multifunctional parts, and that those remaining should be easy to put together. Before proceeding further, the reader is invited to review the concept of assembly as presented in Chapter 4. There are mechanical, cohesive, and adhesive systems—some permanent, others semi-

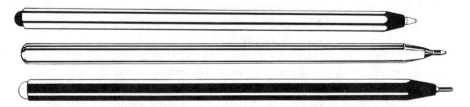

Figure 8.10 Examples of typical throw-away, ballpoint pens. Courtesy of Mikron Corp. Monroe.

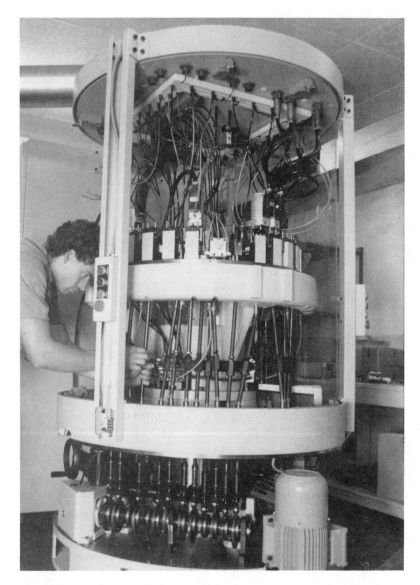

Figure 8.11 An automatic rotary transfer machine. Courtesy of Mikron Corp. Monroe.

permanent—presented for consideration by designers faced with the problem of how to join things together. A knowledge of these systems will be beneficial in DFA.

As stated earlier in this chapter, the majority of the direct costs of product manufacture is incurred in making parts. Although the cost of assembling these parts is relatively small, the issue of DFA is important because of its potential for reducing

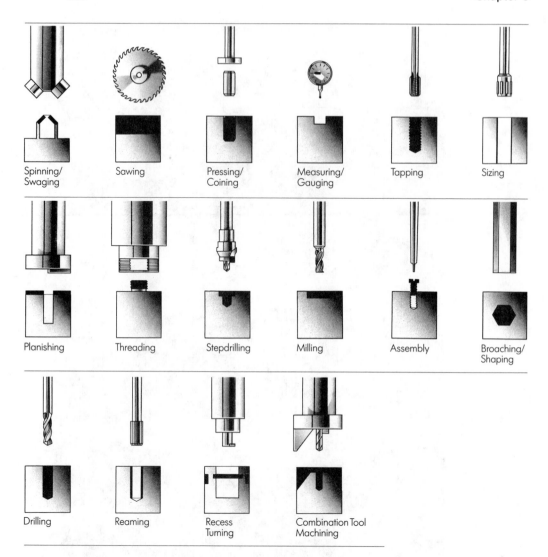

Figure 8.12 Typical operations performed on the rotary transfer machine. Courtesy of Mikron Corp. Monroe.

the numbers of product parts. A part removed is a part that need not be designed, tooled, made, handled, assembled, insured, or stored, and so a considerable reduction of overhead costs can result.

Two leading figures in assembly design theory are Geoffrey Boothroyd and Peter Dewhurst of the University of Rhode Island. Their pioneering work has resulted in a methodology for quantifying the techniques for manual assembly, special-purpose machine assembly, and programmable machine assembly. Their

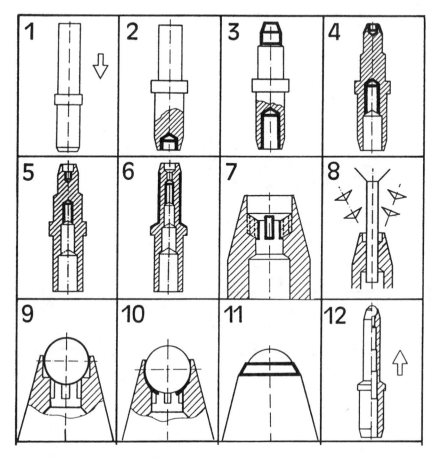

Figure 8.13 Twelve operations are necessary to produce the ballpoint pen tip. Courtesy of Mikron Corp. Monroe.

1. upper half — loading in collet chuck
2. upper half — chuck closing
2. lower half — center drilling
3. upper half — outside diameter turning
3. lower half — first drilling
4. upper half — ball seat drilling, cone turning
4. lower half — second drilling
5. upper half — capillary hole drilling
5. lower half — third drilling

6. upper half — outer diameter turning
6. lower half — fourth drilling
7. upper half — ink channel broaching (inside shaping)
8. upper half — ink channel deburring
9. upper half — ball loading
10. upper half — ball sleeve hammering
11. upper half — ball sleeve spinning
12. lower half — collet chuck opening
13. upper half — completed pen tip ejecting

Figure 8.14 Steps in the pen tip manufacturing cycle. Courtesy of Mikron Corp. Monroe.

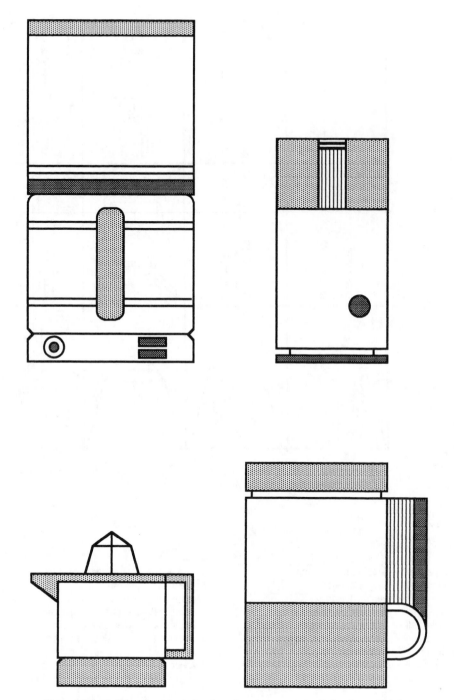

Figure 8.15 Coffee pots and grinders based on the common cylinder shape. This is a
familiar German style and is very producible.

Figure 8.16 An easy-to-make surgical staple extractor—minimal parts, simple assembly, usable, and hygienic. Courtesy of Richard-Allen Medical Industries.

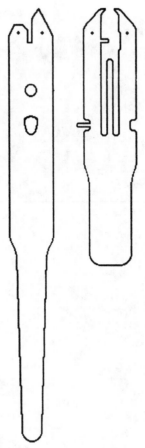

Figure 8.17 A stretchout of the extractor body parts before forming.

methods are widely used in modern industry and have contributed to its improvement. Examples and descriptions of their system are to be found on pages 358 to 368, reprinted with the kind permission of the *Machine Design* magazine.

To summarize the Boothroyd-Dewhurst work, their aim is to illustrate the advantages of redesigning product components to reduce the number of parts, and then to display the saved assembly time. For example, the newly designed reciprocating mechanism for a hand-held power saw resulted in a part reduction from 41 to 28, and a saved assembly time from 409 to 215 seconds. Much can be learned by studying the old and new power saw designs, for one can see the thinking behind the design changes.

Designers also should learn that assembly simplification so often involves only minor part modifications. Inserting a rod into a hole can be easy or difficult depending on component shape. (See Figure 8.19.) A blunt rod point is difficult to insert in a blunt hole, as in "A." Insertion is enhanced by applying a chamfer to the hole

Figure 8.18   The upper example of the beach chair has a structure made of welded tube members. The chair at the bottom has a more producible bent-tube frame.

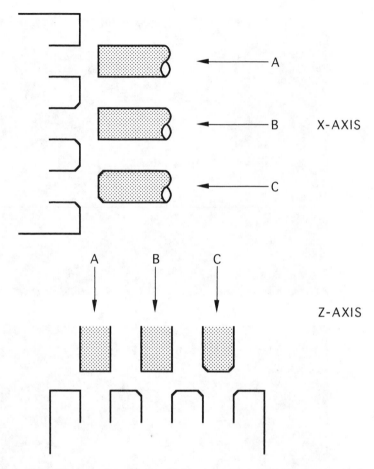

Figure 8.19 The logic of good design thinking is demonstrated in these rod and hole insertion examples. Note also that Z-axis assembly simplifies the task.

opening, as in "B." The best arrangement is to chamfer both the rod point and the hole opening, as in "C." Chamfering is a simple operation and in this case a simple solution. This same line of thinking would hold if a machine screw were to be engaged with a tapped hole: Chamfering would expedite either a manual or automatic assembly process. Imagine robot fingers trying to engage a blunt screw with a blunt hole opening, and the efficacy of chamfering becomes obvious. Look for the easy solution first.

Refer again to Figure 8.19. Note that the rod and hole assembly operation is much easier if the parts are oriented to a downward or Z-axis assembly, utilizing gravity, rather than a horizontal or X-axis position. An upward orientation would be most difficult. This matter of axis orientation is important in general assembly and very important in automatic assembly where the force of gravity is significant.

The subject of screw fasteners invariably emerges in discussions of assembly theory, and rightfully so. Threaded fastening is a chore to be avoided if possible, and many mechanical alternatives are available. Any number of ingenious push-pins, metal tacks, snap-fits, and expandable locks have been devised for use by designers. However, nuts and bolts and screws are very often indispensable for appliance panels that must be opened for servicing, automobile wheels that must be removed for tire changes, and furniture components that are shipped knocked-down and assembled upon arrival. The need for such semi-permanent fasteners will never go away, but they must be used judiciously.

Assembly was simplified by redesigning the metal housing in Figure 8.20. The original unit was comprised of a cap fastened to a body with four machine screws. The product requirements were such that a screw assembly was necessary for both security and ease of disassembly for servicing. The four screw holes in the body had to be drilled and tapped, and the assembly was complicated by having to blind-locate the four screws into these holes. Manual assembly was difficult, automatic assembly was impossible. The redesigned unit featured an L-bend on one edge of the cap, two milled screw slots in the cap, and only two drilled and tapped body holes. The manual assembly of the redesigned unit is considerably easier because it occurs in the open (one can see the screw holes), and there is sufficient play in the assembly for convenient alignment. Joining the two parts could be at least partially automated if both the holes and the screw-points were chamfered. (Refer to Figure 8.19.) If the designer had studied the component assembly problems early in the planning stage, the better solution may well have emerged.

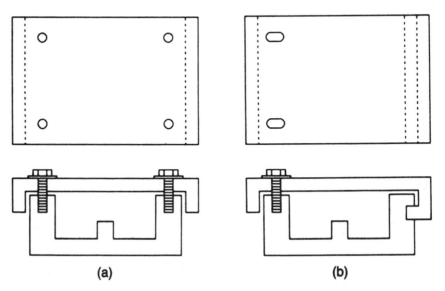

(a)   **(b)**

Figure 8.20 The original (a) and the redesigned (b) housing unit. The redesigned part facilitates the product assembly process, as well as the servicing of the units.

Other approaches to rationalizing the assembly process include modifying part features to provide symmetry. (See Figure 8.21.) The subtle changes reduce the number of different ways to position a part during assembly. Human assemblers resort to trial and error before they arrive at the correct orientation—a problem which is absent if the part has symmetry.

The familiar metal electric outlet box in Figure 8.22 is comprised of the 13 parts shown in Figure 8.23. This product requires numerous progressive sheet metal shearing and bending operations to produce the parts, as well as a considerable assembly effort. In contrast, the plastic box in Figure 8.24 is a one-shot injection molded product, fabricated with the attachment nails in place, and obviously resulting in a more producible item. Similarly, the plastic soap dish is molded in one operation to form an efficient container with top and bottom joined by an integral hinge. (See Figure 8.25.) These examples illustrate one of the basic tenets of design for assembly—namely, reducing the number of parts for any single product.

Injecting heated plastic or molten metal under pressure between two or more components is called *injection assembly*, which is a very economical way to perma-

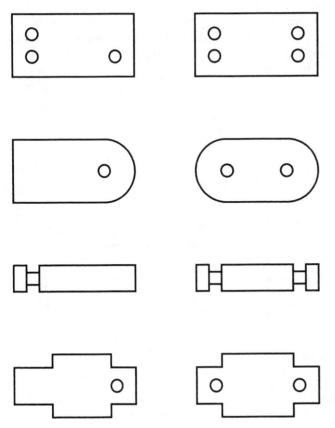

Figure 8.21 Minor changes in part geometry can simplify the assembly process and also reduce assembly errors.

Figure 8.22   A common electric outlet box.

nently join parts. The use of molten metal is an application of die-cast technology applied to the assembly operation. Employing plasticized resin involves injection molding technology in part assembly. In either situation, the parts to be joined are held in special assembly fixtures, which in turn are an integral part of the closed mold cavity. The heated metal or plastic is then injected into the mold cavity and allowed to cool or set. The mold is then opened to eject the completed part assembly. These results are called *insert-molded products*. A typical application of this technique is shown in Figure 8.26. It is evident that with this molten metal injection process, the steel shaft and cast gear have been joined with a newly formed cast gear and hub. This illustrates one of the many applications of the process. (See Figure 8.27.) Other industrial component examples include electric motor rotors, braided cable end nubs, and motor armatures. Many consumer products, such as food strainers, are also made using this method.

For example, utility brushes such as in Figure 8.28 can be made of plastic (or metal) tufts of bristles secured in a fixture, with polypropylene injection molded around the tufts to lock them tightly into the structure. This is a common and efficient way of producing such brushes, and results in a durable, functional, and usable tool.

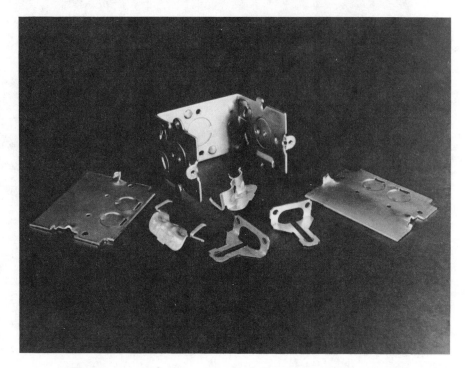

Figure 8.23   The box in Figure 8.22 is made of 13 parts, including screws, and must be assembled by tedious hand methods.

The brush shown has a striking appearance and refutes the all-too-common criticism that well-designed products seldom find their way into the ordinary consumer marketplace. The designer of this product knew how such brushes should be used and created a form to follow this function. The arching handle delivers a downward force, while allowing the user to position the free hand on the curved tip of the brush to provide a controlled, two-handed stroke. This solution adjusts the whole action of the wrist to minimize the scrubbing effort, especially for the older user. The designer also knew something about injection assembly.

Another excellent injected product is the clever, award-winning electrical plug in Figure 8.29, whose ultra-thin profile allows it to sit flat against the wall. This attribute not only results in a safer plug but also one that is less intrusive in home decor. A convenient folding safety-grip ring permits easy removal of the plug from a wall outlet, especially for the elderly. Also note the ingenious offset of the cord boss to permit clearance for plugging in another cord. (See Figure 8.30.) All internal parts are fixtured and then injection molded, as just described.

The three different products discussed are representative of the range of devices that can be produced by the efficient injection assembly process—a process

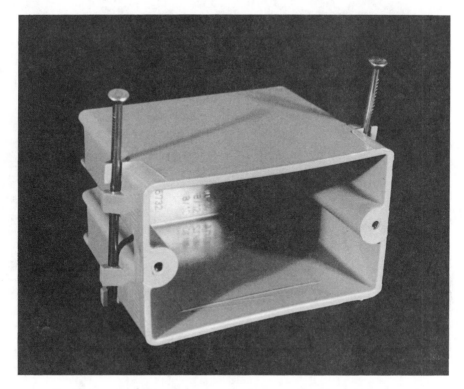

Figure 8.24   A one-piece plastic electric box is injection molded with the nails in place, and requires no assembly.

that eliminates assembly, even though several parts are involved. On the other hand, the garden tools in Figure 8.31 are one piece die-cast aluminum structures that obviously need no assembling. They are elegant organic shapes, comfortable to use and very durable. A more common, more traditional style appears in Figure 8.32— harder to make, harder to assemble, and apt to fail.

Unlike the examples above, a bicycle frame is a product requiring the assembly of parts, which can be difficult or easy, depending on the predesign research that gives direction to the designing. Presented in Figure 8.33 are two familiar frame assemblies. One is constructed of several metal tubes joined by welding. The quality of this product depends on the accuracy of the fixtures used to position the members for welding, and on the skill of the human welder. The other example utilizes tough metal lugs into which the frame members are secured. This joint can be permanently fastened either by an engineering adhesive (preferable) or by a much simpler weldment. Assembly in this latter instance is far less costly and far more reliable. There are several bicycle frames currently on the market that are one-piece carbon filament molded structures, tough, attractive, durable, and very producible.

Many types of bottles must be oriented in a package-labeling line to present the

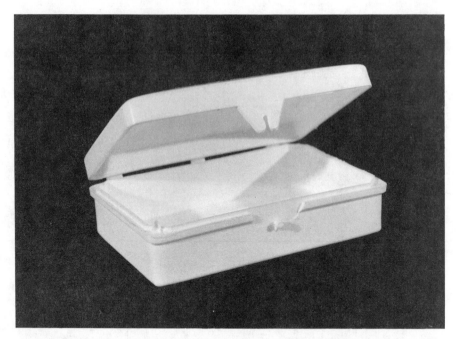

Figure 8.25 This plastic soap box is an example of high-producibility design. The top and bottom are injection molded, joined by an integral hinge in a single, fast operation. It is durable and convenient.

Figure 8.26 A steel shaft and die-cast gear are shown before and after injected metal assembly.

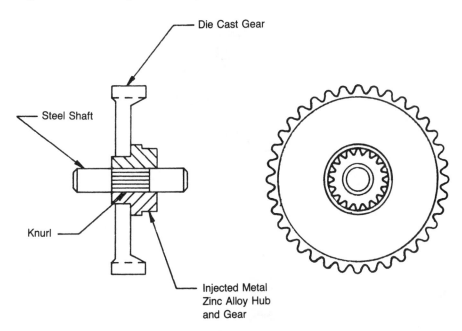

Figure 8.27   Note how the zinc alloy hub and gear become an integral part of the injection assembly. The knurl on the shaft aids in securing it to the cast hub and gear.

Figure 8.28   Brushes such as these can be made by injection molding polypropylene plastic around the fixtured bristle tufts. On larger brushes, the bristle tufts can be stapled in place. This brush is an attractive, ergonomically correct product. Courtesy of Empire Brushes, Incorporated.

Figure 8.29 A new concept in electrical plugs is obtained with this novel design, which is yet another example of an economic injection assembly. Its slim profile hugs the wall safely and improves the appearance of all visible wiring. Courtesy of Paige Manufacturing Corp., Canada.

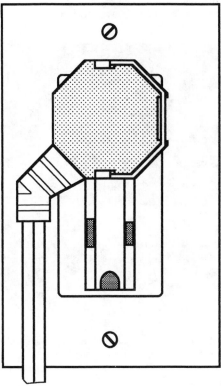

Figure 8.30 The usability of the Flat-Plug is enhanced by its offset design feature.

Figure 8.31 These attractive Plus-Four garden tools are durable, usable, and producible. Courtesy of Allen Simpson Marketing and Design, Canada; design by Todd Wood; photography by Doug Millar.

correct bottle face for label application, which is a special kind of assembly operation. Clever machines are designed to perform this function by tactically recognizing the front of the bottle and turning that face toward the labeling device. (See Figure 8.34.) Other machines employ optical scanners to read molded-in decorations to achieve proper orientation. This is yet another example of automatic part directing and positioning during assembly operations.

Design for disassembly. Reference was made in Chapter 6 to the issue of environmental concerns that must be addressed by product designers. One impor-

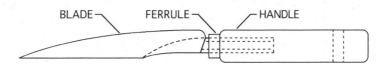

Figure 8.32 This familiar garden tool is more complicated than the tools in Figure 8.31.

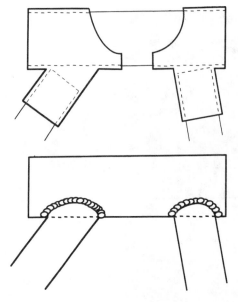

Figure 8.33 Two bicycle frame designs, two approaches to product assembly. The welded frame poses more assembly problems.

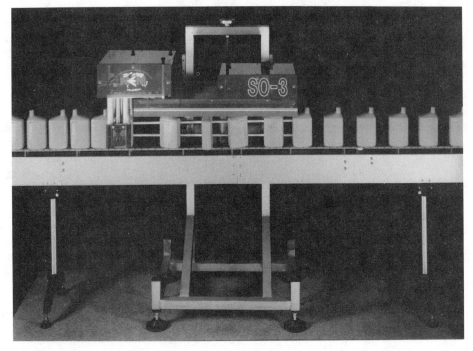

Figure 8.34 This directionalizer machine orients bottles correctly for subsequent label applications. Courtesy of New England Machinery, Inc.

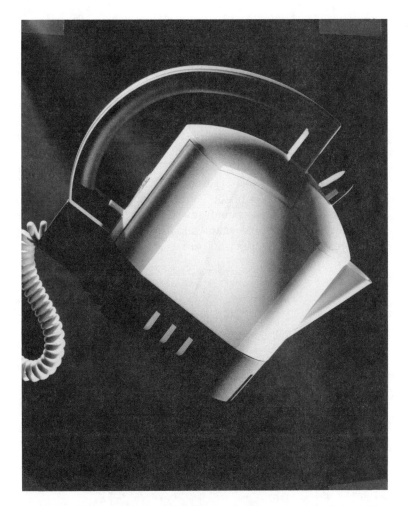

Figure 8.35 The revolutionary, designed-for-disassembly, energy-efficient UKettle. Courtesy of Polymer Solutions Inc.

tant emerging technology relates to *design for disassembly*. Here, the designer is confronted with the matter of creating products that are easy to take apart, so that the product components can be conveniently recycled. This feature, incidentally, also lends itself to easier maintenance. Such products have five common features:

1. They are designed so that all components can be easily separated, handled, and cleaned to permit economical recycling.
2. They include two-way snap-fits, or have break points on the snap-fits, which simplifies both assembly and disassembly.

3. They clearly identify separation points, which also eases disassembly.
4. They reflect tight tolerance design principles so that parts interlock, thereby substantially reducing the need for fasteners such as screws, bolts, and adhesives.
5. They have fewer parts, which saves energy, time, and money.

An excellent example of a product designed for disassembly is shown in Figure 8.35. This electric kettle meets the need for an attractive, versatile appliance used to boil small quantities of water for household or office use. The water reservoir, base, lid, rear cover, cross brace, and toggle are injection molded of a modified polyphenylene oxide plastic. (See Figure 8.36.) The kettle handle grip and lid grip are molded of a thermoplastic elastomer. By restricting the kinds of plastic materials used in the construction of the kettle, product recyclability is enhanced.

Another feature is its innovative construction. To aid the disassembler (and the assembler), the kettle has snap fists instead of the more traditional fasteners. Six thermoplastic parts interlock and are secured by snap fits in the base of the unit. (See

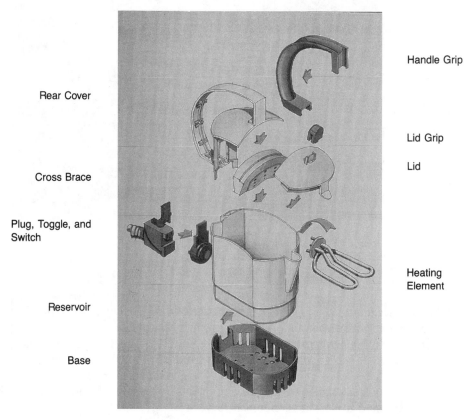

Figure 8.36 Exploded assembly diagram of the parts of the UKettle. Courtesy of Polymer Solutions Inc.

Figure 8.37.) These snaps are designed with molded-in breakpoints that are tough enough to withstand the rigors of daily use, but that can be broken apart at the joints by a disassembler. Injection molded logos clearly identify the materials and shown part separations points. It should be stressed that the innovative designers of the UKettle addressed the problems of disassembly at the outset of the design process in order to create this remarkable product.

Many other companies are beginning to see the "greening" of products as a marketing tool. To give consumers easy access to recycling information, one company is putting an 800 telephone number on all its products. One foreign automobile maker has a car model with an all-plastic shell of body parts connected to the zinc-plated unibody chassis with special elastic joints. The "skin" can be disassembled from the chassis in 20 minutes for simplified recycling. The keys to this approach are to limit the types of plastics used and to substitute "pop-in, pop-out" snap fasteners for threaded fasteners and adhesives. A considerable amount of engineering must go into the design of these fastenings to assure safe and reliable joint systems, and yet facilitate the speedy disassembly of the product. This theory of design for disassembly is in its infancy, especially in the United States. It requires a special mind-set without a doubt, and is yet another challenge for corporate design teams.

A creative knock-down home freezer conceptual model utilizes plastic panels that can be stored, packaged, and shipped disassembled at great cost savings and efficiency. (See Figures 8.38 and 8.39.) The consumer, or the merchandiser, can assemble the components quickly and easily with common tools, as each end panel is equipped with four fasteners to which the side panels can be attached and locked.

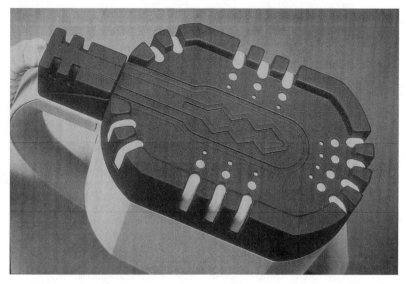

Figure 8.37   The six thermoplastic parts interlock tightly and are secured by snapfits in the base of the UKettle. Courtesy of Polymer Solutions Inc.

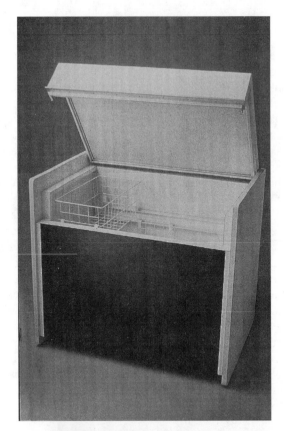

Figure 8.38 Made of individual panels of ABS plastic and polyurethane foam, this knock-down freezer exemplifies the reduction of merchandizing and shipping costs by creatively using plastics technology. Courtesy of The Dow Chemical Company.

Figure 8.39 This knock-down freezer can be stacked and shipped in an 18′ high box by using good assembly technology techniques. Courtesy of the Dow Chemical Company.

Additionally, the side panels are available in 3-, 4-, and 5-foot lengths to accommodate differing freezer capacity requirements. As an added bonus, the unit can be taken apart for eventual recycling, in keeping with the best design for disassembly practices. As such, it combines the theories of both assembly and disassembly in a most satisfactory way.

MANUFACTURING IN SPACE

The National Aeronautics and Space Administration (NASA) has for many years been engaged in advanced research to support its many missions into outer space. This has led to numerous technology spin-offs for use by U.S. industry such as freeze-dried foods and drinks; self-contained breathing and working apparatus; self-propelled robotic devices; space-age materials, such as carbon fiber structures; fiber optics; and production schemes, among many others. One of the more fascinating of these ventures is the whole concept of building huge space structures. (See Figure 8.40.) A key factor in such constructions are the truss members that must first be fabricated and then assembled, and several experimental schemes have been devised. One employs telescoping coils of sheet metal, tightly rolled around a central

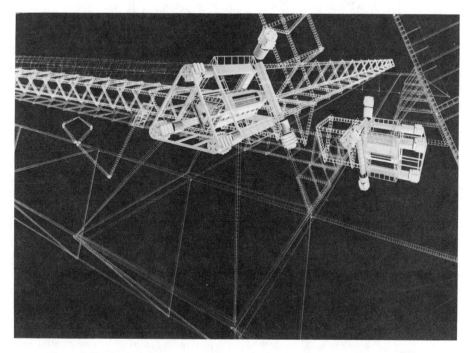

Figure 8.40 An artist's concept of a giant space construction. Courtesy of NASA.

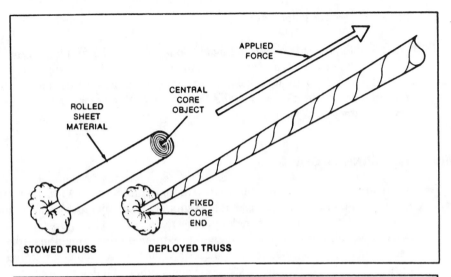

STOWED TRUSS   DEPLOYED TRUSS

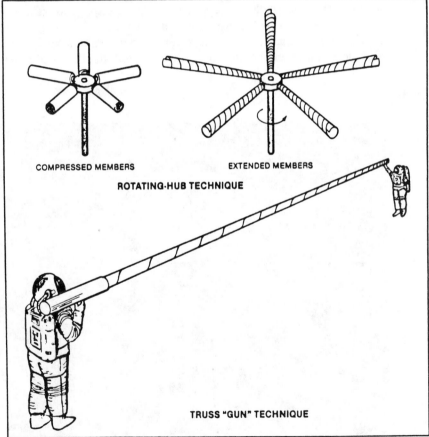

Figure 8.41 Three methods for deploying truss members in space by using applied forces to form the truss cones. Courtesy of NASA.

core, which can be transported in large quantities by the space shuttle. Three coil deployment techniques have been devised. (See Figure 8.41.) By one method, the coil can be extended by centrifugal force to form a cone. With both ends fastened to prevent unrolling, a rigid truss member can be made. Another method involves fixing several coils to a rotating hub that generates centrifugal force to create the truss cones. In the third method, a hand-held gun is fired to propel the core object, thereby forming the truss cone. The ingenuity shown in these subsystems derives from the need for easily stowed and transported compact coils that can be deployed as long trusses, usually without human intervention, and then assembled at some future space station.

An equally creative method is the hoop/column concept, the details of which are shown in Figure 8.42, and where the relationship to the common umbrella is obvious. The compact unit is carried into space where, upon radio signal, it gradually unfurls to form an opened "umbrella." (See Figure 8.43.) Two of these units become the satellite dishes of a Tracking and Data Relay Station (TDRS) used to transmit information to a terminal on Earth. (See Figure 8.44.)

Space research has resulted in some unique concepts for lunar mining operations, where robots and automated equipment are used to mine and process ores without direct human involvement. The operation employs lunar robots (see Figure 8.45) to mine, grade, and haul ores to a material processing plant according to the scheme in Figure 8.46. The processed ingots are subsequently transported to Earth.

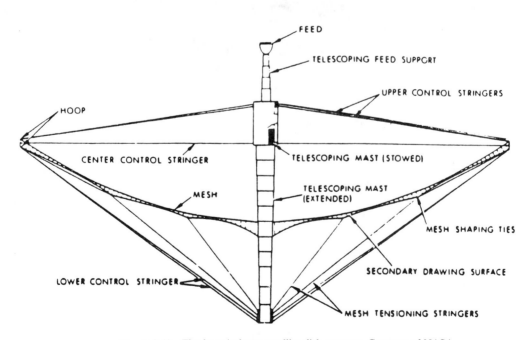

Figure 8.42   The hoop/column satellite dish concept. Courtesy of NASA.

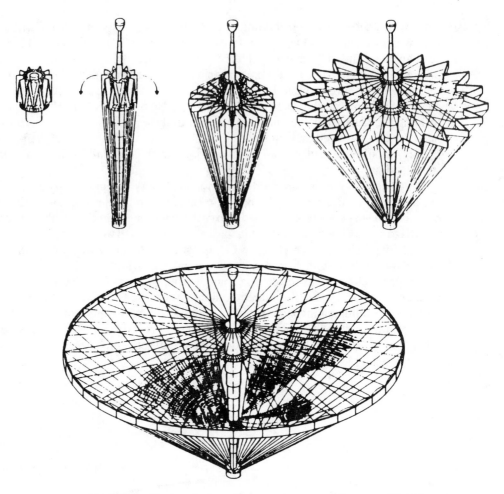

Figure 8.43   The hoop/column deployment sequence. Courtesy of NASA.

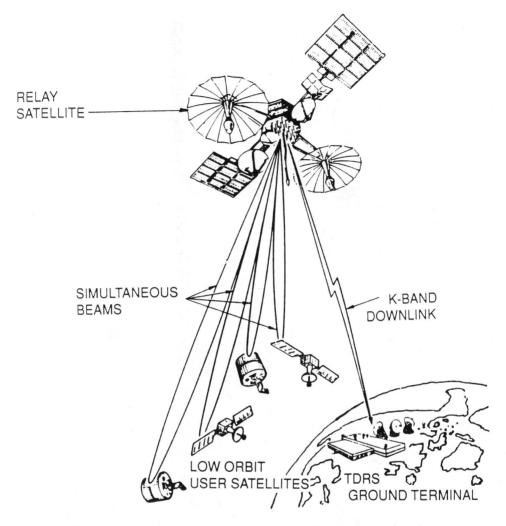

RELAY
SATELLITE

SIMULTANEOUS
BEAMS

K-BAND
DOWNLINK

LOW ORBIT
USER SATELLITES

TDRS
GROUND TERMINAL

Figure 8.44 An application of the hoop/column satellite dish in the Tracking and
Data Relay Station. Courtesy of NASA.

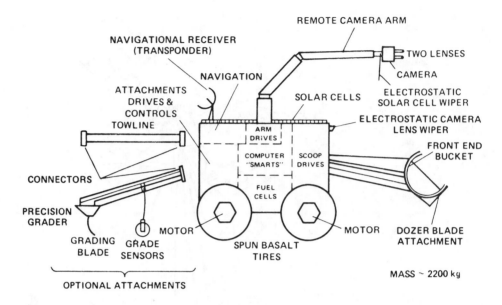

Figure 8.45 A conceptual design of a lunar mining robot. Courtesy of NASA.

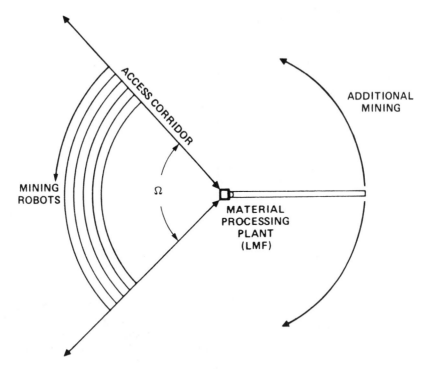

Figure 8.46 A lunar surface strip mining scheme. Courtesy of NASA.

One proposed use of this technology has been the design of the bidirectional, automatic coal-mining machine for use on Earth. (See Figure 8.47.) According to the concept, the machine uses two drums that cut into the coal seam as they advance on crawler tracks. The cut coal is fed to the crusher screws, water is added, and the coal slurry is conveyed to a haulage tube where it is pumped to the surface. When the machine reaches the end of a pass, the positions of the two drums are reversed so that the cutting can continue, uninterrupted, in the opposite direction. This maneuver could be controlled automatically or remotely by human operators. Coal mining is dangerous, unhealthy, and arduous. People were not meant to be picking at coal deep underground; this machine can prevent such practices.

Several conclusions can be drawn from this exposition on space research and development. First, the researchers must have had a great time engaging in such wild speculations, and thrilled at the potential of some of their workable theories. Second, ground-based humans should study these research results carefully to ascertain what uses they can make of them in their product development. Automatic manufacturing schemes devised for the far reaches of space surely must be more attainable on Earth. And finally, designers should routinely engage in wild speculation to tax their imaginations toward the creation of perfect artifacts.

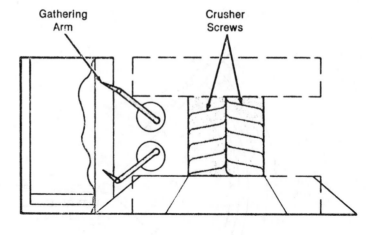

Gathering Arm

Crusher Screws

Coal-Slurry Flow

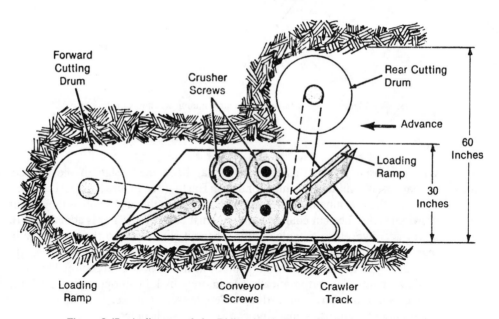

Forward Cutting Drum

Crusher Screws

Rear Cutting Drum

Advance

60 Inches

30 Inches

Loading Ramp

Loading Ramp

Conveyor Screws

Crawler Track

Figure 8.47 A diagram of the Bidirectional Automatic Coal Mining machine. Based on space technology concepts, this unit could be used on Earth. Courtesy of NASA.

QUESTIONS AND ACTIVITIES

1. Write your definition of *producibility* and describe its importance in design for manufacturing.

2. Study the guides to design for producibility. Select several of the guidelines and describe how they can facilitate product manufacture.

3. Describe how process design factors can influence the efficiency and quality of part production.

4. Select some product examples that reflect good and poor applications of process design theory.

5. The kitchen appliances shown in Figure 8.15 exhibit a knowledge of machine process characteristics. Can you find some similar examples that emerge as a product style?

6. Write your definition of *design for assembly*, and describe its importance in design for manufacturing.

7. What specifically is the Boothroyd-Dewhurst approach to design for assembly?

8. Examine the literature dealing with design for assembly and identify some examples of creating a new design by reducing the numbers of parts in an old design.

9. Cite some differences in designing a product for manual assembly as opposed to designing that product for automatic assembly.

10. Discuss the significance of part symmetry as an aid to product assembly.

11. Describe the process of injection assembly and find some examples of injection-molded products.

12. Describe the theory of designing products for disassembly, and summarize the five common features of such products.

13. One important rationale for disassembly design is to simplify product recycling. Does this in any way influence the producibility of the product?

14. Consult the NASA reports on "space technology spin-offs" and identify some important consumer product applications.

DESIGN FOR ASSEMBLY: SELECTING THE RIGHT METHOD

Surprisingly, the least costly assembly method can be identified early in the design stage. If the product is then designed for that process, manufacturing cost can drop 20 to 40% and assembly productivity rise 100 to 200%.

GEOFFREY BOOTHROYD
Professor
PETER DEWHURST
Professor
University of Massachusetts
Amherst, MA

DESIGN is the first stage of manufacturing. It is here that manufacturing costs are largely determined. In addition, the assembly process is usually the single most important process contributing to both manufacturing costs and labor requirements.

When productivity improvements are sought, design for ease of assembly must be given the highest priority. Indeed, even when automated assembly is considered, one must determine whether the design of the product lends itself to such automation. This is particularly true in such high-technology batch-production industries as computer hardware.

Recent studies of computer-related products have shown that reductions of 20 to 40% in manufacturing cost and increases of 100 to 200% in assembly productivity are readily obtainable through proper consideration of assembly at the design stage. For example use of design-for-assembly techniques could save Xerox Corp. an estimated $150 million per year.

Savings like these can be realized with simple techniques for analyzing even rough designs and predicting assembly costs. The first step in these techniques is to identify the assembly process that is most likely to be economic for a particular product. Then the product itself can be designed for that particular process.

The reason that early process selection is important is that manual assembly differs widely from automatic assembly in the requirements it imposes on product design. An operation that is easy for a person may be impossible for a robot or special-purpose workhead, and operations that are easy for machines may be difficult for people.

Surprisingly, detailed knowledge of product design is not required to make a good estimate of the most economical assembly process. Essentially, what must be known is projected market life, number of parts, projected production volume, and company investment policy.

Process characteristics

The cost of assembling a product is related both to the design of the product and to the assembly process used for its production. Assembly cost is lowest when the product is designed so that it can be economically assembled by the most appropriate process. The three basic processes are manual assembly, special-

Cutting parts and cost

The twin objectives of design-for-assembly studies are to reduce the number of parts in a product and to increase the ease of assembling the remaining parts. In the example shown here, a reciprocating mechanism from a hand power saw, the number of parts was reduced from 41 to 28 and assembly time from 409 to 215 seconds. The time savings cut assembly cost by $0.95, but what

Pages 358 to 368 reprinted from *Machine Design*, 1984. Copyright 1984 by Penton Publishing Inc., Cleveland, OH. Reprinted by permission.

purpose machine assembly, and programmable-machine assembly.

In manual assembly (MA on the accompanying chart), the tools required are generally simpler and less costly than those employed on automatic assembly machines, and the downtime caused by defective parts is usually negligible. Cost of manual assembly is relatively constant and independent of production volume. Manual processes also have considerable flexibility and adaptability. In some instances it is economical to provide the assembler with mechanical assistance (MM) in order to reduce assembly time.

Special-purpose assembly machines are those that have been built to assemble a specific product. They consist of transfer devices with single-purpose workheads and parts feeders at the various workstations. The transfer devices can operate on an indexing (synchronous) principle (AI) or on a free-transfer (non-synchronous) principle (AF).

These special-purpose machines are costly and require considerable engineering development before they can be put into service. Downtime caused by defective parts can be a serious problem unless the parts have

surprises most engineers is that parts cost was reduced as well — by $1.28 in this instance. Equal or greater savings in parts costs are typical of the design specifications achieved by deliberate design for assembly.

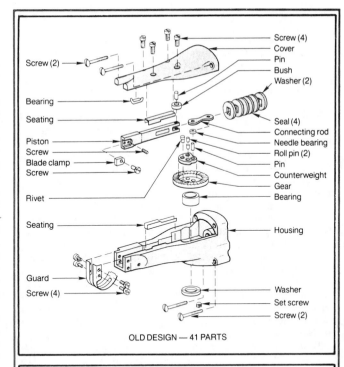

OLD DESIGN — 41 PARTS

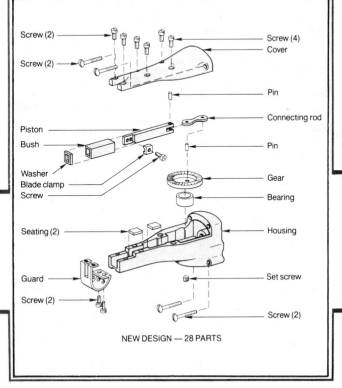

NEW DESIGN — 28 PARTS

relatively high quality. Also, special-purpose machines work on a fixed cycle time, with a fixed rate of production. If they are underutilized they cannot be used for any other purpose, resulting in a marked increase in assembly cost.

Programmable assembly machines are similar to the nonsynchronous special-purpose machines except that the workheads are general-purpose and programmable. This arrangement (AP) allows more than one assembly operation to be performed at each workstation. It also provides for considerable flexibility in production volume and greater adaptability to design changes and different product styles.

For lower production volumes, robotic assembly with a single robot workstation may be preferable. Here, two robot arms normally work interactively at the same work fixture (AR).

For the latter two systems, parts are normally made available at workstations in manually loaded magazines. Parts feeders like those employed for machine assembly are usually too costly.

Selecting a process

The method for assessing the available processes is summarized in the accompanying chart. The chart is based on an analysis of mathematical models of the various assembly processes. Using the chart requires that six basic facts be known:

• Production volume per shift
• Number of parts in an assembly
• Single product vs variety of products
• Number of parts required for different styles of the product
• Number of major design changes expected during product life
• Company policy on investment in labor-saving machinery

The production volume per shift and number of parts in an assembly determine the correct row in the chart. The other factors determine the correct column.

The box at the intersection of row and column describes the most economical assembly process. Processes in parentheses are no more than 10% less economical than the primary process in the box.

White boxes indicate low-cost assembly, light-colored boxes medium-cost assembly, and dark boxes expensive assembly.

By varying the basic information about the product and observing how each variation affects the box selected, one can see which factors have the greatest influence on the assembly process.

Several cautions should be observed in using this chart. The numbered comments that follow correspond to the "notes" called out in the chart.

1. Defective parts can cause severe problems in automatic assembly machines by jamming feeding devices, preventing workhead operation, or spoiling an otherwise acceptable assembly. These defective parts could be screws without threads, chipped or discolored parts, parts out of tolerance, pieces of swarf, or any foreign items in the feeding devices of the assembly machine. Automation is unlikely to be successful if the proportion of defective parts is greater than about 2%.

2. Automatic assembly machines produce a steady output of assemblies. Therefore, significant fluctuations in demand, such as may occur with sports equipment, must be accommodated by stockpiling. The cost of stockpiling may rule out automatic assembly for some products.

3. On automatic assembly machines, different product styles can be accommodated by arranging alternative parts at the workstation. Instructions are then given to the machine as to which part should be inserted. For example, in a three-part assembly with two alternatives for each part, eight product styles could be produced.

4. For the purposes of this analysis, one design change means a change that will require a new feeding device and workhead on an automatic assembly machine.

5. An important factor in considering investment in automation equipment is the company investment factor, R_i. The larger the number of shifts worked and the higher the figure for economical investment to replace each operator, the greater the opportunity for automation.

Using the chart

To see how this chart works in practice, consider a product assembled from 35 parts. It is to be manufactured in ten different styles, obtained by having one alternative for each of ten of the parts in the assembly.

Ten major design changes will probably take place during the first three years of product life. Each design change will require a new feeding and orienting device and a new workhead if automatic assembly is used. Expected annual production is 1,000,000 units, 500,000 per shift.

As a matter of company policy, the amount that can be spent on an item of automation equipment that will do the work of one operator on one shift is $40,000. This figure allows for the purchase of the equipment and all engineering and debugging necessary before it is fully operational in the plant. The annual cost of one assembly operator is estimated to be $20,000, including overhead.

In this example, the annual volume per shift, V_s, is 500,000 and the number of parts in the assembly, N_a, is 35, so Row 3 is selected.

Because this is a single product with a market life greater than three years, the choice of columns is restricted to Columns 0

This article is largely based on the *Design for Assembly Handbook*, which has been developed by Boothroyd and Dewhurst at the University of Massachusetts. The handbook is available from Automatic Assembly Program, Department of Mechanical Engineering, U. of Mass., Amherst, MA 01003, tel. (413) 545-0054.

CHOICE OF ASSEMBLY METHOD

$N_p = 1$ Single product has a market life of three years or more without significant variations in demand; manual fitting of any of the parts is not necessary and the proportion of defective parts is less than 2%. See notes 1 and 2.	
$N_t < 1.5 N_a$ Number of parts needed to build different product styles less than 1.5 times the number of parts in the assembly (3). AND $N_d < 0.5 N_a$ Fewer than half of the parts will be subjected to major redesign during the product market life (4).	$N_t \geq 1.5 N_a$ More than 50% extra parts are needed to build the range of different product styles (3). OR $N_d \geq 0.5 N_a$ More than half of the parts are likely to be affected by design changes during the product life (4).

Column 8: A variety of different but similar products, no manual fitting required and less than 2% defective parts.

Column 9: Variety of products, manual fitting of some parts necessary, fluctuations in demand or low investment potential.

Company investment $R_i = S_h Q_r / W_a$ (5)

$R_i \geq 5$	$5 > R_i > 2$	$2 \geq R_i \geq 1$	$R_i < 1$	$R_i \geq 5$	$5 > R_i > 2$	$2 \geq R_i \geq 1$	$R_i < 1$		

V_s group	N_a	#	0	1	2	3	4	5	6	7	8	9
$V_s > 0.65$ Annual production volume per shift greater than 0.65 million assemblies.	$N_a \geq 16$ 16 or more parts in the assembly.	0	AF	AF	AF	MM (AF)	AP	AP	AP (MM)	MM	MA (AP)	MA
	$15 \geq N_a \geq 7$ Between 7 and 15 parts in the assembly.	1	AF	AF (AI)	AI (AF)	MM (AI)	AP	AP	MM (AP)	MM	MA	MA
	$N_a \leq 6$ 6 or fewer parts in the assembly.	2	AI	AI	AI	AI	AI	AI (AP)	MM	MM	MA	MA
$0.65 \geq V_s > 0.4$ Annual production volume per shift between 0.4 and 0.65 million assemblies.	$N_a \geq 16$ 16 or more parts in the assembly.	3	AP	AP	MM (AP)	MM	AP	AP	AP	MA (MM)	MA	MA
	$15 \geq N_a \geq 7$ Between 7 and 15 parts in the assembly.	4	AI	AI	AI	MM	AP	AP	MM (AP)	MA (MM)	MA	MA
	$N_a \leq 6$ 6 or fewer parts in the assembly.	5	AI	AI	MM (AI)	MM	AI (MM)	MM	MM	MA (MM)	MA	MA
$0.4 \geq V_s > 0.2$ Annual production volume per shift between 0.2 and 0.4 million assemblies.	$N_a \geq 16$ 16 or more parts in the assembly.	6	AP	AP	MM	MM	AP	AP	AP	MA	MA	MA
	$15 \geq N_a \geq 7$ Between 7 and 15 parts in the assembly.	7	AI (MM)	MM	MM	MM	AP	MM	MA (MM)	MA	MA	MA
	$N_a \leq 6$ 6 or fewer parts in the assembly.	8	MM	MM	MM	MM	MM	MM	MA (MM)	MA	MA	MA
$V_s \leq 0.2$ Annual production volume per shift less than or equal to 0.2 million assemblies.		9	MM	MM	MM (MA)	MA	MM	MA	MA	MA	MA	MA

How assembly processes differ

Early in the design of any product it is important to decide which type of assembly process is likely to yield the lowest cost. This decision has a major bearing on the design because manual assembly differs widely from automatic assembly. Each type of process has its own advantages and limitations. For analytical purposes, the six basic assembly processes are:

AI: Automatic assembly using special-purpose indexing machines, workheads, and automatic feeders. One supervisor for the machine when $N_a < 6$ (rotary indexing machine) and one supervisor together with one assembly operator when $N_a > 6$ (in-line indexing machine).

AF: Automatic assembly using special-purpose free-transfer machines, workheads, and automatic feeders. One supervisor and one assembly operator for the machine.

AP: Automatic assembly using manually loaded part magazines and a free-transfer machine with programmable workheads capable of performing several assembly tasks. One supervisor and one assembly operator for the machine.

AR: Automatic assembly using manually loaded part magazines and a sophisticated two-arm robot with a special-purpose gripper that can handle all the parts for one assembly. One supervisor needed for the robot.

MA: Manual assembly on a multistation assembly line. The transfer device is a free-transfer machine with one buffer space between each operator.

MM: Manual assembly with mechanical assistance. This system is the same as MA, but feeders or other devices are provided and the assembly time per part thereby reduced.

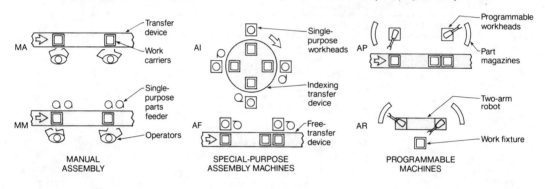

| MANUAL ASSEMBLY | SPECIAL-PURPOSE ASSEMBLY MACHINES | PROGRAMMABLE MACHINES |

through 7. A shorter market life, parts of poor quality, or manual fitting would have suggested Columns 8 or 9.

The total number of parts required to build the different styles is 45, which is less than 1.5 times the number of parts in one assembly. Also, fewer than half of the parts will be subject to major redesign during the life of the product, because only ten redesigns are expected. These factors restrict the choice to Columns 0 through 3.

To select among these columns, calculate the investment factor R_i:

$$R_i = \frac{S_h Q_e}{W_a}$$
$$= \frac{2(40,000)}{20,000}$$
$$= 4$$

Therefore Column 1 is selected. The box at the intersection of Column 1 and Row 3 is labeled AP, for automatic ass-

embly with programmable workheads.

If the production volume were higher, moving the selected box to Row 0 Column 1, special-purpose automatic assembly machines could be used. Such machines could also be used if the product could be broken down into smaller assemblies, shifting the selected box to Row 4 Column 1.

If the company's investment policy were not so generous, moving the selected box to Row 3 Column 2, then manual assembly with mechanical assistance would be the most appropriate method.

For the conditions specified, Row 3 Column 1 indicates that relatively low assembly cost can be achieved with automation, and that design for ease of automation should be pursued.

Succeeding articles will cover design for manual assembly, de-

sign for automatic assembly, and current developments in software that aids these design procedures in much the same way that "Work Factor" and "MTM" programs now assist production engineers.

Nomenclature

N_a = Number of parts in the completed assembly

N_d = Number of design changes during the first three years that would necessitate a new feeder or workhead on an automatic assembly machine

N_p = Number of different products to be assembled using the same basic assembly system during the first three years

N_t = Total number of parts required for building different product styles

Q_e = Capital expenditure allowance to replace one operator on one shift

R_i = Investment factor

S_h = Number of shifts

V_s = Annual production volume per shift

W_a = Annual cost of one assembly operator

DESIGN FOR ASSEMBLY: MANUAL ASSEMBLY

A a technique that quantifies the difficulties of manual assembly can cut production cost dramatically. The analysis spotlights not only parts that require excessive time for assembly, but also those that can be eliminated or combined with other parts.

GEOFFREY BOOTHROYD
Professor

PETER DEWHURST
Professor
University of Massachusetts
Amherst, MA

INDUSTRIAL productivity depends in large part on ease of assembly. In fact, the way a product is assembled is usually the single most important control over both manufacturing cost and labor requirements. Minimum production cost for a product is achieved when the most economical assembly process is selected early in the design stage, and the product is then designed for that process.

Contrary to the conventional wisdom, final product designs are not needed to make a reasonable selection of the most economical assembly process. Selection is based on projected market life, number of parts, projected production volume, and company investment policy.

Products with half the parts

Manual assembly requires workers to select parts, orient them, insert them, and often to fasten them with other parts that must also be selected, oriented, and inserted. Many of these time-consuming steps can be eliminated if preliminary designs are examined with an eye both to the assembly requirements they impose and also to the potential for combining parts with each other and with integral fasteners.

Substantial reductions in parts count and assembly time are possible, even when the original design has already been subjected to value analysis. Traditional value analysis tends to overlook small parts such as fasteners. Each fastener has a relatively small initial price, yet the assembly of such seemingly insignificant parts often adds substantially to the total cost of the product.

One example is the ribbon base plate from a Diablo printer. The "Old Design" shown here had already undergone value analysis, yet when the assembly was redesigned for ease of manual assembly the number of parts was cut from 77 to 36 and the cost was reduced by $4.90.

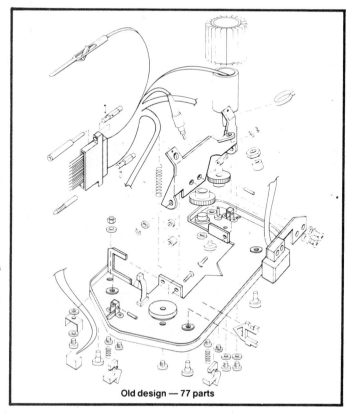

Old design — 77 parts

A detailed procedure for selecting the least costly assembly process was presented in "Design for Assembly: Selecting the Right Method," MD, Nov. 10, 1983, p. 94. The present article assumes that manual assembly has been selected and shows how to analyze designs for ease of manual assembly. Succeeding articles will deal with automatic assembly and software systems that assist in analysis and redesign.

Once it has been decided that a product is to be assembled manually, features of the design are examined systematically, and a "design efficiency" is calculated. This efficiency allows different designs to be compared for ease of assembly.

Examination of the preliminary design answers two important questions for each part in the assembly: Can this part be eliminated or combined with other parts in the assembly? And how long will it take a worker to grasp, manipulate, and insert this part?

With this information it is possible to estimate the total assembly time, compare it to the assembly time for an ideal design, and identify design features that result in high assembly cost.

Design analysis

The first step in the analysis is to obtain the best information available on the product. Useful items are engineering drawings, exploded three-dimensional views, an existing version, or a prototype.

The second step is to take the assembly apart, or imagine how it might be done. The complete assembly is assigned identification number 1. As each part is removed from the assembly, it is assigned an identification number in sequence. If the assembly contains subassemblies, they should be treated as "parts" at first, then analyzed separately later.

The third step is to reassemble the product. First assemble the part with the highest identification number to the work fixture, then add the remaining parts one by one. As the product is reassembled, data on handling, assembly, and operation cost are entered on a worksheet. Also entered is an estimate of the potential for eliminating the part or combining it with another part.

When reassembly is complete, data from the work sheet are summed to give the total estimated manual-assembly time and cost, as well as the theoretical minimum number of parts in the product.

Finally, the manual-assembly design efficiency is obtained from

$$E_m = 3N_m/T_m$$

where E_m = design efficiency; N_m = minimum number of parts, and T_m = total assembly time.

This equation compares the estimated assembly time for an assembly containing the theoretical minimum number of parts, each of which can be assembled in the "ideal" time of 3 s. This ideal time is based on the assumption that each part is easy to handle and insert, and that about one third of the parts are secured immediately on insertion with well-designed snap-fit fasteners.

Unfortunately, there is no broadly applicable guideline for a satisfactory design efficiency.

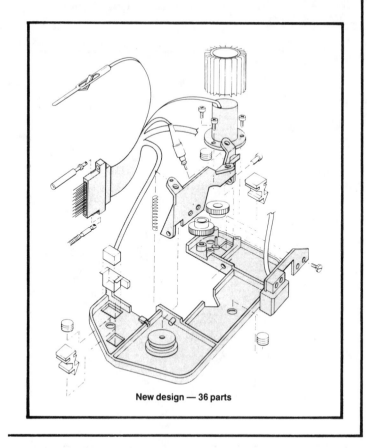

New design — 36 parts

The assumption that all parts in the assembly are easy to handle and insert is impossible to meet in many products.

At one extreme, complex electromechanical products that require extensive wiring and gasketing tend to have low design efficiencies, even when well designed. Many companies making such products have decided that design efficiencies around 20 to 30% are quite acceptable.

At the other extreme, simple products such as pneumatic piston assemblies with few parts can have design efficiencies as high as 90%. Ultimately, experience with a range of similar products is the only way to decide on an acceptable efficiency.

Generating data

The most critical step in the design analysis is the reassembly, where data are generated for possible redesign. The easiest way to understand this process is to consider the example of a riser-panel assembly.

For purposes of reassembly and data generation, never assume that one part is grasped in

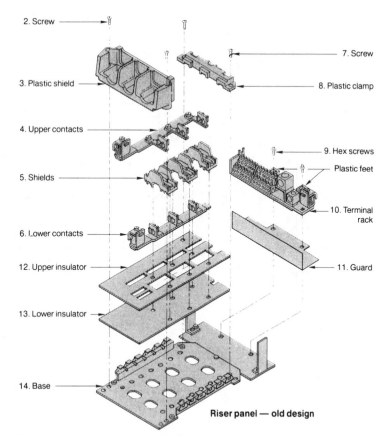

2. Screw
3. Plastic shield
4. Upper contacts
5. Shields
6. Lower contacts
12. Upper insulator
13. Lower insulator
14. Base

7. Screw
8. Plastic clamp
9. Hex screws
Plastic feet
10. Terminal rack
11. Guard

Riser panel — old design

Manual handling times

			Parts are easy to grasp and manipulate					Parts present handling difficulties				
			Thickness > 2 mm			Thickness ≤ 2 mm		Thickness > 2 mm			Thickness ≤ 2 mm	
			Size > 15 mm	6 mm ≤ Size ≤ 15 mm	Size < 6 mm	Size > 6 mm	Size ≤ 6 mm	Size > 15 mm	6 mm ≤ Size ≤ 15 mm	Size < 6 mm	Size > 6 mm	Size ≤ 6 mm
			0	1	2	3	4	5	6	7	8	9
Parts can be grasped and manipulated by one hand without the aid of grasping tools	$(\alpha + \beta) < 360°$	0	1.13	1.43	1.88	1.69	2.18	1.84	2.17	2.65	2.45	2.98
	$360° \le (\alpha + \beta) < 540°$	1	1.5	1.8	2.25	2.06	2.55	2.25	2.57	3.06	3	3.38
		2	1.8	2.1	2.55	2.36	2.85	2.57	2.9	3.38	3.18	3.7
	$540° \le (\alpha + \beta) < 720°$	3	1.95	2.25	2.7	2.51	3	2.73	3.06	3.55	3.34	4
	$(\alpha + \beta) = 720°$											

☐ ONE HAND ASSEMBLY

Charts in the Design for Assembly Handbook give standard handling codes and times. α and β are the required rotations for end-to-end and side-to-side symmetry. "Size" is the largest orthogonal dimension of the part.

365

Design for manual assembly worksheet

1	2	3	4	5	6	7	8	9	Name of assembly
Part I.D. no.	Number of times the operation is carried out consecutively	Two-digit manual handling code	Manual handling time per part	Two-digit manual insertion code	Manual insertion time per part	Operation time, sec $(2) \times [(4) + (6)]$	Operation cost, ¢ $0.4 \times (7)$	Figures for estimation of theoretical minimum parts	Riser panel (old design)
14	1	30	1.95	00	1.5	3.45	1.38	1	Base
13	1	33	2.51	00	1.5	4.01	1.60	1	Lower insulator (< 2 mm)
12	1	33	2.51	06	5.5	8.01	3.20	0	Upper insulator (< 2 mm)
11	1	30	1.95	08	6.5	8.45	3.38	0	Guard
10	1	30	1.95	08	6.5	8.45	3.38	1	Terminal rack
9	2	10	1.50	49	10.5	24.00	9.60	0	Hex. screws (8 x 16 mm)
8	1	30	1.95	00	1.5	3.45	1.38	0	Plastic clamp
7	2	11	1.80	49	10.5	24.60	9.84	0	Screw (9 x 14 mm)
6	1	30	1.95	08	6.5	8.45	3.38	1	Lower contacts
5	3	20	1.80	00	1.5	9.90	3.96	0	Shields
4	1	30	1.95	02	2.5	4.45	1.78	1	Upper contacts
3	1	30	1.95	02	2.5	4.45	1.78	1	Plastic shield
2	2	11	1.80	49	10.5	24.60	9.84	0	Screw (9 x 14 mm)
						136.3	54.5	6	Design efficiency $= \dfrac{3\,N_m}{T_m} = 0.13$
						T_m	C_m	N_m	

The old design of the riser panel required 18 parts and 136.3 s for assembly. Analysis shows that the number of parts can theoretically be reduced to six (Column 9 on the worksheet). High assembly times in Column 7 indicate opportunities for improved design.

Manual insertion times

			After assembly no holding down required to maintain orientation and location.				Holding down required during subsequent processes to maintain orientation or location.			
			Easy to align and position during assembly.		Not easy to align or position during assembly.		Easy to align and position during assembly.		Not easy to align or position during assembly.	
			No resistance to insertion.	Resistance to insertion.	No resistance to insertion.	Resistance to insertion.	No resistance to insertion.	Resistance to insertion.	No resistance to insertion.	Resistance to insertion.
			0	1	2	3	6	7	8	9
Addition of any part where neither the part itself nor any other part is finally secured immediately.	Part and associated tool (including hands) can easily reach the desired location.	0	1.5	2.5	2.5	3.5	5.5	6.5	6.5	7.5
	Part and associated tool (including hands) cannot easily reach the desired location. Due to obstructed-access or restricted vision.	1	4	5	5	6	8	9	9	10
	Due to obstructed access and restricted vision.	2	5.5	6.5	6.5	7.5	9.5	10.5	10.5	11.5

☐ Part added but not secured

Manual insertion times depend on ease of handling parts and tools, need for holding parts down after insertion, and ease of alignment, positioning, and insertion.

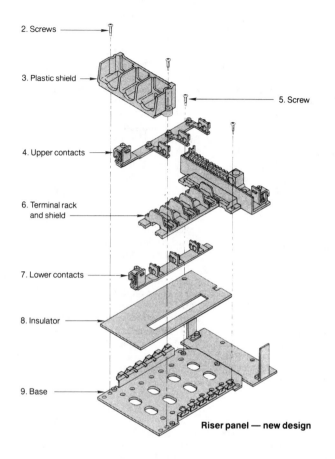

2. Screws

3. Plastic shield

5. Screw

4. Upper contacts

6. Terminal rack
   and shield

7. Lower contacts

8. Insulator

9. Base

Riser panel — new design

each hand and joined, and then placed in the assembly. One row on the worksheet must be completed for each part as it is added in turn to the assembly.

Reassembly is started with the part having the highest identification number, 14 for the old design of the riser panel. The base is inserted in the fixture, and the identification number entered in Column 1 of the worksheet. The operation is carried out once, so "1" is entered in Column 2.

A two-digit handling process code is found on a chart, which also shows an estimated time for manual handling. The process code is entered in Column 3 of the worksheet and the corresponding handling time in Column 4.

The assembly code and insertion time are found from a second chart. The assembly code is entered in Column 5 and insertion time in Column 6. The total operation time in secoonds is calculated by adding the times in Columns 4 and 6 and multi-

The new design of the riser panel has only 10 parts compared to 18 in the old design, reducing assembly time by 84.6 s. Further reductions would be possible if the screws (Parts 5 and 2) could be replaced with integral fasteners.

Design for manual assembly worksheet

1	2	3	4	5	6	7	8	9	Name of assembly
Part I.D. no.	Number of times the operation is carried out consecutively	Two-digit manual handling code	Manual handling time per part	Two-digit manual insertion code	Manual insertion time per part	Operation time, sec (2) x [(4) + (6)]	Operation cost, ¢ 0.4 x (7)	Figures for estimation of theoretical minimum parts	Riser panel (new design)
9	1	30	1.95	00	1.5	3.45	1.38	1	Base
8	1	30	1.95	00	1.5	3.45	1.38	1	Insulator (> 2 mm)
7	1	30	1.95	00	1.5	3.45	1.38	1	Lower contacts
6	1	30	1.95	00	1.5	3.45	1.38	1	Terminal rack and shield
5	2	10	1.5	38	6.0	15.00	6.00	0	Screw (9 x 20 mm)
4	1	30	1.95	02	2.5	4.45	1.78	1	Upper contacts
3	1	30	1.95	00	1.5	3.45	1.38	1	Plastic shield
2	2	10	1.5	38	6.0	15.00	6.00	0	Screw (9 x 20 mm)
						51.7 T_m	20.68 C_m	6 N_m	Design efficiency $= \dfrac{3\,N_m}{T_m} =$ 0.35

plying by the number of repeat operations in Column 2. This result is entered in Column 7.

Total operation cost in cents is obtained by multiplying the operation time in Column 7 by 0.4 and entering the result in Column 8. The value of 0.4 ¢/s is reasonably representative for many companies, and if designs from different companies are to be compared it is useful to use such a standard value. However, a more accurate value for a given company can be used in place of 0.4.

Probably the most critical entry on the worksheet is the theoretical minimum number of parts, entered in Column 9. Three criteria determine this number:

1. Must this part be separate because it moves with respect to all other parts already assembled? (Sometimes parts can be combined by manufacturing them from flexible material, so that limited relative motion is possible.)

2. Must this part be made of a different material than or be isolated from all other parts already assembled? (Only fundamental reasons concerned with material properties may be considered here.)

3. Must this part be separate from all other parts already assembled because necessary assembly or disassembly would otherwise be impossible? (Fasteners are rarely counted as essentially separate parts because integral fasteners can be employed, at least in theory.)

If the answer to any of these questions is yes, a "1" is placed in Column 9 unless multiple identical operations are indicated in Column 2. In that case, the number of parts that must be separate is placed in Column 9.

In the riser panel, the base (Part 14) must be separate because it is the first part in the assembly. The lower insulator (Part 13) must be separate because it insulates the metal base from the metal contacts, satis-

fying the second criterion.

The upper insulator (Part 12), guard (Part 11), plastic feet on the terminal rack (Part 10), three shields (Part 5), and the plastic clamp (Part 8) do not move and need not be made of a different material than the lower insulator (Part 13). Hence a "0" has been placed in Column 9 of the worksheet for all these parts.

The lower contacts (Part 6) and the upper contacts (Part 4) must be separate for isolation purposes, and the plastic shield (Part 3) must be separate for reasons of assembly. Finally, the screws need not be separate; integral fasteners could be employed.

Redesign

The design analysis and the data it generates provide useful design information in two areas. First, the criteria for separate parts point out where the number of parts can be reduced. In many instances parts cannot be eliminated because of other constraints, such as the economics of manufacture or unavailability of specialized equipment needed to manufacture the combined parts. However, the areas for possible improvement are clearly spelled out in Column 9 of the worksheet.

Second, the areas for improvement of handling and assembly can be seen by reviewing the figures in Columns 4 and 6. Any operations resulting in excessive times should be examined critically.

In the riser panel, the theoretical minimum number of parts is six and the actual number in the old design is 18. Column 9 of the worksheet indicated that all the plastic parts except the plastic shield (Part 3) could theoretically be manufactured as one component, integral with the plastic feet on the terminal rack. However, the lower contacts must be sandwiched between the insulator and the three shields, so the insulator that results from combining Parts 12 and 13 must be

separate for reasons of assembly. The three shields, the plastic feet on the terminal rack, the guard, and the plastic clamp can be combined. This combination also eliminates the need for the two hex screws (Part 9). Because integral fasteners are not feasible for this assembly, the number of parts can be reduced from the original 18 to 10.

The completed worksheet for the new design shows that the estimated manual assembly time is 51.7 s, less than half the original time of 136.3 s. The theoretical minimum number of parts is 6 in both old and new designs, so design efficiency is increased from 13 to 35% and estimated assembly cost is reduced from 54.5 to 20.7¢.

Because the analysis procedure must use standard times and costs to produce comparable data for different designs, the figures it produces do not necessarily reflect actual industrial experience. For example . the analysis assumes that parts are added to the assembly one at a time. This assumption is valid for assembly lines where workers add only one part at each station, but for bench assembly and on most assembly lines, workers often handle two parts simultaneously. This practice reduces overall assembly time by about one-third. Thus, a more accurate bench assembly time is obtained if the theoretical time is divided by 1.5. This procedure does not affect design efficiency because "ideal" assembly time is reduced in the same proportion.

The analysis also assumes that parts are presented in bulk and randomly oriented. However, some parts are available in magazines or special containers. If accurate estimates of assembly time are needed and appropriate data are available, they can be included in the analysis. MD

This article is largely based on the *Design for Assembly Handbook*, which has been developed by Boothroyd and Dewhurst at the University of Massachusetts. The handbook is available from Professor Boothroyd, Automatic Assembly Program, Department of Mechanical Engineering, Univ. of Mass., Amherst, MA 01003, (413) 545-0054.

CHAPTER 9

Design for Graphics and Packaging

Graphic design is essentially a matter of organizing visual elements on a flat plane or curved surface; its function is to convey a message or a mood to the beholder. This is true whether the item is an outdoor advertising billboard, a book cover, or a toothpaste carton. By way of example, a package is a three-dimensional construction that contains a product and to which graphics are applied. The graphic display on the successful package identifies the product, and often visually depicts the contents in order to entice a potential purchaser. A selection of familiar products in Figure 9.1 exemplifies good package graphic design. Because the package influences product choice at the point of sale, quality, consistency, and integrity in graphic design and printing is of paramount importance.

Problems in graphics can be approached as one would approach any problem in design. The *function* of a graphics work is simply that it must communicate some message effectively. In addition, packages must house their contents safely and economically, and enhance storage and display. The *material* requirement is important, for the graphics media are varied indeed. One can choose from substrates such as paper, paperboard, plastic, or metal in many colors, textures, and configurations. (See Figure 9.2.) Processes enter in and can impose certain restrictions on the work of the designer. Gravure, flexography, and lithography are among the several graphic reproduction methods with potentials and limitations to be accounted for by knowledgeable designers. Also significant are factors related to numbers of product; types of substrates; characteristics of inks, package shapes, and materials; and environmental impact. Additionally, federal legislation regarding package labels can impose certain limitations on the designer who must comply with the required product

Figure 9.1 Polypropylene packaging materials are attractive, safe, durable, and have many product applications. Courtesy of Amoco Chemical Company.

information and warnings. The *visual* aspect is of special significance, for only through a sensitive organization of elements—balanced shapes, good proportions, proper type faces, emphasis through contrast or bold colors—can the communication succeed.

All these requirements must be addressed in order to produce the sensitive design. Graphic design can be divided into a number of categories, each identifying a

Figure 9.2 Metalized and clear plastic films lend themselves to striking package graphics. Courtesy of Amoco Chemical Company.

specific area of attention. *Packaging design* is concerned with both the carton and the information displayed on it. *Advertising design*—such as posters, magazine and newspaper advertisements, banners, and billboards—is essentially a two-dimensional expression of a product promotion. *Display design* involves the creation of three-dimensional structures that exhibit and describe specific products. *Fabric design* includes wall coverings, wrapping papers, and textiles. And finally, there is an omnibus category that includes such items as book covers, greeting cards, printing on candies and vitamin tablets, and postage stamps.

The precepts of design for manufacturing also must be considered by graphic designers. The creation of superb graphics for a package may be so difficult to apply to it as to render this creative effort worthless. Graphic reproduction technologists must be consulted by graphic designers in the same manner as production engineers are by artifact designers. A package design that is difficult to produce carries the same economic constraints as does a chair that is difficult to produce. In both situations, it is imperative that designers and makers work closely early in the design process.

GRAPHIC REPRODUCTION TECHNOLOGY

Printing technology is a vast and complex field that encompasses a range of reproduction methods. Modern technology makes considerable use of computers in designing packages and their graphic layouts, and in planning and monitoring the printing process. The theory of printing, however, is simple: (1) an image is prepared on a printing surface called a plate; (2) ink is applied to the plate; and (3) the substrate is pressed against the printing surface to receive the inked image. This cycle can be repeated until the required number of impressions have been obtained. Printing is applicable to hundreds of different products, ranging from packages, bottles and cartons, to greeting cards, newspapers, and posters.

The primary printing methods currently in use, along with their approximate distributions, are offset lithography (55 percent), gravure (18 percent), flexography (17 percent), screen printing (6 percent) and letterpress (4 percent). Note that these methods differ in the way that the image is carried to the print surface. (See Figure 9.3.) With the *planographic* (lithography) method, printing occurs from a plane surface; *intaglio* (gravure) uses a depressed surface for the image; *porous* (screen) employs a mesh stencil; and *relief* (flexography and letterpress) features raised printing areas. There also are many electrostatic and other office duplicating methods that are but modifications of the primary systems. Descriptions of these printing processes follow.

Offset Lithography

Lithography, a form of planographic printing, is based on the principle that oil and moisture do not mix. The term *offset* derives from the fact that the plate cylinder does

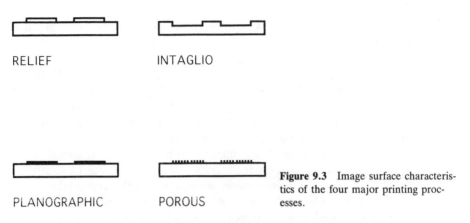

RELIEF   INTAGLIO

PLANOGRAPHIC   POROUS

Figure 9.3 Image surface characteristics of the four major printing processes.

not contact the paper, but instead transfers (offsets) the tacky ink image to an intermediate rubber blanket cylinder. (See Figure 9.4.) The process typically employs a photographic image etched on the plate cylinder. In operation, the ink travels via a series of ink rolls to the plate cylinder. At the same time, separate rolls carry and deposit moisture on the plate cylinder. The plate accepts ink and repels moisture in the etched image areas, and attracts moisture and repels ink in the nonimage areas. The plate image next offsets to the blanket cylinder where it transfers to the substrate by mutual rotary contact over an impression roller. One reason for the popularity of lithography is that the compressed blanket cylinder allows for a clearer impression on a wide variety of smooth and textured surfaces. It is commonly used for printing

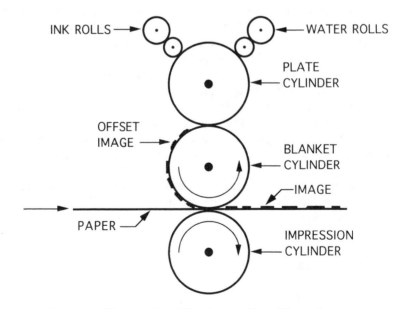

Figure 9.4 The essentials of the planographic or lithography process.

Figure 9.5 Graphics such as this striking book layout are products of lithographic printing. Design and Style is a paper promotion produced by Mohawk Paper Mills, Inc. Editor: Steven Heller; Designer: Seymour Chwast.

newspapers, advertising inserts, catalogs, business forms, books, and magazines. (See Figure 9.5.)

Gravure Printing

The gravure technique, called *intaglio*, a word implying a countersunk incision, involves direct printing from a depressed surface image engraved with a diamond stylus into a copper-plated cylinder. The gravure cylinder is rotated in a pan of ink to

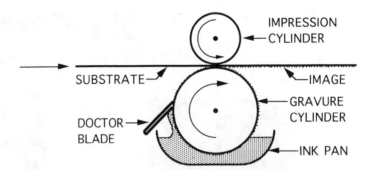

Figure 9.6 The essentials of the intaglio or gravure process.

fill the recessed crevices. The excess ink is wiped from the cylinder by a flexible steel doctor blade, leaving ink only in the etched depressions. This imprint then transfers to the print surface by a heavy rotary contact between the gravure and impression cylinders. (See Figure 9.6.) The gravure method is used to print on paper sheets or rolls (hence, rotogravure), the latter being the most common. Gravure notably reproduces color pictures of exceptional quality, but because of high cylinder costs, it is restricted to large volume runs, such as the familiar Sunday newspaper advertising inserts. Other products include glossy catalogs and magazines, plastic laminate paper sheets, wallpaper, labels, and foil bags. Commemorative postage stamps honor the people, arts, crafts, animals, birds, plants, and events of nations, and quality products of gravure printing. (See Figure 9.7.)

Gravure printing permits fine detail and excellent ink coverage because of the

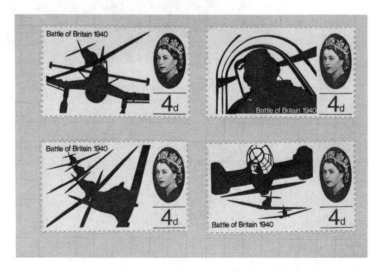

Figure 9.7 Postage stamps such as these attractive symbolic examples are typical of gravure printing. Reproduced by permission of Royal Mail.

extremely tiny engraved, three-dimensional cells, up to 22,000 per square inch. The depth of these recesses determines the quality and intensity of the color. Shallow cells result in less intense ink deposition, while deeper cells provide thicker ink deposits and better opacity. This variation of cell size and depth makes possible precise replication of continuous tone photography—excellent but expensive.

Flexography

Flexography is special rotary web relief printing method. It is an inexpensive and simple reproduction process whose use is steadily increasing, largely because of its applicability to package printing. By this process, a rubber fountain roll picks up ink from a pan and passes it to a steel anilox or metering roll. The anilox roll delivers a controlled ink film to the plate cylinder, where the image is transferred to the substrate by the impression cylinder. (See Figure 9.8.) Flexible rubber plates and quick-drying water-based inks are used, which save energy and are more environmentally benign. Flexographic printing is used for decorative wrappings and tissues, foil and plastic bags, milk cartons and boxes, and bar code and copy labels. (See Figures 9.1 and 9.2.) Most paperback books and many newspapers also use this technique. High-quality, high-speed printing is possible with this method, as shown in Figure 9.9.

Screen Printing

Screen printing is a kind of stenciling where an insoluble woven mesh partially covered with a stencil design is stretched over the surface to be decorated. A moving

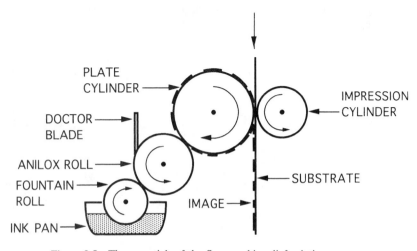

Figure 9.8 The essentials of the flexographic relief printing process.

Figure 9.9 A flexographic rotary web imprinting maching. Courtesy of Adolf Gottscho, Inc.

rubber squeegee forces ink through the fine mesh openings and onto the print surface. (See Figure 9.10.) With this method, products of many sizes and shapes—including bottles, cartons, decals, sheets, specialty items, and fabrics—can be decorated economically and with good results. The familiar and very user-friendly battery tester is a screen-printed product. The innovative tester strip (Figure 9.11) utilizes temperature-sensitive, color-changing inks to visually indicate battery strength. Screen printing also is particularly adaptable to large, multicolored, intricate designs,

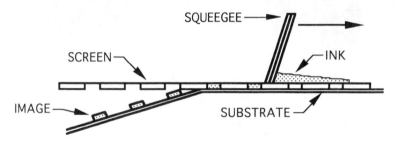

Figure 9.10 The essentials of the porous or screen process.

Figure 9.11 This familiar Copper Top flashlight battery tester strip is screen printed by the Murfin Division of the Menasha Corporation. Copper Top is a registered trademark of Duracell Inc.

such as the graphics on the sides of semi-trailers and the striking banners in Figure 9.12. Manual screen techniques are capable of producing pieces at the rate of 30 to 60 per hour. (See Figure 9.13.) If higher production is required, semi-automatic methods can be used. With automatic machinery, it is possible to produce up to 5,000 parts per hour.

Figure 9.12 These attractive banners were screen printed. Courtesy of Kalamazoo Banner Works, Inc.

Figure 9.13 This large banner is being manually screen printed. Courtesy of Kalamazoo Banner Works, Inc.

This form of printing is often referred to as *silk screening* or *serigraphy* because early work was done with screens made of silk. Modern screens are made of synthetic fabrics such as nylon, and stainless steel mesh is used for long runs or when screens are subjected to abuse. Fine image definition is achieved by using fine meshed screens. Photomechanical stencil-making techniques generally are employed, but hand-cut ones also are used on larger pieces.

An interesting example of stencil printing is found with the familiar gasoline-dispensing pumps. These pumps are subject to considerable user abuse, and also become unsightly through normal weathering. They deteriorate to the point that the lettering on the pump displays are unreadable. As a result, the sheet metal graphic panels either must be removed for restoration or replaced entirely about every five years. To avoid this problem, one manufacturer has converted to a transparent composite film to make display panels for its electronically controlled gasoline dispensers. (See Figure 9.14.) The thin glass-clear film is screen printed on the reverse side, so that the panel graphics are not exposed to harmful sunlight or caustic chemicals. The composite film combines the toughness, durability, and impact strength of polycarbonate, with the chemical, abrasion, and ultraviolet resistance of polyvinyl chloride. The two films are securely bonded by proprietary adhesion and extrusion technology to produce the rugged film.

Figure 9.14   A special two-film plastic composite was used to create this durable and very readable screen-printed gasoline pump display panel. Courtesy of Gilbarco, Inc.

Letterpress Printing

Letterpress printing is a direct ink transfer relief method where the image areas are raised above the nonprinting surface. Anyone who has carved a linoleum block to produce a printed greeting card has practiced this form of printing. Woodcuts also are examples of relief work. Both flatbed and rotary kinds of presses can be used. Ink applied to the raised carrier areas of the image cylinder is transferred to the substrate by the squeeze pressure of the impression cylinder. (See Figure 9.15.) Because letterpress is the only method where printing is done directly from type, it is a very usable process in that on-the-job changes are easily accomplished. This feature makes it especially economical for price and parts lists, as well as schedules and time

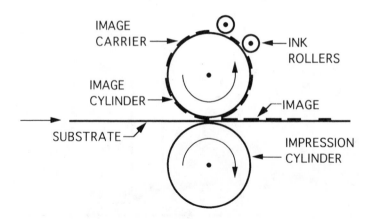

Figure 9.15 The essentials of the relief or letterpress process.

tables that are subject to regular change. The type plate can be kept standing and modified as necessary. Individual metal or wood letters also are hand-set for artistic prints. The process is used in magazine and newspaper printing with extensive production runs, corrugated and folding boxes, grocery bags, labels, and "blister" pack cards. A rubber wrap-around plate is shown in use in Figure 9.16. Letterpress has declined in popularity due to the emergence of other printing methods of higher quality and efficiency.

There are several other image reproduction processes available to the graphics designer. *Ink-jet printing* is a noncontact plateless method where charged, electrostatically directed ink droplets are sprayed to form characters on a carton. (See Figure 9.17.) The inks can adhere to recessed, nonabsorbent surfaces at high speeds and low costs. This method is mainly used in the beverage industry. *Laser printing* is another noncontact system where a computer-controlled laser beam produces an image pattern on a material. The laser beam prints by changing the color of the material or of the ink pigment, or by burning off a precoating. (See Figure 9.18.)

A process known as *sublimation printing* is used to make keytops for computer and typewriter keyboards, as shown in Figure 9.19. This is a heat transfer system based on the technology by which images are produced for polyester textiles and carpets. The process utilizes solid dye crystals on a film transfer, which convert into gas under heat and pressure. The gas penetrates into the polyester keytop material and then reconstitutes as a dye to produce the visible characters. (See Figure 9.20.) The result is a durable product with high visibility. *Hot stamp printing* is used to apply crisp, legible, smudge-proof images to any flexible material, and is widely used in package imprinting and labeling. (See Figure 9.21.) Hot stamping is often called *foil printing* because it uses dry inks bonded to a thin carrier film. When the foil is sandwiched between heated metal type and the object to be printed, the ink is transferred to the object. This process was used to produce the sharp graphics on the utility knives in Figure 9.22.

Figure 9.16   This carton imprinter is an example of a letterpress printing machine. Courtesy of Adolf Gottscho, Inc.

Pad printing is an indirect, contact gravure process, where ink is deposited on a photo-etched flat metal plate (called a *cliché*) to fill the image depressions, and a doctor blade removes excess ink from the nonetched plate areas. The ink-filled portions of the plate now represent the print image. The ink is picked up by a flexible silicone pad, the pad is positioned over the package to be printed, and then is pressed against the package to transfer the image. (See Figure 9.23.) The soft pad, which is the equivalent of a printing plate, will conform to almost any shape and can print close to edges and corners, making pad printing the method of choice for imprinting contoured containers. This method originated in Europe to imprint watch faces.

The broad range of measuring tools—including scales, rules, micrometers, protractors, and calipers—require accurate, durable, and easily readable markings. Such markings are engraved, etched, or imprinted, depending on the tool material (i.e., plastic, stainless steel, or wood.) In the micrometer in Figure 9.24, the corporate name, tool number, and button controls are printed on separate metal tabs that are permanently adhered to the tool. The graduations on the barrel are photo-engraved.

The familiar membrane or touch switches are used on the control panels of

Figure 9.17 This ink-jet print system is applying carton graphics. Courtesy of Diagraph Corporation, St. Louis, MO.

Figure 9.18 This is one of several types of laser printers. Courtesy of Diagraph Corporation, St. Louis, MO.

Figure 9.19 Computer keyboard keytops must withstand continual tactile abuse.
Courtesy of Hoechst Celanese Plastics Ltd.

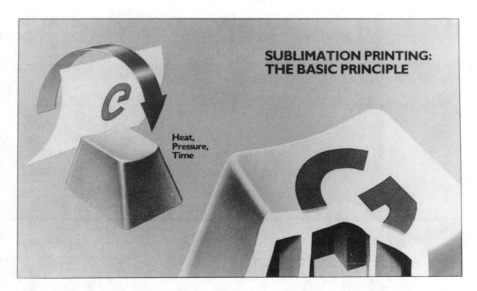

Figure 9.20   An illustration of the sublimation printing process. Courtesy of Hoe-chst Celanese Plastics Ltd.

many electronic devices and production machines. To produce this switch, two flexible polyester sheets are screen printed with electrically conductive ink to provide the switch circuit. A plastic spacer, die-cut at the contact points, is sandwiched between the two printed circuit sheets. The die-cut holes allow a touch of the finger to close the contacts to complete the circuit and cause some function to occur. (See Figure 9.25.) A tough colorful plastic overlay is added so the user can identify the different switch operations.

Every computer operator uses some type of printer to reproduce the output of the computer, be it words or pictures. Dot matrix, ink-jet, and laser printers are common. A *pen-plotter* is often used to produce high-quality, high-speed, single- or multi-color drawings. (See Figure 9.26.) The model shown comes with an eight-pen turret with the capability of mixing liquid ink, liquid ball, or fiber pens for plotting on roll or sheet materials. Other models feature direct imaging technology, where the imaging process occurs entirely in the thermosensitive media the device plots on. No pens are used.

To summarize this treatment of graphic reproduction technology, remember that its basic purpose is to decorate or graphically enhance objects or to supply written information using a range of printing processes. In general, metals are lithographed; glass bottles are screen-printed; plastic containers are imprinted by hot stamping, pad and screen printing, lithographing, and sublimation or heat transfer printing; and paper cartons are letterpressed, lithographed, or ink-jet printed. And separate printed labels can be used on all containers and packages. (See Figure 9.27.)

Figure 9.21 A hot stamp or foil imprinter. Courtesy of Adolf Gottscho, Inc.

Figure 9.22 The lettering on these soft-grip knives was applied by hot stamp printing. Courtesy of Hunt Manufacturing Co.

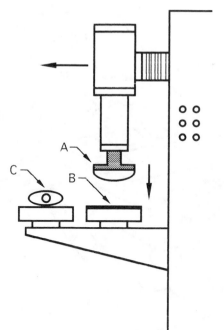

Figure 9.23 Pad printing is especially applicable to imprinting contoured containers, as illustrated here. The pad (A) picks up the ink image from the plate (B) and deposits it on the package (C).

Figure 9.24 Measuring instruments have many graphics requirements. This lightweight and readable digital electronic micrometer is a classic example. Courtesy of The L. S. Starrett Company.

The problem here is obvious: Which process should be used for any given product application? The answer is that process selection is based on cost, size of printing run, type of material, type of container, and the visual requirements, among others. Most product designers are not graphics technology specialists, and so they should very prudently consult one before they proceed too far with their graphic design efforts.

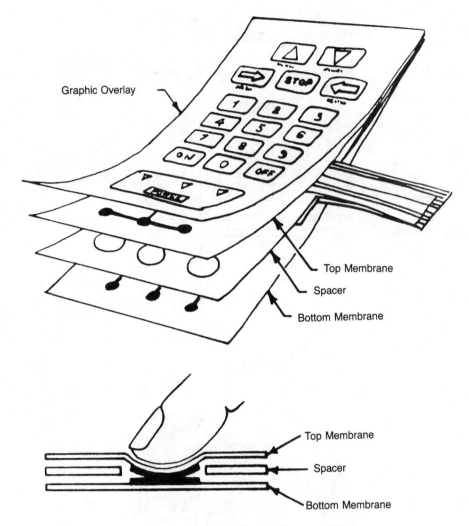

Graphic Overlay

Top Membrane

Spacer

Bottom Membrane

Top Membrane

Spacer

Bottom Membrane

Figure 9.25 Membrane switches are comprised of screen-printed plastic sheets and spacers. As shown in the inset, the finger depresses the top membrane to contact the bottom membrane, thereby completing the electric circuit. Courtesy of Memtron Technologies, Inc.

GRAPHIC DESIGN

As suggested at the beginning of this chapter, the challenge of the graphic designer is to present information in an effective and visually pleasing manner. In the next few pages, the reader will be able to follow the creative efforts of designer Patricia Hennessy at work preparing a new calendar concept.

Figure 9.26 A typical pen plotter printer used in computer graphics printing applications. Courtesy of CalComp.

For 1/3 Wrap Around

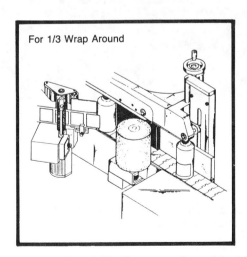

For Full Wrap Around

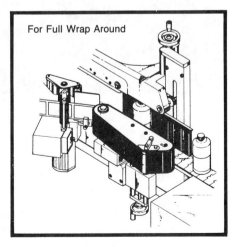

Figure 9.27 Two systems for applying labels to containers. Courtesy of KK Power Impressions, Inc.

January

Sunday		7	14	21	28
Monday	1	8	15	22	29
Tuesday	2	9	16	23	30
Wednesday	3	10	17	24	
Thursday	4	11	18	25	
Friday	5	12	19	26	
Saturday	6	13	20	27	

Page 390: Basic type arrangement
Page 391: Basic horizontal, vertical, and diagonal grids
Page 392: Basic horizontal, vertical, and diagonal grids superimposed
Page 393: Removal of grid parts activating the negative space
Pages 394–395: First full calendar year
Pages 396–397: Second full calendar year
Pages 398–399: Third full calendar year
Pages 400–401: Integration of circle/sphere shape

This project documents a design process involving the calendar. A calendar system defines the days and months of the year. Used to organize and communicate a sense of time visually as well as conceptually, a calendar is based on a grid structure. Printed calendars, as we know them today, are generally composed of images separated from the calendar grid. In this calendar, the grid itself becomes the image. Documented here are examples representing the many possible solutions to the challenge of designing the calendar grid as image.

After thorough sketches, one type arrangement became basic (with variations limited to locations of the numbers of days as they change in individual months of the year). This typographic constant allows for stronger visual solutions to the image/grid design problem. Several basic grids combine with each other to form various starting points. The variations of the grids are created simply by taking away parts in a clear systematic manner. The image created by the absence of parts becomes as important as the image created by the grid itself. The development of figure/ground relationships result from the use of the negative space in many of the solutions.

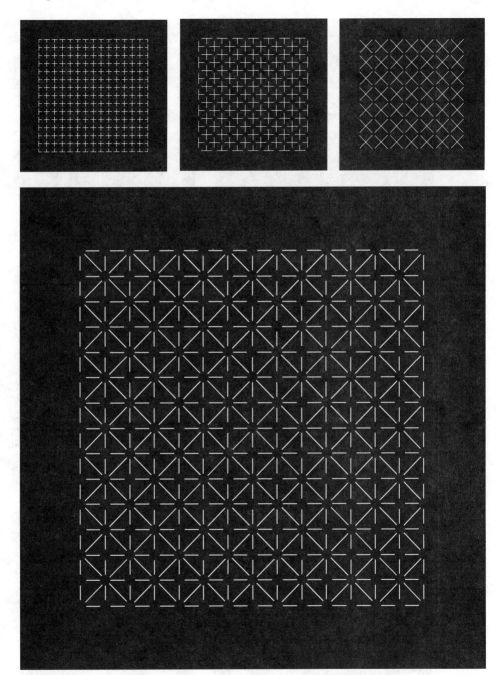

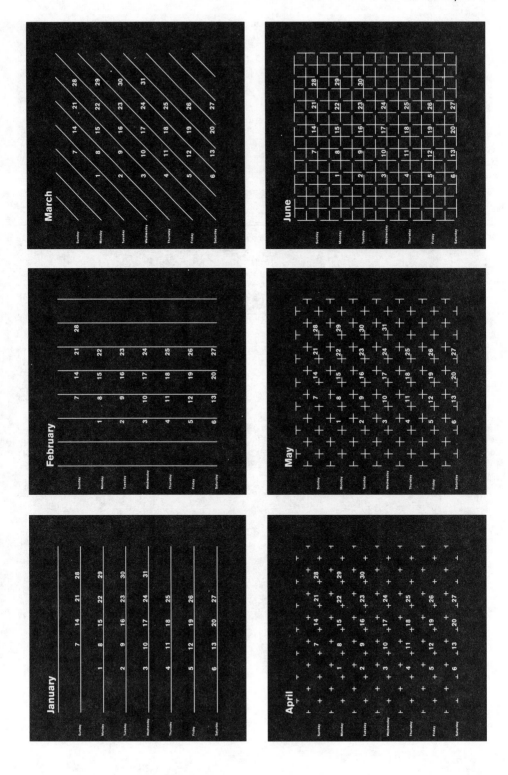

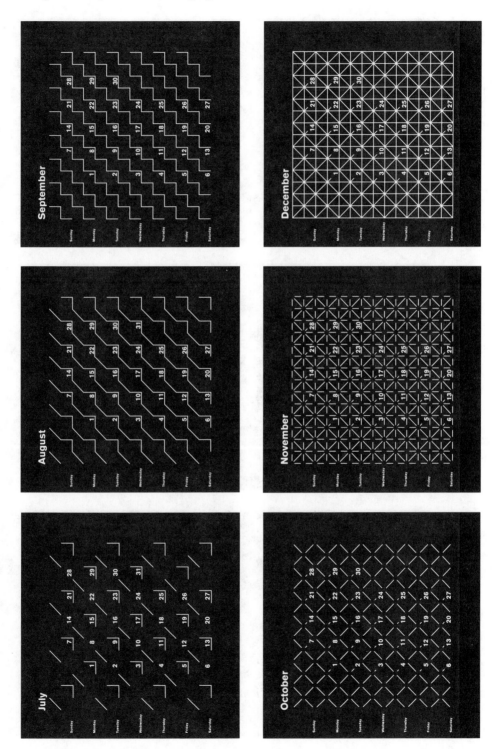

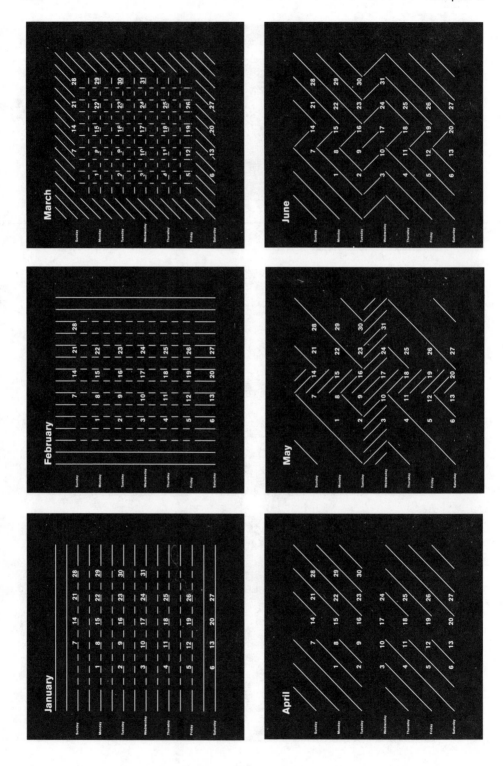

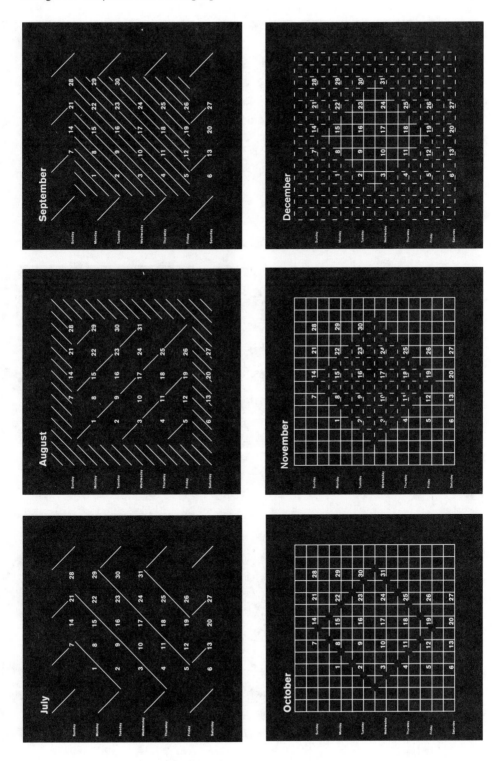

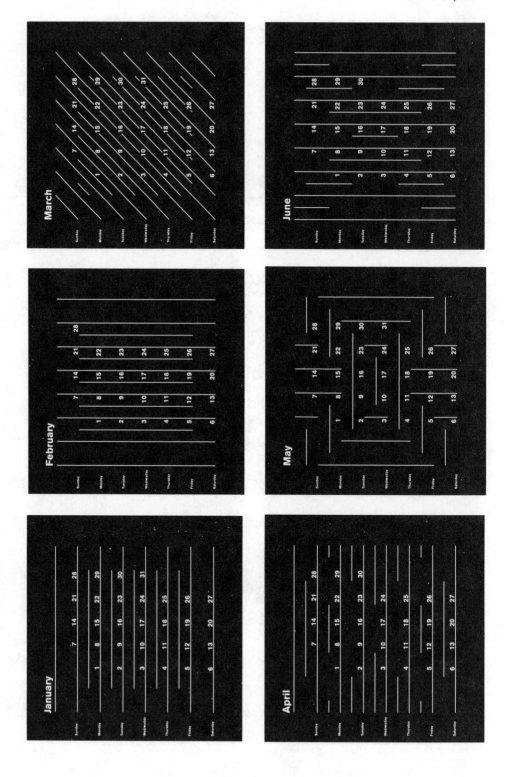

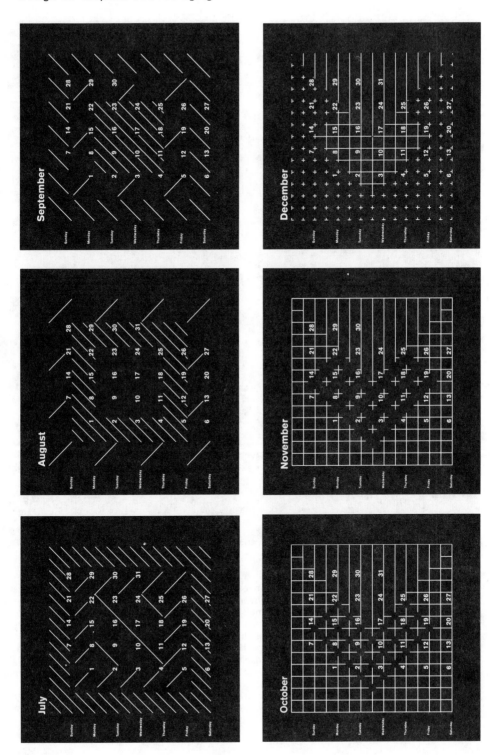

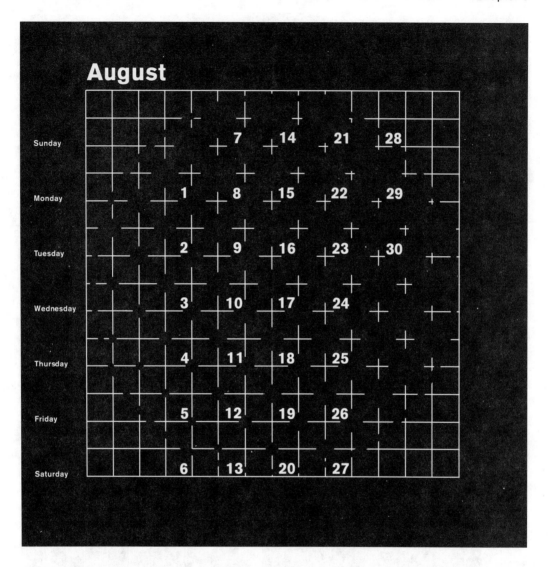

In the design process the first grid investigations began separate from the typographic calendar investigations. As the visual design investigations progressed, common ground became evident in both systematic structures. It seemed both logical and interesting to bring them together. The study of systems, series, and sequence is evident in the visual search. Clarity in progression of various levels of communication was the main foci in the study.

Ideas grew from the search as specific criteria and limitations became evident. The design concept pushed the limitations for results with evidence of discovery and elements of surprise. The results vary from the most simplistic, straightforward, and even obvious to the more complex, experimental, playful, subtle, and intellectual. These ideas combine to give the calendar a person-ality of its own. The initial investigations and solutions were formulated using a type arrangement printed letterpress and all grid studies were hand drawn. The working method has changed and is now completed on the Macintosh computer.

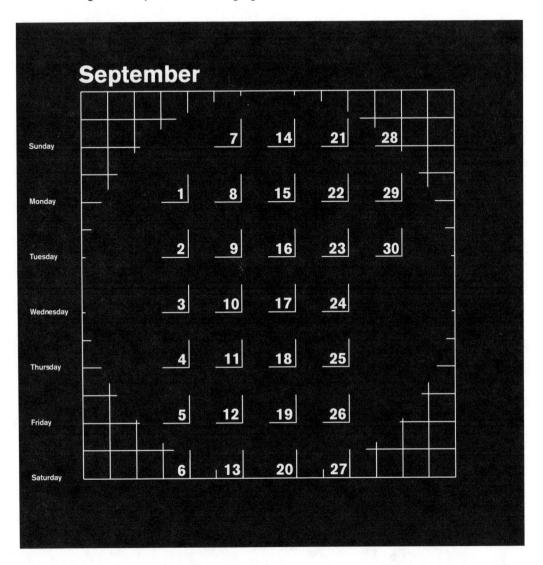

September

Sunday		7	14	21	28
Monday	1	8	15	22	29
Tuesday	2	9	16	23	30
Wednesday	3	10	17	24	
Thursday	4	11	18	25	
Friday	5	12	19	26	
Saturday	6	13	20	27	

The documentation of this work, which was partially completed at the Basle School of Design, Switzerland, under the direction of Wolfgang Weingart, was made possible by a College of Fine Arts Faculty Development Grant, Western Michigan University. Copy editor: Michael Barrett   Production assistant: Michael Kohnke.

It should be mentioned here that graphic designers regularly make use of symbols in their work. The familiar road signs are a good example. (See Figure 9.28.) These are wordless pictorial messages and are effective communicators. They save space, as with the many symbolic messages on the instrument panel of an automobile. The various buttons for operating the windshield washers and wipers, horn, lights, defroster, heater, and air conditioner are good examples, as are the indicator lights for the battery, fuel level, engine temperature, and oil pressure. They provide needed information in a very limited amount of space. Such symbols are becoming international, so that U.S.-made cars can be sold and used abroad.

Product and corporate identification problems are challenging tasks for graphic designers. The packages and the familiar trademarks or logotypes are examples of efforts by organizations to promote a consumer awareness of a product. It is important that this facet of marketing keep pace with product changes and with graphics style changes, as illustrated in the sequence of logotype changes for a pharmaceutical company. (See Figure 9.29.)

One of the early selling points of the tablets manufactured by this company was that they could be reduced to powder under the thumb. This was important then, for a great number of the first pills and tablets prepared for human consumption were so hard that they could be driven into a pine board with a hammer. Obviously, these did not dissolve in the stomach and were of little value as a medication. The Upjohn Company developed what they called a "friable pill," which meant simply that it could be easily crushed and was therefore readily soluble when taken internally. This

Figure 9.28 Symbols convey messages without using words. These shown could be easily understood if the word messages were erased.

was an important aspect of the company's beginning and was therefore incorporated into their trademark.

A second trademark introduced in 1907 was merely a variation of the original, and the one developed in 1913 was simpler, with all the shading omitted. In 1938, it was simplified still more and the emphasis moved from noting the friability of the product to the company name, Upjohn. The two following logotypes featured only the company name in a new typeface with no visual references to the friable pill.

It is evident that the modifications of this logotype have been characterized by a move toward simplicity because its applications have changed. Originally, the mark was affixed to boxes and packages containing the various company products, and the details of the friable pill were visible. Later, the applications became more varied and ranged in size from large markings on company vehicles to smaller markings that appeared on packages and tablets. It was obvious that the more intricate earlier logotypes were unsatisfactory, whereas the current rendition meets all of the corporate identification and advertising needs. Another reason for simplification was a desire to have the public become aware of the company name first and foremost. Upjohn was no longer only a manufacturer of pills and tablets, but had moved into a broader area of pharmaceutical manufacturing and had a logotype that more accurately reflected its products and services.

Advertising literature provides graphic designers with challenging opportunities to portray products visually and to provide the necessary descriptive information. The strikingly simple and direct brochure cover for the Ford Taurus automobile (Figure 9.30) is a notable example. It attracts attention and invites the readers to investigate the booklet contents. Inside, the pages feature high-quality visuals and just enough data to satisfy the readers and hold their interest. The layout of the page shown in Figure 9.31 offers clear illustrations and explanations of some of the notable automotive details.

APPLIED GRAPHICS

An extensive discussion of packaging and the environment was presented in Chapter 6. Such issues as excessive packaging, recyclability, and energy were examined, and further examples are described here.

Container production technology is a challenging and interesting field, and graphics designers should be familiar with some of its methods in order to simplify their work. Plastic bottles generally are formed and filled and in a series of line operations utilizing sophisticated and inventive equipment. Shown in Figure 9.32 is a diagram of such a process, where a plastic parison is blow-molded to form a container, filled with a metered dose of some medical liquid, and then sealed, all under the most sanitary conditions. This is one of many similar systems. Soft-drink plastic and glass bottles are formed, filled, capped, and then labeled. In-mold labeling techniques can be used to eliminate the separate labeling operation for some plastic bottles. Here, a paper label coated with a heat-activated adhesive is placed in the

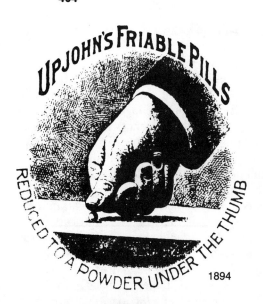

1894

1907

1913

1938

1945

Present

Figure 9.29 The evolution of a corporate trademark. Courtesy of The Upjohn Company.

Figure 9.30 The clean, crisp layout of this Taurus automobile advertising brochure cover exemplifies good graphic design. Courtesy of Ford Motor Company.

mold at the start of the blow-molding cycle. The parison is inflated against the label, the heat activates the adhesive, and the label sticks to the plastic bottle.

Modern plastic bottles and tubes are comprised of several plastic layers or barriers to provide a container that will not impart a undesirable taste to the contents. Toothpaste tubes are a good example, where the thin plastic shell is in fact a series of laminations, each of which has material qualities required to create the most satisfactory container. This was a problem with early bottles designed to hold potable substances—a problem that fortunately has been resolved.

Aluminum drink cans are deep-drawn, imprinted with colored graphics, sterilized, and then filled and end-capped. The can-making process is shown in Figure 9.33. Note that this high-speed method involves forming, cleaning, coating, and printing in a series of efficient, automated processes.

The wax-coated paperboard gabletop carton was developed in 1934 by the American Paper Bottle Company and was used exclusively for packaging fresh pasteurized milk. This was a significant achievement that provided a lightweight, shatter-proof container—an improvement over the common milk bottle. The wax

TAURUS EQUIPMENT

There are a lot of ways to enjoy the Taurus driving experience. It's just a matter of choosing and equipping the Taurus model that best suits your transportation needs and style.

There are three models, each with its own standard features list and at least one Preferred Equipment Package. Packages make ordering easy and offer a savings on select features.

Compared with the prices of the items if purchased separately, the savings may be equivalent to getting certain items at no extra charge.

To further personalize your Taurus, choose from the selection of appearance, comfort and convenience options that are available separately, some of which are shown here.

Preferred Equipment Package features are subject to change over the course of the year. For up-to-date information, see your Ford Dealer.

The optional power moonroof lets in the sun or the stars for a great open-air feeling.

Above Left: The rear window wiper washer is included on LX wagons with package 208A.

Included on wagons with the LX Preferred Equipment Package is the "picnic table" load floor extension.

The optional compact disc player provides superb clarity and includes a convenient random play feature.

The Ford JBL Sound System option (sedans only) is Bi-amplified for a total of 145 watts of power.

Left to right: wheel cover (GL); deluxe wheel cover (GL with Preferred Equipment Package 204A); optional sparkle cast aluminum wheel (GL); sparkle cast aluminum wheel (LX); unidirectional sparkle spoke cast aluminum wheel (SHO).

SHO COLORS

Exterior Paint Colors	Interior Trim Colors		
	Opal Grey	Mocha	Black
Oxford White	▪	▪	▪
Royal Blue Clearcoat Metallic	▪		▪
Silver Clearcoat Metallic	▪		▪
Black Clearcoat (non-metallic)	▪	▪	▪
Deep Emerald Green Clearcoat Metallic	▪	▪	▪
Crimson Clearcoat (non-metallic)	▪	▪	▪

LX & GL COLORS

Exterior Paint Colors	Interior Trim Colors				
	Opal Grey	Crystal Blue	Mocha	Cranberry	Black(1)
Oxford White	▪	▪	▪	▪	▪
Royal Blue Clearcoat Metallic	▪	▪			▪
Silver Clearcoat Metallic	▪	▪		▪	▪
Black Clearcoat (non-metallic)	▪	▪	▪	▪	▪
Mocha Frost Clearcoat Metallic		✓	▪		▪
Medium Cranberry Clearcoat Metallic	▪			▪	▪
Dark Plum Clearcoat Metallic	▪		▪		▪
Crystal Blue Frost Clearcoat Metallic	▪	▪			▪
Caribbean Green Clearcoat Metallic	▪		▪		▪
Opal Grey Clearcoat Metallic	▪			▪	▪

(1) Not available on GL/LX wagon, or on GL without Preferred Equipment Package 204A.

Figure 9.31 This page from the Taurus brochure is an effective combination of graphics and copy. Courtesy of Ford Motor Company.

Extruding. The plastic parison, extruded from polymer, is accepted by the opened blow mould and cut below the die of the parison head.

Moulding. The main mould closes and simultaneously seals the bottom. The special mandrel unit settles onto the neck area and forms the parison into a container, using compressed air. Small containers are formed by vacuum.

Filling. By way of the special mandrel unit, the product, precisely measured by the dosing unit, is filled into the container.

Sealing. After the special mandrel unit retracts, the head mould closes and forms the required seal by vacuum.

Mould opening. With the opening of the blow mould, the container exits from the machine and the cycle repeats itself. Transfer for further processing is achieved by means of a conveying system.

Figure 9.32 An illustration of excellence in packaging technology. Courtesy of rommelag blow, fill, seal system.

How All-Aluminum Beverage Cans Are Made

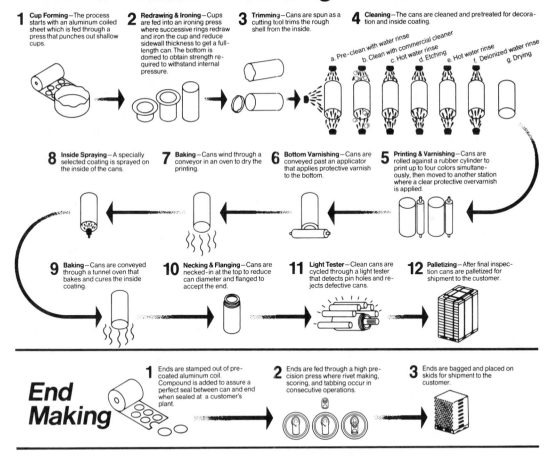

1 Cup Forming—The process starts with an aluminum coiled sheet which is fed through a press that punches out shallow cups.

2 Redrawing & Ironing—Cups are fed into an ironing press where successive rings redraw and iron the cup and reduce sidewall thickness to get a full-length can. The bottom is domed to obtain strength required to withstand internal pressure.

3 Trimming—Cans are spun as a cutting tool trims the rough shell from the inside.

4 Cleaning—The cans are cleaned and pretreated for decoration and inside coating.
 a. Pre-clean with water rinse
 b. Clean with commercial cleaner
 c. Hot water rinse
 d. Etching
 e. Hot water rinse
 f. Deionized water rinse
 g. Drying

8 Inside Spraying—A specially selected coating is sprayed on the inside of the cans.

7 Baking—Cans wind through a conveyor in an oven to dry the printing.

6 Bottom Varnishing—Cans are conveyed past an applicator that applies protective varnish to the bottom.

5 Printing & Varnishing—Cans are rolled against a rubber cylinder to print up to four colors simultaneously, then moved to another station where a clear protective overvarnish is applied.

9 Baking—Cans are conveyed through a tunnel oven that bakes and cures the inside coating.

10 Necking & Flanging—Cans are necked-in at the top to reduce can diameter and flanged to accept the end.

11 Light Tester—Clean cans are cycled through a light tester that detects pin holes and rejects defective cans.

12 Palletizing—After final inspection cans are palletized for shipment to the customer.

End Making

1 Ends are stamped out of pre-coated aluminum coil. Compound is added to assure a perfect seal between can and end when sealed at a customer's plant.

2 Ends are fed through a high precision press where rivet making, scoring, and tabbing occur in consecutive operations.

3 Ends are bagged and placed on skids for shipment to the customer.

Figure 9.33 Metal cans are formed and graphically imprinted as shown here. Courtesy of Reynolds Metals Company.

coating was replaced by a heat-sealable polyethylene lamination in 1961. Since then, this container has undergone a number of additional refinements such as foil and other high-barrier coatings. Also, the natural opacity of paperboard coatings protect the package contents from deterioration caused by direct exposure to light. They are of simple construction, which allows both ease of graphics printing and ease of forming and filling. (See Figure 9.34.) Aside from their traditional uses for dairy and juice products, these cartons hold wines, candies, dry cereals, frozen concentrates, and pet foods, as well as dry cements, plasters, and grouts. Some other advantages of this paper container include a tamper-evident design, built-in pour spout, rigidity, environmental acceptability, and ease of disposal.

Plastic toothpaste tubes are extruded sleeves with a head injection molded on one end to produce an open tube, which is then filled with dentifrice and crimped at the bottom to seal in the contents. The graphics are imprinted on the flat sleeve prior to heading. Plastic container materials are laminations of plastic, sealants, adhesives,

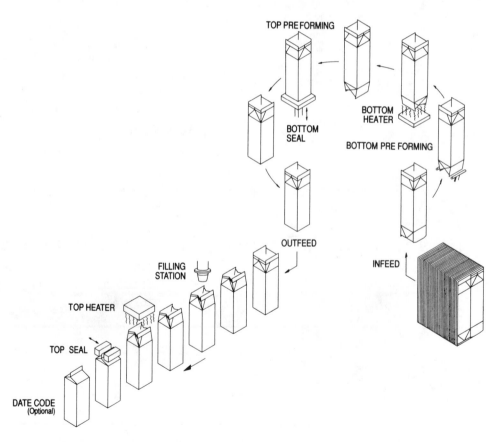

Figure 9.34 A diagram of the forming, filling, and sealing process for gabletop cartons. Courtesy of Nimco Corporation.

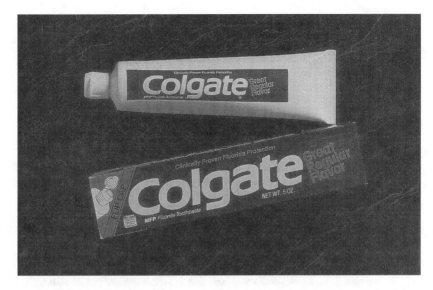

Figure 9.35 The regular Colgate toothpaste tube. Courtesy of Colgate-Palmolive Company.

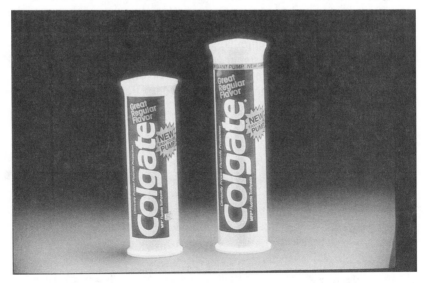

Figure 9.36 The Colgate Pump toothpaste container. Courtesy of Colgate-Palmolive Company.

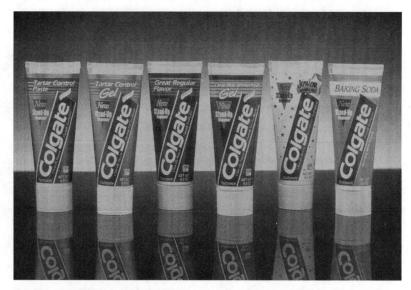

Figure 9.37 The Colgate Stand-Up toothpaste tube. Note the pleasing graphics on each of the types of Colgate packages. Courtesy of Colgate-Palmolive Company.

barriers, and covers, depending on the product it is to contain. Packages for edible products such as toothpastes, decorative cake icings, or condiments must have barrier properties to protect the contents from any objectionable tastes or odors that might be imparted to them. Glues and sealers may dissolve a plastic tube not otherwise protected from solvent action. As can be seen, each type of product has differing package requirements, and these special needs will affect the work of the package, the graphics designer, and the printer.

To elaborate, there are a number of different package styles available to contain and deliver the familiar tooth powders, liquids, and pastes. Powder packages

Figure 9.38 The handle lettering on this ergonomically correct adjustable wrench was imprinted as part of the forging process. It is economical and durable. Courtesy of AB Sandvik Bahco.

Figure 9.39 The graphics on this antique cast-iron medallion were part of the original casting pattern. Many current products have cast-in-place graphics.

must be designed to permit a measured amount of the dentifrice to be deposited on the brush by tapping or squeezing, without undue waste or mess. The opening in the tip of a container for tooth liquid cannot be so large that the user has trouble in controlling the drop size applied to the brush. Pastes are perhaps the most popular dentifrice form because of easier control and the fact that they cling to the bristles better.

The standard toothpaste tube, Figure 9.35, is probably the most common style of paste container. The example shown requires that the user remove the cap and gently squeeze to extrude a ribbon of dental cream onto the brush. As the tube empties, it is necessary to roll the bottom of the tube to force the toothpaste to the nozzle end. This tube should be packaged in a separate cardboard carton to protect the product.

A second type of paste container is the "pump-type" tube, where the paste ribbon is forced through the tip opening by thumb-pressing a plunger button located

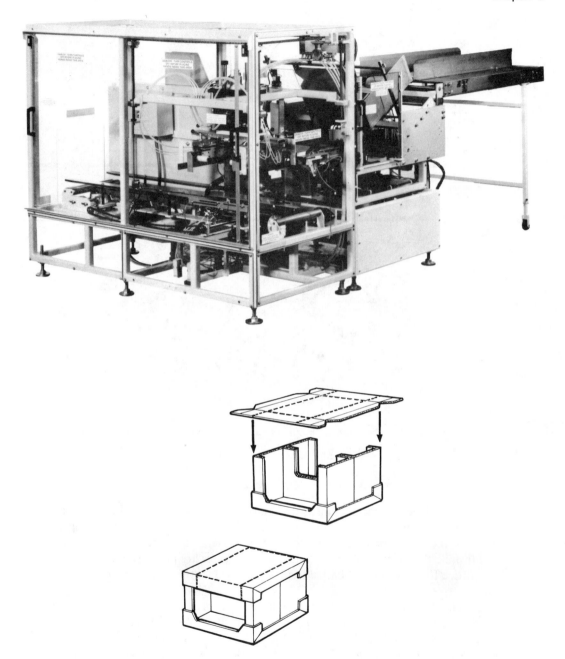

Figure 9.40   The machine illustrated here is used to form packaging cartons and lids. Courtesy
of Moen Industries.

at the top of the plastic cylinder. (See Figure 9.36.) This tube has the added convenience of easy shelving because it sits on its end.

A newer package style is the "stand-up" tube, as shown in Figure 9.37. This actually is an improved version of the standard tube, with the addition of an oversized cap to permit the tube to stand firmly on its cap in an upright position. This unique feature causes the paste to be pulled back into the tube by gravity so that it is always ready for easy use. The convenient cap can be either flipped up or unscrewed, and the

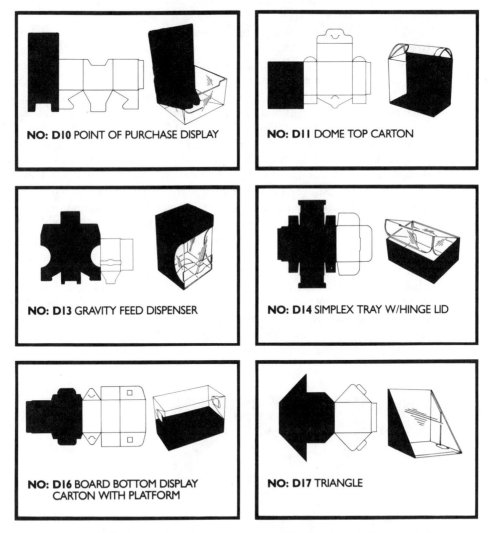

NO: D10 POINT OF PURCHASE DISPLAY

NO: D11 DOME TOP CARTON

NO: D13 GRAVITY FEED DISPENSER

NO: D14 SIMPLEX TRAY W/HINGE LID

NO: D16 BOARD BOTTOM DISPLAY CARTON WITH PLATFORM

NO: D17 TRIANGLE

Figure 9.41 Several designs for one-piece, combination clear plastic and paperboard cartons are shown in both the flat, stretchout and folded configurations. Courtesy of Klearfold.

tube squeezed for paste delivery. The package also has a view window along the side to indicate the amount of paste remaining in the tube. An additional attribute of this container, as well as the pump model, is that no other packaging is required, resulting in both environmental and economic savings.

All three of these plastic packages display striking red, white, and blue graphics—colorful, attractive, and easy to read. The flexible tubes generally are printed flat and then assembled and filled. The stiff pump cylinder graphics are typically printed on a flat transparent plastic film, and then wrapped around the cylinder to form a permanent sleeve. Although there are many other printing methods available, these are common and effective.

Metalworking technology makes available several processes with applications to graphics design. Wrenches and pliers often have identifying letters forged onto the tool handle. This is an efficient and economic method, where the graphics become part of the forging dies themselves. (See Figure 9.38.) In the example shown, the wrench jaw opening sizes are acid etched into the machined surfaces. Most hand tools have some identifying graphics printed by the common printing techniques. The antique cast-iron fire medallion (Figure 9.39) features a logo and letters that were part of the casting pattern. This device was affixed to buildings whose owners had purchased insurance and were therefore eligible for the services of the local fire department. Cast automotive and marine engine blocks frequently have graphics included on the casting. In a like manner, early beverage bottles had their graphics molded onto the container surface as part of the glass bottle forming process. Each of these examples illustrates a graphics application that is an inherent element of the forming process. They are efficient, economic, and require nothing extra.

Ingenious machines created to fold and assemble paperboard cartons are conceived in response to the requirements of carton construction. A designer studies a line of such cartons and plans an apparatus to cut, mark, fold, and glue, and perhaps imprint them in an efficient manner. (See Figure 9.40.) The equipment design engineer and the carton designer must work side by side in these efforts, for the carton should be designed for ease of producibility and assembly. It might be that a self-locking fold would be more economical than integral gluing, or that more

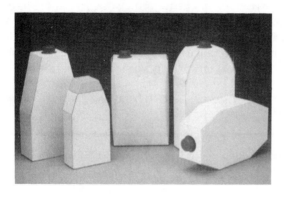

Figure 9.42 The Cebox Packaging Systems unique designs feature plastic film bags within a paper carton. These can contain powders or liquids. Courtesy of Jefferson Smurfit Corporation.

Figure 9.43 These clever dropper bottles feature convenient captive closures. Reprinted by permission of Nalge Company, a Subsidiary of Sybron Corp.

positive lidding could occur if the lid is formed to the box. The concurrency of design operations always will assure a more successful product.

Carton innovations can attract attention and enhance product quality image. The series of one-piece, transparent plastic and paperboard cartons in Figure 9.41 are attention getters—applicable to a range of special packaging requirements, easy to produce, and acceptable to almost any graphics scheme. With this product, the manufacturer also has produced a "green box," whose rigid plastic window material is made of at least 50 percent of recycled materials. This carton style is yet another example of the endless variety of package designs. The package in Figure 9.42 is another type of green box that combines a folded paper carton with a co-extruded plastic film bag to accommodate any liquid or powder product. When empty, the paper carton can be recycled and the bag disposed of (or recycled) without unduly taxing sanitary landfills.

The clever captive closure on the dropper bottle (Figure 9.43) resulted from a careful study of the problem of closures (or caps) for sterile liquids such as ophthalmic solutions. The term *captive* implies that the cap and tip stopper are a single unit. The one-piece, flip-top closure permits one-handed use for quick and easy access, eliminates the stocking of conventional tips and caps, allows convenient one-step capping on conventional production lines, and provides a leak-proof seal.

QUESTIONS AND ACTIVITIES

1. Describe the functional, material, and visual requirements as they pertain to graphic design.
2. Write descriptions of the terms *planographic, intaglio, porous*, and *relief* as they relate to printing technology.
3. What are the special attributes of the above four processes?
4. Collect and discuss some examples of work produced by the offset lithography, gravure, screen, and lithography processes.
5. Collect data on the current percent distributions of the primary printing methods.
6. Why is gravure printing so widely used in advertising inserts?
7. Write comparative explanations of the ink-jet, laser, sublimation, hot stamp, and pad printing methods. Collect some examples of each.
8. Explain the preparation of the familiar membrane switch.
9. Examine the graphic design example of the calender. Prepare a similar layout for a product of your choice.
10. Collect some examples of corporate trademarks or logos, and discuss their appropriateness of the company or product.
11. Investigate the different methods used to apply graphics to metal cans, paper cartons, and plastic bottles used as beverage containers.
12. Three different toothpaste containers were described in this chapter. Identify another product that is marketed in a variety of packages and analyze its graphics printing methods and problems.
13. How does packaging machinery influence the design of packages and their graphics?

CHAPTER 10

Legal Issues in Product Design

At some early stage in the development of new artifacts or techniques, designers must attend to the legal aspects of their craft, such as guarding the rights to their inventions and considering the matter of product safety. Certain results of human creative efforts are called *intellectual property*. In general, much like other kinds of property, rights in intellectual property can be protected against unauthorized appropriation or use by others, and can be sold (assigned) and rented (licensed) to others. Manufacturer concerns for product safety include both protecting the buyer from possible injury and protecting the company from possible product liability claims. These issues are the subject of this chapter.

INTELLECTUAL PROPERTY

The United States Constitution gives Congress the power to (1) regulate commerce with foreign nations and among the several states and (2) promote the progress of science and the useful arts, by securing for limited times to authors and inventors the exclusive rights to their respective writings and discoveries. Any product of the human mind that has potential value in the marketplace should be protected by intellectual property law for two reasons. First, it rewards the originator for expended time, energy, and money, thereby encouraging further creative work. Second, by requiring that the results of this new work be made known, it increases the amount of technical and artistic knowledge available to the public. Everyone gains by such an arrangement.

The four basic categories of intellectual property protection are patents, copyrights, trademarks, and trade secrets. Patents and trademark registrations are granted by the U.S. Patent and Trademark Office in Arlington, Virginia, and copyrights are registered by the Copyright Office, part of the Library of Congress in Washington, DC. Trade secrets are protectable under state law, as are some aspects of copyright and trademark related matters not covered by federal law.

Patents

A *patent* gives its owner (inventor or his or her successor) the right granted by the United States Patent and Trademark Office to exclude others for a limited time from making, using, or selling the patented invention in this country. A patent is a published legal document in which the invention is clearly disclosed to the public, and the scope of the patentee's exclusionary rights is clearly defined. The patent right does not mean that the Patent Office will protect the patent owner against infringers. However, the patent right does give the patentee "offensive rights" to pursue claims against infringers through the federal courts.

Two types of U.S. patents exist. *Design* patents cover the new, original, and ornamental form or appearance of an article of manufacture, such as a toy doll or a watch face. Neither the concept "doll" nor the watch workings are being patented—only their unique appearances are protected. Tire tread patterns are other examples of original forms covered by design patents. These patents are granted for a term of 14 years and are not renewable.

More important to product designers are *utility* (or *product*) patents, which cover new and useful processes, machines, manufactured articles, compositions, and improvements of any of these. A typical patentable *process* involves the steps performed on materials to produce a desired result, such as a method for making plastic laminates or for removing stains from cloth. A *machine* is a mechanical device that functions to produce a result, such as dental drills, typewriters, and lawn mowers. *Manufactured articles* include toothbrushes, pliers, and electronic circuits. A *composition* is a substance formed by combining ingredients to create, for example, new drugs, adhesives, or insecticides. Most patentable inventions are *improvements*, such as a variation or enhancement of some existing process, machine, manufactured article, or composition of matter. A utility patent is granted for a term of 17 years, subject to payment of government maintenance fees at required intervals after issuance of the patent. At the end of its patented term, the patented invention becomes public property and anyone can make, use, or sell the invention without the permission of the patent owner.

Charles Eames, known especially for his chair designs, developed a method for making laminated articles and was granted a utility patent for it in 1946. The complete patent text and and drawings are shown in the exhibits in Figure 10.1. This patent expired in 1963, but serves as a good example of the kinds of disclosures and claims to be found in a utility patent covering a process of manufacture.

UNITED STATES PATENT OFFICE

2,395,468

METHOD OF MAKING LAMINATED ARTICLES

Charles Eames, West Los Angeles, Calif., assignor to Evans Products Company, Detroit, Mich., a corporation of Delaware

Application May 28, 1942, Serial No. 444,774

3 Claims. (Cl. 144—309)

This invention relates to methods of forming laminated articles, to apparatus for forming laminated articles, and to laminated articles themselves. More particularly it involves laminated articles having surfaces conforming to compound 5 curves, along with the manner and means of providing such conformations. (Note: In the following, by a compoundly curved surface is meant a surface presenting a two dimensional curve, such as a spherically shaped surface, as distin- 10 guished from an unidimensional curve, such as a cylindrically shaped surface.)

An object of this invention is to provide a novel and improved method of molding laminated materials into shapes having compoundly curved 15 surfaces by utilizing low molding pressures. Another object is to set forth novel and improved apparatus for fabricating shaped laminated sheets. A further object is to provide a novel and improved surgical splint. A still further object is 20 to provide a novel and improved chair. Yet another object is the provision of a novel and improved stretcher. Additional objects will be in part obvious and in part pointed out as the description proceeds below.

One example of the laminated materials above referred to is plywood. Plywood has been used in varying thicknesses in furniture, boxes, houses, and certain other structures. In some instances it has been bent into curved configurations for 30 use, for example, in the seats and backs of folding wooden chairs. Heretofore, the use of curved sheets of plywood has been limited primarily to articles wherein, at any point on the surface, one element of the surface was a straight line or, in 35 other words, wherein the surface of the article conformed to a surface generated by a straight line moving along a curve. Recently it has become known that plywood could in some instances be molded to conform to a limited extent to more 40 irregular, compoundly curved surfaces, such as those generated by a curved line moving through space along a curve. Such compound curves were achieved in sheets of plywood only through the use of expensive dies operating under relatively 45 great pressures and then usually only with some sacrifice in the quality of the finish of the wood in the molded surfaces.

The present invention comprehends methods and means whereby compoundly curved surfaces 50 in plywood articles are made possible without the use of high pressures or expensive dies and without deleterious effects on the wood surfaces. It includes methods and means whereby a laminated article may be fabricated conforming to a 55 great variety of configurations and having almost any specified strength. As will appear, the materials required are easily obtained, highly skilled labor is not essential, and mass production is readily achieved.

The basic method of the invention stems from an appreciation inculcated through experience of the characteristics and possibilities of the materials that go together to make a sheet of plywood. A single sheet of veneer, properly cut from suitable wood, was highly flexible and easily distorted and it was discovered that, by cutting a small opening in a veneer and removing a dart extending from an edge of the veneer to the opening, the pliant sheet could easily be formed to give a compound surface (the precise shape depending upon the nature of the dart and the forces directed on the sheet) with the salient portion of the surface coextensive with the point of the dart and the opening. It was further discovered that a plurality of veneers, when properly cut, could easily be simultaneously shaped to give a compound surface, with each intermediate sheet creeping slightly with respect to the adjacent upper and lower sheets as the laminae were molded. By cutting the veneers so that no cut directly overlay an adjacent cut, by permanently bonding the veneers one to another after they had been individually shifted with respect to one another during the actual molding, and by combinations of openings and darts to give combinations of salient portions, it was finally discovered that a sheet of plywood could be formed over a range of thicknesses to almost any given compound configuration and without the application of excessive power.

The invention opens a new field for wood veneers, plies and other sheet material. Articles fabricated in accordance with the invention may be given the desired configurations through combinations of compoundly shaped portions, each one of which portions is more or less similar to a portion of the surface of a sphere. The resulting article has the advantage of possessing exceptional structural strength. Even as a plane-surfaced support is, theoretically, the least rigid type of structural support so, on the other hand, is a spherically shaped supporting structure the ideal rigid support because of the perfect force distribution which it provides. Each of the curved portions of an article made in accordance with the invention approaches a theoretically ideal spherically shaped structure with the result that the strength of the article as a whole approaches a maximum.

Figure 10.1 A copy of the Charles Eames patent for a method of producing laminated articles.

The discovery in its early stages dealt primarily with plywood and it will be illustrated here as it may be applied to plywood articles but it is not to be construed as limited to plywood or to sheet-like materials built up wholly from wood veneers. It may be practiced in the fabrication of other materials built up from a plurality of laminae. In so far as the invention is concerned it does not matter whether each of the laminae be of the same composition, thickness, or physical properties.

In the drawings, in which exemplary structures are illustrated for attaining the objectives set forth and in which like reference characters refer to like parts:

Figure 1 is a perspective view of a surgical splint embodying and made in accordance with the process of the invention;

Figure 2 is a plan view of a veneer for the splint showing the general pattern of the laminae from which the splint is fabricated;

Figure 3 is a perspective view of a chair embodying and made in accordance with the process of the invention;

Figure 4 is a plan view of a veneer for the chair showing the general pattern of the laminae from which the chair is fabricated;

Figure 5 is a perspective view of a stretcher embodying and made in accordance with the process of the invention;

Figures 6, 7, and 8 are sectional views of the stretcher of Figure 5 taken in the direction of the arrows along lines 6—6, 7—7, and 8—8, respectively;

Figure 9 is a plan view of a reinforcing member;

Figure 10 is a perspective view of a series of splint veneers in registration with an aligning device;

Figure 11 is a perspective view of a die member for forming a splint in accordance with the invention;

Figure 12 is a perspective view of the veneers of Figure 10 applied to the die of Figure 11;

Figure 13 is an enlarged view of a detail of Figure 12 taken in the direction of the arrows along the line 13—13; and

Figure 14 is a perspective view, with parts broken away and parts in section, of the veneers and die of Figure 12 located within a pressure chamber, in what might be termed a molding position, in accordance with the invention.

The splint

The splint, indicated by 12 (Figure 1), is fabricated from layers of wood veneer. The veneers are cut according to a pattern to form blanks, such as the blank 14 shown in Figure 2, and each blank is so cut from the veneer sheeting that the grain of the wood in each runs in a different direction. The blank 14 (Figure 2) is provided with cut-out portions as indicated by P. These portions are located and proportioned in relation to the location and configuration of the desired compoundly curved shapes of the finished article. Each portion P is adapted to provide a dome-like contour, the top of the dome corresponding generally to the location of an opening 20 and the character of the dome depending upon the shape of a dart-shaped part 22. If the height of the dome is to be slight, the dart removed is quite narrow at its broadest point; if the height of the dome is to be not so slight, the dart removed is wider at its base. In the form illustrated the dart-shaped parts removed have been given generally straight side lines. These lines may in some instances be arcuate or irregular, depending upon the shape of the surface desired. Some of the cut-out portions include specially formed openings 21 adapted to cooperate with an aligning device, to be described, to align the veneers in registration with one another, and are connected by cuts with the outer edges of the veneers as shown at 21a, Fig. 2.

For any given compoundly curved portion the dart in each veneer is angularly displaced somewhat with respect to the corresponding dart of the next adjacent veneer (as indicated in dotted lines by 23) so that, when the edges of a given dart are brought into juxtaposition during the molding to form a seam, no seam directly overlies another seam. The overlapping joint thus formed is substantially as strong as any other portion of the splint. Similarly, ends 24 and 26 of each veneer are cut to give an interlocking, overlapping joint, as at 25 (Figure 1), when they are combined during the molding to form stirrup 27 of the splint. For a detailed view of this type of joint see Figure 13. Dotted lines 18 and 19 indicate where portions 24 and 26 might be cut in alternate blanks to provide for joint 25. A patch 16 allows for expansion of the upper portion 17. The patch may be varied through the several layers to avoid bulkiness.

The splint illustrated combines strength with light weight in a member which will afford support, including traction, to an entire leg. The embodiment described has a weight of a scant 700 grams and a series of splints may be nested together for compact storage or easy transportation.

The chair

Figure 3 illustrates a molded plywood chair 28. The pattern 29 of the blanks used in making the chair appears in Figure 4 and shows how cut-out portions P may be combined. In this instance, three portions P have been combined to form a veneer which may be accurately shaped to conform to the contours of a person who is to occupy the chair. Sections 30 and 32 of the blanks, for a given chair, would be displaced somewhat with respect to one another, as shown in dotted lines, so that they would interfit to make a solid frame over the top of the chair back, with the opposing edges of sections 30 and 32 of any one veneer coming together in off-set relationship to that of the corresponding sections of adjacent veneers to provide a strong, smooth joint, as at 31 (Figure 3).

The chair is very light in weight and at the same time, because of the almost unlimited configurations to which such chairs may be conformed, it gives the maximum comfort and freedom for movement necessary to use in such locations as, for example, in aeroplanes.

The stretcher

The stretcher, shown in Figures 5, 6, 7, and 8 and indicated generally by numeral 38, is another example of a novel, improved product made in accordance with the invention. It may be fabricated from veneers cut in accordance with a pattern. A fitted body section 40 and fitted leg sections 42 and 44 are all made possible through the skillful location and combination of cut-out portions in the blanks.

The stretcher is strong, light in weight, and, because of its scientifically proportioned surfaces, provides proper support for each member of the body without additional pads, rubber sheets, and the like, previously necessary in the field with known stretchers, and any number of them may be nested together with the result that they enable economies in space and one man can carry a number of them. Furthermore, the stretcher is waterproof and can float while bearing a man. It acts as a complete splint in itself to immobilize any part of the body, conserves body heat, and, when covered by a second, inverted stretcher (as indicated by dotted lines in Figure 8) serves to keep out rain, snow or insects and affords protection generally from exposure and from the rough handling which inevitably accompanies movements such as from one ship to another or through heavy underbrush.

The reinforcement

Figure 9 depicts a means for strengthening an isolated section of an article made in the practice of the invention. Piece 36 is adapted to be inserted between layers of an article to be reinforced. Holes, such as indicated at 16, extend through all the laminae of the portion strengthened, as well as through piece 36, and serve to cooperate with an aligning device, to be described, to maintain piece 36 in position during the formation of the article.

The strength of an article may be increased over a desired area by the insertion of reinforcing pieces such as piece 36. The reinforcements may occur in a compound surface, in which event they may include cut-out portions such as P (Figure 2) so that they may be easily shaped along with the other laminae. Several reinforcements may be inserted over the same area and they may even be used to give slight changes in the contour of the surface of the article. Best results are obtained by tapering the edge portions of the reinforcing piece so that the adjacent laminae may flow smoothly over the piece.

The aligning device

Figure 10 illustrates an aligning device 45 consisting of a base plate 46 carrying pins 48. The pins are arranged to conform to the location of specially shaped openings in the blanks to be aligned. By way of example the device is shown aligning the specially shaped openings 21 of the splint blanks 14. It may be used with other blanks, however, or in connection with reinforcing pieces. The blanks may be assembled by slipping them over pins 48 and onto the base plate 46, which supports them in registration with each other while enough blanks are being assembled into a sheaf to form a splint.

The die member

Figure 11 represents a die member 49 consisting of a base 50, a die portion 52, an electric heating element 54 and a conventional thermal control 56. The die member may be made of concrete, for example, since it is not required to withstand great pressure or heat. For the purposes of illustration the die portion 52 shown is the one used to form a splint. Embedded in base 50 and closely surrounding die portion 52 is a plurality of U-shaped members 58, for a purpose to be described.

The pressure chamber

Figure 14 shows a pressure chamber 60 consisting of a die bed 62 for supporting interchangeable dies, a housing 64, a rubber bladder or diaphragm 66, and, by way of illustration, the die member 49 for the splint resting on the die bed.

An air pump 68 is connected by a conduit 70 to the interior of the bladder. The numeral 72 indicates an air gauge, 74 indicates a one-way check valve, 76 indicates a relief valve and 78 indicates a door to the pressure chamber.

In operation, a die member carrying a sheaf of blanks to be molded is placed on die bed 62 underlying bladder 66, and door 78 is closed and locked. By pumping air into the bladder, the sheaf of blanks may be brought into accurate conformation with the matrix presented by the die member. Simultaneously, the article being processed may be heated under the control of the thermostat associated with the heating element of the die member.

The process

The process of the invention comprises a series of relatively simple steps. Success is achieved in a practicable manner which does not require expensive equipment or high pressures or temperatures.

A pattern is planned in accordance with the article to be produced. Portions of it are cut out in relation to the curved surfaces desired. By way of illustration, reference is made to the splint illustrated in Figure 1, the pattern for which is shown by the blank in Figure 2. Blanks are cut from a veneer sheet in such a manner that the grain in each blank runs in a different direction. The various blanks of a given splint display the further differences necessary for the location of the cut out portions and the tab portions which will comprise the overlapping joints. The blanks should preferably be designed so that no one seam will directly overlie another seam.

A thermo-setting, phenolic resin binder material is applied to the blanks and the blanks are placed in registration with one another, as, for example, with the aid of an aligning device such as illustrated in Figure 10. The sheaf of registering blanks is then applied to a die, as illustrated in Figure 12, by binding the laminae of the sheaf into conformation with the configurations of the die matrix and by assuring that the edges of the joint portions are properly interfitted. The U-shaped members 58 of die member 49 aid in the binding procedure.

The die member carrying the sheaf of blanks is put into a pressure chamber, such as illustrated in Figure 14. The thermal control of the heating element in the die member is set for about 165° C., the door of the pressure chamber is closed, air at a pressure of about 30 pounds per square inch is pumped into the bladder and the whole is left for about an hour, during which time the thermo-setting binder is permanently fixed to cement the laminae together. Thereafter the finished article is removed.

None of the equipment involved is complicated. The specially cut blanks of the invention lend themselves peculiarly well to being molded so that only a low pressure is required to make the laminae conform to the die while the binder is setting. The old concept of male and female die surfaces is eliminated entirely and the one die that remains is not subjected to hard wear and hence can be made, for example, of cast concrete. The low pressure used has the further advantage of eliminating the impaired exterior surfaces heretofore accompanying the attempts made at molding compoundly curved surfaces in plywood articles.

Since many embodiments of the invention are possible and since many changes might be made

4   2,395,468

in the embodiments set forth, protection is not to be limited to anything described or presented in the above specification and drawings but only to the scope of the hereinafter attached claims.

I claim:

1. In the art of fabricating a laminated article having compound curved surfaces the steps of preparing a plurality of blanks, removing from the blanks predetermined portions related to areas to be subsequently shaped into compound curves, removing from the blanks portions to provide in each blank a plurality of registry openings, connecting each of said openings with the outer edges of its blanks by cuts in the blank, treating the surfaces of the blanks with a cement moldable when subjected to pressure and elevated temperature, stacking the blanks in a sheaf, positively aligning the blanks by holding said registry openings in alignment, placing said sheaf thus registered in juxtaposition to a matrix, conforming the sheaf generally to the matrix, and subjecting the sheaf to pressure and temperature while on said matrix to give the sheaf the desired shape and to cause the cement to flow and bind the blanks together and hold the said shape.

2. In the art of fabricating a laminated article having compound curved surfaces the steps of preparing a plurality of blanks, removing from the blanks predetermined portions related to areas to be subsequently shaped into compound curves, removing from the blanks portions to provide in each blank a plurality of registry openings, connecting each of said openings with the outer edges of its blanks by cuts in the blank, removing from the blanks predetermined dart-shaped portions related to areas to be subsequently shaped into compound curves, angularly offsetting each dart-shaped portion with respect to the corresponding dart-shaped portion of any adjacent blank whereby no one dart-shaped portion directly overlies any other dart-shaped portion, treating the surfaces of the blanks with a

cement moldable when subjected to pressure and elevated temperature, stacking the blanks in a sheaf, positively aligning the blanks by holding said registry openings in alignment, placing said sheaf thus registered in juxtaposition to a matrix, conforming the sheaf generally to the matrix, and subjecting the sheaf to pressure and temperature while on said matrix to give the sheaf the desired shape and to cause the cement to flow and bind the blanks together and hold the said shape.

3. In the art of fabricating a laminated article having compound curved surfaces the steps of preparing a plurality of blanks, removing from the blanks predetermined portions related to areas to be subsequently shaped into compound curves to provide openings in the blanks and locating said openings in such manner that they register one with the other in a subsequent stacking and conforming operation, whereby edges of said openings may be used for registering purposes, removing from the blanks predetermined dart-shaped portions related to areas to be subsequently shaped into compound curves, angularly offsetting each dart-shaped portion with respect to the corresponding dart-shaped portion of any adjacent blank whereby no one dart-shaped portion directly overlies any other dart-shaped portion, treating the surfaces of the blanks with a cement moldable when subjected to pressure and elevated temperature, stacking the blanks in a sheaf, positively aligning the blanks by holding said edges in alignment, placing said sheaf thus registered in juxtaposition to a matrix, conforming the sheaf generally to the matrix, and subjecting the sheaf to pressure and temperature while on said matrix to give the sheaf the desired shape and to cause the cement to flow and bind the blanks together to hold the said shape.

                                    CHARLES EAMES.

Feb. 26, 1946.   C. EAMES   2,395,468

METHOD OF MAKING LAMINATED ARTICLES

Filed May 28, 1942   5 Sheets—Sheet 1

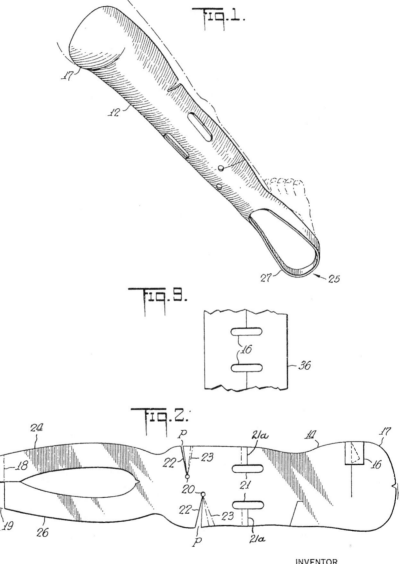

Fig.1.

Fig.3.

Fig.2.

INVENTOR
Charles Eames
BY
Blair, Curtis + Hayward
ATTORNEYS

Feb. 26, 1946.   C. EAMES   2,395,468

METHOD OF MAKING LAMINATED ARTICLES

Filed May 28, 1942   5 Sheets—Sheet 2

Fig.3.

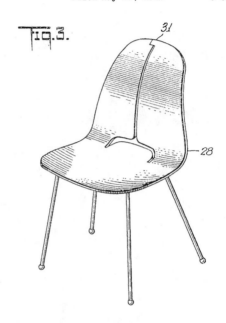

Fig.4.

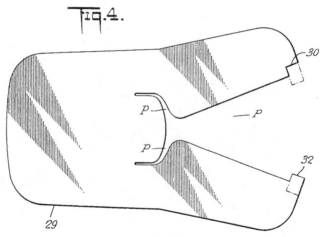

INVENTOR
Charles Eames
BY
Blair, Curtis & Hayward
ATTORNEYS

Feb. 26, 1946.   C. EAMES   2,395,468

METHOD OF MAKING LAMINATED ARTICLES

Filed May 28, 1942   5 Sheets—Sheet 3

Fig.5.

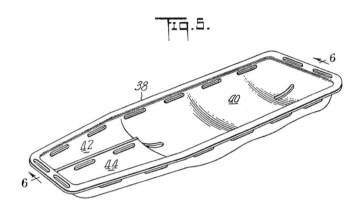

Fig.6.

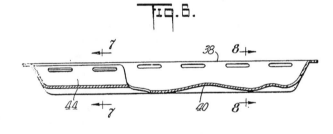

Fig.7.

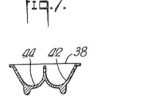

Fig.8.

INVENTOR
Charles Eames
BY
Blair, Curtis & Hayward
ATTORNEYS

Feb. 26, 1946.   C. EAMES   2,395,468

METHOD OF MAKING LAMINATED ARTICLES

Filed May 28, 1942   5 Sheets—Sheet 4

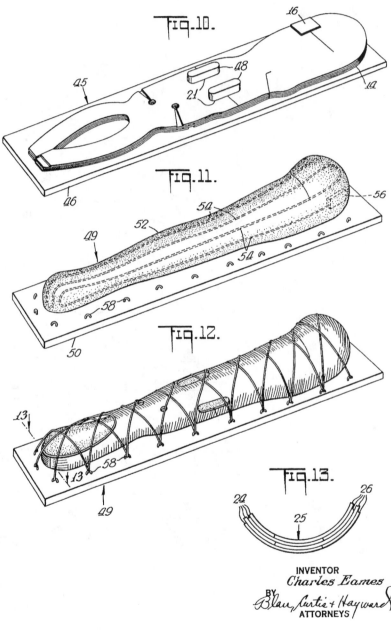

INVENTOR
Charles Eames
BY
Blau, Curtis + Hayward
ATTORNEYS

Feb. 26, 1946.   C. EAMES   2,395,468

METHOD OF MAKING LAMINATED ARTICLES

Filed May 28, 1942   5 Sheets—Sheet 5

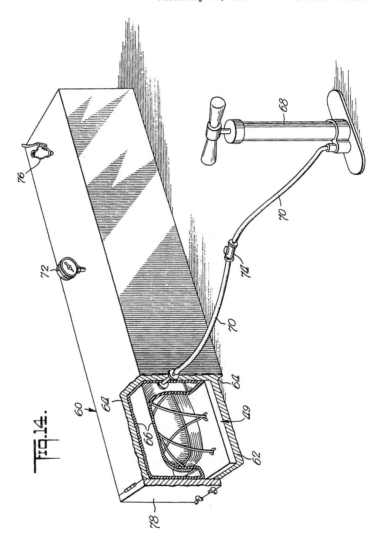

INVENTOR
Charles Eames
BY
Blair, Curtis & Hayward
ATTORNEYS

An invention must fulfill a number of requirements for it to be patentable, three of which follow. First, the invention must be useful. Examples of inventions deemed not useful and hence not patentable are inoperative devices such as perpetual motion machines, inventions detrimental to health or morality, and inventions useful only for illegal purposes, such as a technique for counterfeiting currency. Methods of doing business and printed matter also are considered to be not patentable.

Second, the invention must be new and "novel." In general, an invention is not novel if it is identical to something already available to the public, or in legal terms, something in the "prior art." Listed in Section 102 of the federal patent statute are a number of conditions under which an invention may be unpatentable, or patentability may be lost. For example, the right to file a patent application is lost irretrievably by failure to be timely in filing the patent application. Under European patent law, making the invention available to the public any where and in any way instantly bars filing of a patent application. Many other countries have similarly stringent filing deadlines. Under the more liberal United States patent laws, filing of a patent application is barred not at the instant, but one year after, a unit embodying the invention is sold, offered for sale, publicly used in the United States, or disclosed in a printed publication anywhere in the world. In any event, it is unwise to release an invention from secrecy without competent counsel.

Third, an invention, though novel, also must be nonobvious to a person possessing ordinary skills in the field (or art) to which the invention pertains at the time of the invention. This mythical person is deemed to have access to all the available pertinent prior art. In rejecting an application as being obvious, the Patent Office often combines the teachings of several prior art patents or publications. Needless to say, this can be a confusing issue and is one of a number of reasons for the existence of patent attorneys.

In addition to the design and utility patent categories, the *plant* patent extends to plants that are asexually reproducible (e.g. grafting or rooting) and are of a distinct and new variety, such as Delicious apples and American Beauty roses. Plant patents are beyond the scope of this present book and require no further explanation.

The phrases *Patent Applied For* and *Patent Pending* provide no legal protection against the copying of an invention. However, they do usefully serve as public warning that a patent application has been filed in the Patent Office, and that if the patent is granted, any unauthorized infringer may be liable for infringement.

Copyrights

A *copyright* is, in general, the right to exclude others from unauthorized copying of original works such as writing, musical and dramatic works, motion pictures, graphics and sculptures, computer programs, and sound recordings. Usually, a work must include some artistic component and be original in order to be copyrighted. Copyrights are protected under state "common law" and under federal copyright law. Copyrights are different and separate from trademarks or patents. For example, a product name cannot be copyrighted, but may be protected under trademark law. An invention cannot be copyrighted, but may be protected under federal patent law. The text of a book cannot be protected under patent or trademark law, but may be copyrightable.

In general, the protection for copyrighted work lasts for the lifetime of the author plus an additional 50 years after the author's death. The copyright owner has the exclusive right to reproduce, distribute, condense, and translate the work, to present displays and performances as appropriate, or to authorize someone else to do it. Other persons can make limited fair use of a copyrighted work for purposes of research, teaching, criticism, or news reporting without infringing the copyright. Whether the use is indeed "fair" depends on the effect of that use on the potential market, and profit, for the copyrighted work.

Trademarks

A *trademark* is a name, word, symbol, or device used by a manufacturer to identify the source of a product and thereby distinguish it from other product sources in the marketplace. Similarly, a *service mark* may be used to identify a source of services offered to the public. The words *Public Broadcasting System* in connection with the radio network is an example of a service mark. *Trade names*, or company names, may be trademarks or service marks.

As an example, perhaps the world's best-known package—the Coca-Cola bottle—was officially listed by the United States government as a registered trademark of The Coca-Cola Company in 1960. As such, it is one of the few containers ever to be recorded on the Principal Register of the United States Patent and Trademark Office. Further, this most distinctive glass bottle was originated by Alex Samuelson, working under the direction of the Root Glass Company, which first produced the bottle in 1915. The famous form, (Figure 10.2) is easy to hold, even when the bottle is wet, and is so unique that it is recognizable even when held in the dark. This trademarked device cannot be confused with any other bottle.

Similarly, Northwest Airlines uses the phrase NORTHWEST AIRLINES and an associated symbol as its servicemarks. (See Figure 10.3.) The symbol, the word NORTHWEST, and bold color graphics as applied to an airplane distinguish it from any other airline. (See Figure 10.4.) To obtain and retain trademark ownership in the United States, one must be first to use the mark on the goods in commerce in the territory, and such use must be continued if trademark ownership is to be preserved. Federal registration of a trademark provides substantial ancillary advantages.

Trade Secrets

A *trade secret* is any process, formula, device, or business information of value that the owner does not wish to fall in the hands of a competitor. It is guarded, privileged information. Examples include a recipe for a soft drink or a formula for a special

Figure 10.2 This most unique soft drink bottle is a trademarked container. Courtesy of The Coca-Cola Company.

paper. One of the strengths of trade secret law is that there is no limit to the time that the secret may be kept confidential. As long it remains concealed, it will afford a business advantage only to its owner.

Trade secret protection is managed exclusively by state law. Several states have made the theft or unauthorized disclosure of a trade secret an unlawful act. Companies must rely on the integrity of their employees to refrain from revealing a valued procedure or formula. Courts will offer their protection only if a company proves the

NORTHWEST AIRLINES

Figure 10.3 This is an example of a servicemark for a leading airline. Courtesy of Northwest Airlines.

value of the secret, and proves that it has made a reasonable effort to maintain confidentiality.

This discussion of patents, copyrights, trademarks, and trade secrets has aimed to stress the importance of guarding the results of human creative effort. Designers employed by a firm must adhere to policies of confidentiality established by that firm. Independent designers should follow certain basic rules early in the development process: Do not discuss your work with anyone who has not signed a nondisclosure agreement; and maintain a complete, well-documented, illustrated, dated, and

Figure 10.4 Note the servicemark symbol applied to the tail of this jumbo jet. Courtesy of Northwest Airlines.

witnessed record of your work. It also is wise to secure the services of a patent attorney for guidance and advice.

PRODUCT LIABILITY

Product liability is a legal term describing an action whereby a purchaser seeks to recover damages for personal injury or loss of property from a vendor, when it is alleged that the injury or loss resulted from the vendor's defective product. Note that a person need not suffer bodily harm. The person can also/or suffer damage to property. A vendor can be the manufacturer of the goods, the wholesaler, the distributor, or the retailer. A purchaser can be the buyer, the user, or a bystander in some cases. For purposes of the present discussion, the terms *vendor* and *purchaser* will be used.

It becomes immediately obvious that the well-designed product generally is the best defense against product liability claims. In such products, the safety faults will have been "designed out." The issue of design for product safety and buyer protection was described and exemplified in Chapter 6, and some design guidelines were presented. This present discussion centers on the legal ramifications of this serious problem.

While the liability issue is serious, the popular media and some corporations believe that there has been a "litigation explosion" in products liability. This imagined upsurge also purports that the consumer is obtaining an increase in both the percentage of victories and amounts in such cases. However, all evidence indicates the opposite to be true. The actual number of products suits in both state and federal courts has been contracting over the last ten years. In addition, both the amounts paid to the consumer for products claims and the percentage of consumer victories has steadily decreased during this period. Despite this trend, products liability claims can present high costs to both the consumer/purchaser and the vendor/manufacturer in a serious injury situation. When a product proves defective, it can result in a decline in consumer confidence and a loss of corporate reputation.

There are three basic liability theories that can be used by a purchaser to bring suit against a vendor in a product liability action: negligence, breach of warranty, and strict liability.

Negligence

Negligence is simply the failure to exercise due care. A vendor may be liable for a foreseeable purchaser harm or loss caused by a specific product because of the failure to exercise reasonable care to assure that the product would cause no harm or loss. It is not necessary for the purchaser to prove that the vendor acted maliciously or recklessly. Instead, it only is required that the conduct of the vendor be measured against the "reasonable person" test. The vendor's conduct is deemed negligent only if it falls below what a reasonable person would have done in a similar situation. This reasonable person is a hypothetically objective individual who acts with due care under all circumstances. This "mythical" person cannot be identified with any particular group or even a "majority" of people. Nor is it proper to identify such persons with a judge or jury applying the standard. Such persons are merely personifications of a community ideal of reasonable behavior, determined by the jury's social judgment. Negligence is the failure to demonstrate such reasonable behavior.

Warranty

A *warranty* is a vendor assurance that the character or quality of a product is as it is declared to be. If a purchaser elects to buy a product because the vendor assures that the item will function safely and without failure for two years, and the item fails or injury occurs, the warranty has been breached because either the purchaser no longer has the use of her or his property or gets hurt. A *breach of warranty*, therefore, is the failure of a product to perform as represented by its vendor, and the vendor can be held liable. The two kinds of warranties are express and implied.

An *express* warranty is any affirmation made by the seller to the purchaser. This expression is a guarantee that the product will function in a certain way and meet specified standards. Promotional literature, sales agreements, printed guarantees, catalog descriptions, and other publications are examples of express warranty. However, there is no legal requirement that the affirmation must be in writing. Any oral statement made by a sales person also can constitute an express warranty, as can any sample or model of the product. By law, a vendor will be strictly liable if he or she expressly warrants that a product will perform in a certain way and it does not.

By virtue of the fact that a manufacturer makes a product and offers it for sale carries the assumption that the product is safe and fit for use. This is an *implied* warranty for the simple reason that neither an oral nor a written representation of the product is necessary. By the very fact that it appears on a storeowner's shelf and is for sale, the purchaser assumes that the item is merchantable and fits the purposes for which such goods are used. Implicit in the presence of a gallon jug of milk in a supermarket dairy case is the buyer's trust that the milk is safe to drink.

Strict Liability

Strict liability, also referred to *strict liability in tort*, means that a vendor is responsible for harm or loss caused by a product defect, whether or not it can be proved that the defect was caused by vendor negligence. In these situations, fault is not the issue. A vendor can be held liable for product harms or losses regardless of the degree of care that went into the product.

The courts are by no means unanimous in their agreement as to what constitutes a product defect. Most courts require proof that a design defect was responsible for an unreasonably dangerous product. Many courts focus on the consumer, in which cases the vendor is held liable whenever product performance falls below user expectations. Another factor is whether the purchaser should have reasonably anticipated the hazards of the product, or whether from the vendor position the danger was so obvious that no specific product warning was deemed necessary. The current court position on this concept is that unless the awareness of the product hazard is almost universal, a product must be designed to prevent against even the most obvious of dangers.

The defect requirements by law have arbitrarily been divided into manufacturing or production defects; design defects; and a failure to adequately instruct on proper usage or to warn of product dangers. Thus, a product that has an adequate design and that has been perfectly manufactured may still be defective if the vendor/manufacturer fails to instruct or warn. This warnings issue is one to which vendors generally are responding with vigor. Consumer products on the current market are replete with such cautions, and home power tools generally include assembly and user manuals. The general problem with the manuals is that they are not always user-friendly. Too often, they are confusing, misleading, and incorrect. It should be a requirement that all consumer product designers and makers, as well as user-manual writers, personally assemble and use the product in question, by following the manuals they have prepared.

To conclude these brief remarks regarding the serious and costly issue of product liability, vendors should, as matter of first principle, strive for defect-free products. The most effective approach toward this goal is to identify and attack all safety issues in the very early stages of the product design process. A safety problem eliminated from a design is one that cannot cause injury and can never come back to haunt the vendor. Barring the perfect, defect-free product, vendors must endeavor to educate the purchaser in the safe use of the potentially dangerous product. This can best be done by excellent product literature and by clearly identifying danger points on machines.

QUESTIONS AND ACTIVITIES

1. Define the term *intellectual property* and list two reasons for protecting it by law.
2. List and describe the four basic categories of intellectual property protection.
3. What federal agency grants patents and trademark registrations, and which agency registers copyrights?
4. How are trade secrets protected?
5. List and describe the two primary types of U.S. patents.
6. What five types of inventions are covered by utility patents?
7. List three requirements for a patentable invention.
8. What value lies in the phrases *Patent Applied For* and *Patent Pending*?

9. What is meant by the phrase *offensive rights*?
10. What are the terms of duration of design and utility patents?
11. What is the term of duration of a copyright?
12. What is the meaning of the concept of *limited fair use* of a copyrighted work?
13. What is the difference between a trademark and a service mark?
14. What is the limit of time that a trade secret can be kept confidential?
15. Define the terms *product liability, negligence, warranty*, and *strict liability*.
16. What is the difference between an express warranty and an implied warranty?
17. What is the best defense against product liability claims?

CHAPTER 11 | Design Resources

The following lists of books, periodicals, visuals, and design center information may be helpful to the reader for research and familiarization with the broad topic of design in manufacturing. The author found the primary set of books especially helpful in preparing this manuscript. The older books, listed separately, are generally out of print and hard to find, but are excellent historical, theoretical, and philosophical resources, and are well worth the search. The periodicals were selected to give a broad coverage of the many such offerings that will be valuable to designers. There are dozens of videotapes available from industrial sources, as well many quality products from commercial distributors. Teachers can make good use of these to complement lectures and demonstrations. The noted design centers can provide numerous booklets, catalogues, reports, and photographs as evidences of the product design examples available internationally.

BOOKS

ADLER, MORTIMER J. *Aristotle for Everybody*. New York: Macmillan, 1978.

ALDERSEY-WILLIAMS, HUGH. *New American Design*. New York: Rizzoli, 1988.

ALEXANDER, DAVID C. *The Practice and Management of Industrial Ergonomics*. Englewood Cliffs, NJ: Prentice Hall, 1986.

ALLEN, C. WESLEY (ED.). *Simultaneous Engineering*. Dearborn, MI: Society of Manufacturing Engineers, 1990.

AMERICAN MACHINIST. *Computers in Manufacturing*. New York: McGraw-Hill, 1983.

ANDERSON, DAVID M. *Design for Manufacturability*. Lafayette, CA: CIM Press, 1990.

ANDREASEN, M. MYRUP, ET AL. *Design for Assembly*. New York: Springer-Verlag, 1983.

ASFAHL, C. RAY. *Industrial Safety and Health Management*. Englewood Cliffs, NJ: Prentice Hall, 1984.

BAKERJIAN, RAMON (ED.). *Design for Manufacturability. Volume 6: Tool and Manufacturing Engineers Handbook*. Dearborn, MI: Society of Manufacturing Engineers, 1992.

BAKKER, MARILYN (ED.). *The Wiley Encyclopedia of Packaging Technology*. New York: Wiley and Sons, 1986.

BEDWORTH, DAVID D. ET AL. *Computer-Integrated Design and Manufacturing*. New York: McGraw-Hill,1991.

BESTERFIELD, DALE H. *Quality Control*. (3rd edition). Englewood Cliffs, NJ: Prentice Hall, 1989.

BOOTHROYD, G., AND P. DEWHURST. *Design for Assembly*. Wakefield, RI: Boothroyd Dewhurst, Inc., 1983.

BOYCE, CHARLES. *Dictionary of Furniture*. New York: Facts On File Publications, 1985.

BRALLA, JAMES G. (ED.). *Handbook of Product Design for Manufacturing*. New York: McGraw-Hill, 1986.

BROWN, JAMES. *Modern Manufacturing Processes*. New York Industrial Press, 1991.

BUCK, C. HEARN, AND D. M. BUTLER, *Economic Product Design*. London: Collins, 1970.

BURGESS, JOHN H. *Human Factors in Industrial Design*. Blue Ridge Summit, PA: TAB Professional and Reference Books, 1989.

BUSCH, AKIKO (ED.). *Product Design*. New York: PBC International, 1984.

CAMP, ROBERT C. *Benchmarking*: *The Search for Industry Best Practices that Lead to Superior Performance*. Milwaukee: ASQC Quality Press, 1989.

CAPLAN, RALPH. *By Design*. New York: St. Martin's Press, 1982.

CHANG, TIEN-CHIEN, ET AL. *Computer-Aided Manufacturing*. Englewood Cliffs, NJ: Prentice Hall, 1991.

CHOW, WILLIAM WAI-CHUNG. *Cost Reduction in Product Design*. New York: Van Nostrand Reinhold, 1978.

CONSODINE, DOUGLAS M. *Standard Handbook of Industrial Automation*. New York: Chapman and Hall, 1986.

CORBETT, JOHN, ET AL. *Design for Manufacture*. Reading, MA: Addison-Wesley, 1991.

CROSS, NIGEL. *Developments in Design Methodology*. New York: Wiley and Sons, 1985.

DERTOUSOS, MICHAEL L. *Made in America*. Cambridge, MA: MIT Press, 1989.

DESIGN COUNCIL. *Designing Against Vandalism*. New York: Van Nostrand Reinhold, 1979.

DHILLON, BALBIR S. *Human Reliability with Human Factors*. New York: Pergamon Press, 1986.

EALEY, LANCE E. *Quality by Design: Taguchi Methods and U.S. Industry*. Dearborn, MI: American Suppliers Institute, 1988.

EASTMAN KODAK COMPANY. *Ergonomics Design for People at Work*. New York: Van Nostrand Reinhold, 1983.

ELECTA. *From the Spoon to the Town Through the Work of 100 Designers*. Milan: Electa Editrice, 1983.

EL WAKIL, SHERIF D. *Processes and Design for Manufacturing*. Englewood Cliffs, NJ: Prentice Hall, 1989.

ENGHAGEN, LINDA K. *Fundamentals of Product Liability Law for Engineers*. New York: Industrial Press, 1992.

EVANS, BARRIE, ET AL. (EDS.). *Changing Design*. New York: Wiley & Sons, 1982.

FARAG, MAHMOUD M. *Selection of Materials and Manufacturing Processes for Engineering Design*. Englewood Cliffs, NJ: Prentice Hall, 1989.

FELDMAN, EDMUND BURKE. *Varieties of Visual Experience*. (4th edition). Englewood Cliffs, NJ: Prentice Hall, 1992.

FIELL, CHARLOTTE AND PETER. *Modern Furniture Classics*. Washington, DC: American Institute of Architecture Press, 1991.

FLURSCHEIM, CHARLES H. *Industrial Design in Engineering*. New York: Springer-Verlag, 1983.

FORTY, ADRIAN. *Objects of Desire, Design, and Society from Wedgewood to IBM*. New York: Pantheon Books, 1986.

FOSTER, FRANK H., AND ROBERT L. SHOOK. *Patents, Copyrights, & Trademarks*. New York: Wiley & Sons, 1989.

GAUSE, DONALD C., AND GERALD M. WEINBERG. *Exploring Requirements: Quality Before Design*. New York: Dorset House, 1989.

GIGLIOTTI, RICHARD J., AND RONALD C. JASON. *Security Design for Maximum Protection*. Boston: Butterworth, 1985.

GOETSCH, DAVID L. *Advanced Manufacturing Technology*. Albany, NY: Delmar, 1990.

GREENWOOD, DOUGLAS C. (ED.). *Production Engineering Design Manual*. Malabar, FL: Robert E. Krieger, 1982.

GROOVER, MIKELL P. *Automation, Production Systems, and Computer Integrated Manufacturing*. Englewood Cliffs, NJ: Prentice Hall, 1987.

HARRIS, MARVIN. *Why Nothing Works*. New York: Simon and Schuster, 1987.

HEALY, RICHARD J. *Design for Security*. New York: Wiley & Sons, 1983.

HESKETT, JOHN. *Industrial Design*. New York: Oxford University Press, 1980.

JACOBS, P. *Rapid Prototyping & Manufacturing: Fundamentals of StereoLithography*. Dearborn, MI: Society of Manufacturing Engineers, 1992.

JOYCE, MARILYN, AND ULRIKA WALLERSTEINER. *Ergonomics: Humanizing the Automated Office*. Cincinatti, OH: Southwestern Publishing, 1989.

KOCHAN, ANNA. *Implementing CIM*. New York: Springer-Verlag, 1986.

KONCELIK, JOSEPH A. *Aging and the Product Environment*. Stroudsburg, PA: Hutchinson Ross, 1982.

KONZ, STEPHEN. *Facility Design*. New York: Wiley & Sons, 1985.

KROUSE, JOHN K. *What Every Engineer Should Know About Computer-Aided Design and Computer-Aided Manufacturing*. New York: Marcel Dekker, 1982.

LANE, JACK D. (ED.). *Automated Assembly*. (2nd edition). Dearborn, MI: Society of Manufacturing Engineers, 1986.

LEECH, D. J., AND B. T. TURNER. *Engineering Design for Profit*. New York: Halsted, 1985.

LEWALSKI, ZADZISLAW M. *Product Esthetics*. Carson City, NV: Design and Development Engineering Press, 1988.

LINDBECK, JOHN R. ET AL. *Manufacturing Technology*. Englewood Cliffs, NJ: Prentice Hall, 1990.

LUCIE-SMITH, EDWARD. *A History of American Design*. New York: Van Nostrand Reinhold, 1983.

LORENZ, CHRISTOPHER. *The Design Dimension*. New York: Basil Blackwell, Ltd., 1986.

LUGGEN, WILLIAM W. *Flexible Manufacturing Cells and Systems*. Englewood Cliffs, NJ: Prentice Hall, 1991.

LYNN, GARY S., *From Concept to Market*. New York: Wiley & Sons, 1989.

MAULE, H. G., AND J. S. WEINER. (EDS.). *Design for Work and Use*. London: Taylor and Francis, Ltd., 1981.

MEDLAND, A. J., AND PIERS BURNETT. *CAD/CAM in Practice*. New York: Wiley & Sons, 1986.

MILLER, JULE A. *From Idea to Profit: Managing Advanced Manufacturing Technology*. New York: Van Nostrand Reinhold, 1986.

MILLER, LANDON C.G. *Concurrent Engineering Design*. Dearborn, MI: Society of Manufacturing Engineers, 1993.

MILLER, R. CRAIG. *Modern Design in the Metropolitan Museum of Art 1890-1990*. New York: Harry N. Abrams, Inc., 1990.

MOSS, MARVIN A. *Designing for Minimal Maintenance Expense*. New York: Marcel Dekker, 1985.

NATIONAL AERONAUTICS AND SPACE ADMINISTRATION. *Anthropometric Source Book. Volume I: Anthropometry for Designers*. NASA Reference Publication 1024. Washington, DC: National Aeronautics and Space Administration, 1978.

NEVINS, JAMES L., AND DANIEL E. WHITNEY (EDS.). *Concurrent Design of Products and Processes*. New York: McGraw-Hill, 1989.

NIEBEL, BENJAMIN W. "Product Design for Economic Production." *Mechanical Engineer's Handbook*. New York: Wiley & Sons, 1986.

NORMAN, DONALD A. *The Psychology of Everyday Things*. New York: Basic Books, 1988.

OWEN, A. E. *Flexible Assembly Systems*. New York: Plenum, 1984.

PAO, Y. C. *Elements of Computer-Aided Design and Manufacture*. New York: Wiley & Sons, 1984.

PAPANEK, VICTOR. *Design for the Real World*. New York: Van Nostrand Reinhold, 1984.

PARATORE, PHILIP C. *Art and Design*. Englewood Cliffs, NJ: Prentice Hall, 1985.

PETROSKI, HENRY. *Evolution of Useful Things*. New York: Knopf, 1992.

PHADKE, MADAV S. *Quality Engineering Using Robust Design*. Englewood Cliffs, NJ: Prentice Hall, 1989.

PILDITCH, JAMES. *Talk About Design*. London: Barrie and Jenkins, Ltd., 1977.

PRESSMAN, DAVID. *Patent It Yourself*. (2nd edition). Berkley, CA: Nolo Press, 1988.

PRIEST, JOHN W. *Engineering Design for Producibility and Reliability*. New York: Marcel Dekker, 1988.

PULAT, MUSTAFA BABUR, AND DAVID C. ALEXANDER. (EDS.). *Industrial Ergonomics Case Studies*. Norcross, GA: Industrial Engineering and Management Press, 1991.

PULOS, ARTHUR S. *American Design Esthetic*. Cambridge, MA: MIT Press, 1983.

RAND, PAUL. *A Designer's Art*. New Haven: Yale University Press, 1985.

RAY, MARTIN S. *Elements of Engineering Design*. Englewood Cliffs, NJ: Prentice Hall, 1985.

REDFORD, A. H., AND E. K. LO (EDS.). *Robots in Assembly*. New York: Wiley & Sons, 1986.

RICHMOND, WENDY. *Design and Technology*. New York: Van Nostrand Reinhold, 1990.

RILEY, FRANK J. *Assembly Automation: A Management Handbook*. New York: Industrial Press, 1983.

SALVENDY, GAVRIEL. (ED.). *Handbook of Human Factors*. New York: Wiley & Sons, 1987.

SAMUELS, JEFFREY M. (ED.). *Patent Trademark and Copyright Laws*. Washington, DC: The Bureau of National Affairs, Inc., 1991.

SANDERS, MARK S., AND ERNEST J. MCCORMICK. *Human Factors in Engineering and Design*. (7th edition). New York: McGraw-Hill, 1993.

SMITH, CHARLES O. *Introduction to Reliability in Design*. Malabar, FL: Robert E. Krieger, 1983.

SMITH, CHARLES O. *Products Liability: Are You Vulnerable?* Englewood Cliffs, NJ: Prentice-Hall, 1981.

SMITH, DAVID J., AND ALEX J. BABB. *Maintainability Engineering*. New York: Wiley & Sons, 1973.

TANNER, JOHN P. *Manufacturing Engineering*. New York: Marcel Dekker, 1985.

THACKER, ROBERT M. *A New CIM Model*. Dearborn, MI: Society of Manufacturing Engineers, 1989.

THORPE, JAMES F., AND WILLIAM H. MIDDENDORF. *What Every Engineer Should Know About Product Liability*. New York: Marcel Dekker, 1979.

TRUCKS, H. E. *Designing for Economical Production*. (2nd edition). Dearborn, MI: Society of Manufacturing Engineers, 1987.

VALLIERE, DAVID. *Computer-Aided Design in Manufacturing*. Englewood Cliffs, NJ: Prentice Hall, 1990.

WILSON, RICHARD GUY. *The Machine Age in America*. New York: Harry N. Abrams, 1986.

WINGLER, HANS M. *The Bauhaus*. Cambridge, MA: The MIT Press, 1976.

OLDER BOOKS

ALGER, JOHN R., AND CARL V. HAYS. *Creative Synthesis in Design*. Englewood Cliffs, NJ: Prentice Hall, 1964.

BANHAM, REYNER. *Theory and Design in the Machine Age*. New York: Frederick A. Praeger, 1960.

BAYER, HERBERT, AND WALTER AND ISE GROPIUS. *Bauhaus*. Boston: Charles T. Branford Co., 1950.

BAYNES, KEN. *Industrial Design and the Community*. London: Lund Humphries, 1967.

BEVLIN, MARJORIE E. *Design Through Discovery*. New York: Holt, Rinehart, and Winston, 1963.

CAPLAN, RALPH. *Design in America*. New York: McGraw-Hill, 1969.

DE ZURKO, EDWARD R. *Origins of Functionalist Theory*. New York: Columbia University Press, 1957.

DOBLIN, JAY. *One Hundred Great Product Designs*. New York: Van Nostrand Rheinhold, 1970.

DREXLER, ARTHUR, AND GRETA DANIEL. *Introduction to Twentieth Century Design*. New York: The Museum of Modern Art, 1959.

DREYFUSS, HENRY. *Designing for People*. New York: Simon and Schuster, 1955.

FEREBEE. ANN. *A History of Design from the Victorian Era to the Present*. New York: Van Nostrand Rheinhold, 1970.

GILBERT, K. R. *The Portsmouth Block-making Machinery*. London: Her Majesty's Stationery Office, 1965.

HOLLAND, LAURENCE B. (ED.). *Who Designs America?* Garden City, New York: Doubleday, 1966.

ITTEN, JOHANNES. *Design and Form: The Basic Course at the Bauhaus.* London: Thames and Hudson, 1964.

JONES, J. CHRISTOPHER. *Design Methods.* New York: Wiley-Interscience, 1970.

KAUFMANN, EDGAR, JR. *What Is Modern Design?* New York: The Museum of Modern Art, 1950.

KAUFMANN, EDGAR, JR. *What Is Modern Interior Design?* New York: The Museum of Modern Art, 1953.

KEPES, GYORGY. *Language of Vision.* Chicago: Paul Theobald, 1949.

KEPES, GYORGY (ED.). *The Man-Made Object.* New York: George Braziller, 1966.

KOUVENHOVEN, JOHN. *Made in America..* Garden City, New York: Doubleday, 1962.

LINDBECK, JOHN R. *Designing Today's Manufactured Products.* Bloomington, IL: McKnight & McKnight, 1973.

LOEWY, RAYMOND. *Never Leave Well Enough Alone.* New York: Simon and Shuster, 1950.

LUTHRA, S. K. *Applied Art Handbook.* Bombay, India: Kareer Polytechnic, Wandy Road, 1966.

LYNES, RUSSELL. *The Tastemakers.* New York: Harper and Brothers, 1954.

MAYALL, W. H. *Machines and Perception in Industrial Design.* New York: Reinhold, 1968.

MOHOLY-NAGY, LASLO. *Vision in Motion.* Chicago: Paul Theobald, 1947.

MUELLER, ROBERT E. *Inventive Man.* New York: Lancer Books, 1964.

MUMFORD, LEWIS. *Technics and Civilization.* New York: Harcourt, Brace and World, 1963.

NELSON, GEORGE. *Problems of Design.* New York: Whitney Publications, 1957.

NEUTRA, RICHARD. *Survival Through Design.* New York: Oxford University Press, 1954.

PEVSNER, NIKOLAUS. *Pioneers of Modern Design from William Morris to Walter Gropius.* New York: The Museum of Modern Art, 1949.

PYE, DAVID. *The Nature of Design.* New York: Reinhold, 1964.

READ, HERBERT. *Art and Industry.* New York: Horizon Press, 1953.

SCHEIDIG, WALTHER. *Weimar Crafts of the Bauhaus.* New York: Reinhold, 1967.

SCOTT, ROBERT G. *Design Fundamentals.* New York: McGraw-Hill, 1951.

SNEUM, GUNNAR. *Teaching Design and Form.* London: B. T. Bratford Ltd., 1965.

TEAGUE, WALTER D. *Design This Day.* New York: Harcourt, Brace, and World, 1940.

VAN DOREN, HAROLD. *Industrial Design.* New York: McGraw-Hill, 1954.

WALLACE, DON. *Shaping America's Products.* New York: Reinhold, 1956.

VIDEO TAPES

Automated Assembly. Society of Manufacturing Engineers, One SME Drive, Dearborn, MI 48121. 1987. (34 minutes).

Benchmarking Manufacturing Processes. Society of Manufacturing Engineers, One SME Drive, Dearborn, MI 48121. 1994. (90 minutes).

CAD/CAM. Society of Manufacturing Engineers, One SME Drive, Dearborn, MI 48121. 1987. (25 minutes).

CAD/CAM Networking. Society of Manufacturing Engineers, One SME Drive, Dearborn, MI 48121. 1988. (36 minutes).

Competetive Benchmarking and Target Setting. Gordon M. Lewis, CAD/CIM Management Roundtable, Inc., 1050 Commonwealth Avenue, Suite 301, Boston, MA 02215. 1990. (45 minutes).

Cost Estimation Tools. Peter Dewhurst. CAD/CIM Management Roundtable, Inc., 1050 Commonwealth Avenue, Suite 301, Boston, MA 02215. 1990. (45 minutes).

Design for Manufacturing. V1199, 601/2. John R. Lindbeck. Media Services, Western Michigan University, Kalamazoo, MI 49008. 1989. (56 minutes).

Design for Manufacturing. VVC1363. Westinghouse Corporate Video Services. Westinghouse Corporation, 1316 Market, Pittsburgh, PA 06513. 1989. (8 minutes).

DFM:Design for Manufacturing. Society of Manufacturing Engineers, One SME Drive, Dearborn, MI 48121. 1991. (44 minutes).

Ergonomic Economics. Dean Mattox et al. T & S Equipment Company, P.O. Box 496, Angola, IN 46703. 1990. (38 minutes).

Ergonomic Seating. BIOFIT, P.O. Box 109, Waterville, OH 43566. 1994. (21 minutes).

Ergonomics in Manufacturing. Society of Manufacturing Engineers, One SME Drive, Dearborn, MI 48121. 1993. (30 minutes).

Flexible Manufacturing Cells. Society of Manufacturing Engineers, One SME Drive, Dearborn, MI 48121. 1987. (28 minutes).

Injected Metal Assembly Systems. Fishertech, Box 179, Peterborough, Ontario K9J 6Y9 Canada. 1993. (7 minutes).

Key Steps to DFM Implementation, From Beginning to Advanced. William Sprague. CAD/CIM Management Roundtable, Inc., 1050 Commonwealth Avenue, Suite 301, Boston MA 02215. 1990. (45 minutes).

Manufacturing Processes. Preview Tape. Society of Manufacturing Engineers, One SME Drive, Dearborn, MI 48121. 1981. (10 minutes).

Product Liability and the Reasonably Safe Product. Society of Manufacturing Engineers, One SME Drive, Dearborn, MI 48121. 1980. (55 minutes).

Rapid Prototyping for DFM. Society of Manufacturing Engineers, One SME Drive, Dearborn, MI 48121. 1991. (45 minutes).

Robotics Technology: Beyond Basics. Society of Manufacturing Engineers, One SME Drive, Dearborn, MI 48121. 1987. (90 minutes).

Simpler Products, Simpler Methods. William Conlin. CAD/CIM Management Roundtable, Inc., 1050 Commonwealth Avenue, Suite 301, Boston MA 02215. 1990. (30 minutes).

Simultaneous Engineering. Bodine Assembly Systems. The Bodine Corporation, P.O. Box 3245, 317 Mountain Grove Street, Bridgeport, CT 06605. 1991. (9 minutes.)

Simultaneous Engineering. Society of Manufacturing Engineers, One SME Drive, Dearborn, MI 48121. 1989. (45 minutes).

PERIODICALS

Architectural Digest. 5900 Wilshire Boulevard, Los Angeles, CA 90036.

Architectural Review, The. 33-39 Bowling Green Lane, London EC1R ODA, England.

Assembly Engineering. P.O. Box 3001, Wheaton, IL 60189-9925.

AXIS World Design Journal. 5-17-Roppongi Minata-ku, Tokyo 106, Japan.

Business Week. P.O. Box 8829, Boulder, CO 80328-8829.

CADalyst. Aster Publishing, P.O. Box 1965, Marion, OH 43306-4065.

CADENCE Magazine. P.O. Box 202260, Austin, TX 78720-2260.

Computer-Aided Engineering. P.O. Box 985005, Cleveland, OH 44198-5005.

Contract Design. P.O. Box 7615, Riverton, NJ 08077-9115.

Design. 28 Haymarket, London SW1Y 4SU, England.

DesignNet Magazine. P.O. Box 202380, Austin, TX 78720-2380.

Design News. 44 Cook Street, Denver, CO 80208.

Design News. Japan Industrial Design Promotion Organization, P.O. Box 101, World Trade Center Building, Tokyo, Japan.

Design Quarterly. Walker Art Center, Vineland Place, Minneapolis, MN 55403.

Design World. Design Editorial Pty. Ltd., 11 School Road, Ferny Creek, Victoria 3786, Australia.

DOMUS. Centro Domus, Via Manzoni 37, 1-20121 Milan, Italy.

Elle Decor. P.O. Box 51914, Boulder, CO 80323-1914.

European Plastics News. EMAP Maclaren, Ltd., P.O. Box 109, Maclaren House, Croydon CR9 1QH, England.

Form Swedish Design Magazine. Renstiernas Gata 12, S-116 28, Stockholm, Sweden.

Forma. Via Liszt, 21-00144 Roma/Eur, Italy.

Furniture Design & Manufacturing. 650 South Clark Street, Chicago, IL 60605-9936.

Imaging Magazine. 12 West 21st Street, New York, NY 10160-0371.

Industry Week. 1100 Superior Avenue, Cleveland, OH 44197-8020.

Innovation. Industrial Designers Society of America, 1142-E Walker Road, Great Falls, VA 22066.

Interior Design. P.O. Box 173653, Denver, CO 80217-9398.

Interiors. 1515 Broadway, New York, NY 10036.

International Design. 250 West 57th Street, New York, NY 10107.

Machine Design. 1100 Superior Avenue, Cleveland, OH 44197-8024.

Managing Automation. P.O. Box 1145, Skokie, IL 60076-9790.

Manufacturing Engineering. One SME Drive, Dearborn, MI 48121.

Manufacturing Systems. P.O. Box 3008, Wheaton, IL 60189-9972.

Materials Engineering. 1100 Superior Avenue, Cleveland, OH 44197-8089.

Nouvel Economiste, le. 22 Rue de la Tremoille, 75008 Paris, France.

Packaging Digest. 1020 South Wabash Avenue, Chicago, IL 60605-2246.

Plastics World. 44 Cook Street, Denver, CO 80206.

Pollution Engineering. P.O. Box 173329, Denver, CO 80217-3329.

Product Design and Development. P.O. Box 2001, Radnor, PA 79080-9501.

R&D: Research and Development. 44 Cook Street, Denver, CO 80206.

Safety & Health. 650 South Clark Street, Chicago, IL 60605.

Textile World. 29 North Wacker Drive, Chicago, IL 60606-3298.

Wood and Wood Products. Vance Publishing Corporation, 400 Knightsbridge Parkway, Lincolnshire, IL 60069.

DESIGN CENTERS

Accademia delle Arti e del Disegno. Via Orsanmichele 4, 50123 Firenze, Italy.

Alberta Design Works. Renton Belbin Partners, Calgary, Alberta T2R 0A5, Canada.

Associazione per il Disegno Industriale. Via Montenapoleone 18, 20121 Milano, Italy.

Australian Design Council. Amory Gardens, 4-6 Calvill Avenue, Ashfield 2131, Australia.

Barcelona Centre de Disseny. P. de Gracia 90, 08008, Barcelona, Spain.

Danish Design Centre. H. C. Andersens Boulevard 18, DK-1553 Copenhagen V, Denmark.

Design Arts Board. P.O. Box 302, North Sydney, NSW 2060, Australia.

Design Center Budapest. V. Gerloczy, Utca 1. H-1052, Hungary.

Design Centre. 28 Haymarket, London SW1Y 4SU, England.

Design Exchange. 232 Bay Street, Toronto, Ontario M5K 1B2, Canada.

Design Forum Finland. Fabianinkatu 10, SF-00130 Helsinki, Finland.

Design Institute of Australia. Design Editorial Pty. Ltd., 11 School Road, Ferny Creek, Victoria 3786, Australia.

Design Museum. Butlers Wharf, Shad Thames, London SEI 2YD, England.

Design Promotion Center. 340 Tun Hua N. Road, CETRA Exhibition Hall, Sung Shan Airport, Taipei 1059, Taiwan, ROC.

Design Vancouver. 1205-1100 Melville Street, Vancouver, British Columbia V6E 4A6, Canada.

Designers Institute of New Zealand. Wyndham Towers, Albert Street, Aukland 1, New Zealand.

Finnish Society of Crafts and Design. Pohjoisesplanadi 25 A, 00 100 Helsinki, Finland.

Forum Design Montreal. 52, Rey Lawson, Case postale 1122, Etage D. Place Bonaventure, Montreal, Quebec M5G 1G4, Canada.

German Design Council. Rat-Haus, Messegelande, Postfach 970287, D-6000 Frankfurt/Main 97, Germany.

Haus Der Wirtschaft. Willi-Bleicher-Strasse 19, D-7000, Stuttgart 1, Germany.

Industrial Designers Society of America. 1142 Walker Road, Great Falls, VA 22066.

Industrie Forum Design. Messegelande, D-3000 Hanover 82, Germany.

Institute of Design Montreal. 3510 St. Laurent, Suite 300, Montreal, Quebec, H2X 2V2 Canada.

International Design Center. 10-19, Sakae 2-chome, Naka-ku, Nagoya 460, Japan.

Japan Industrial Design Promotion Organization. 4-1, Hamamatsu-cho 2- chome, Minato-ku, Tokyo 105, Japan.

Korea Design and Packaging Center. 128, Yunkun-dong, Chongro-Ku, Seoul 110-460, Korea.

MODO. Ricerche Design Editrice srl, via Roma 21, 20094 Corsico, Milano, Italy.

Norwegian Design Council. Oscarsgate 53, N-0258 Oslo 2, Norway.

Osterreichisches Institut fur Formgebung. Salesianergasse 1, A-1030 Vienna, Austria.

Secretariat to Biennial of Industrial Design. Stari trg 11a, Ljublijana, Yugoslavia 576.

Svensk Form. Renstiernas gata 12, S-116 28, Stockholm, Sweden.

Index